Electronic Sensors for the Evil Genius

TOM PETRUZZELLIS

McGraw-Hill

New York   Chicago   San Francisco   Lisbon
London   Madrid   Mexico City   Milan   New Delhi
San Juan   Seoul   Singapore   Sydney   Toronto

The McGraw-Hill Companies

Cataloging-in-Publication Data is on file with the Library of Congress

Copyright © 2006 by The McGraw-Hill Companies, Inc. All rights reserved. Printed in the United States of America. Except as permitted under the United States Copyright Act of 1976, no part of this publication may be reproduced or distributed in any form or by any means, or stored in a data base or retrieval system, without the prior written permission of the publisher.

3 4 5 6 7 8 9 0   QPD/QPD   0 1 2 1 0 9 8 7 6

ISBN 0-07-147036-0

The sponsoring editor for this book was Judy Bass and the production supervisor was Pamela A. Pelton. It was set in Times Ten by MacAllister Publishing Services, LLC. The art director for the cover was Anthony Landi.

Printed and bound by Quebecor/Dubuque.

This book is printed on acid-free paper.

McGraw-Hill books are available at special quantity discounts to use as premiums and sales promotions, or for use in corporate training programs. For more information, please write to the Director of Special Sales, McGraw-Hill Professional, Two Penn Plaza, New York, NY 10121-2298. Or contact your local bookstore.

The book is dedicated to Josh and Andy—

May inspiration and aspiration carry them successfully through life challenges!

Electronic Sensors for the Evil Genius

Evil Genius Series

Contents

Contents

Introduction

Electronic Sensors for the Evil Genius: 54 Electrifying Projects was created as a voyage of discovery for children, adults, science buffs, and for those curious at heart of all ages. This book was designed to provide a spark for the curiosity seeker, as well as to inspire curious children, students, and adults alike toward experimentation and exploration of the sights, sounds, and smells of the natural world, which may not be detectable by our limited range of human senses. This book was also written for electronics hobbyists, as well as for electronics technicians and engineers who wish to build and experiment with electronics sensing and detection circuits.

Electronic Sensors for the Evil Genius: 54 Electrifying Projects will introduce the reader to how to sense, detect, and monitor sound, light, heat, and gas as well as to vibration, magnetic, electric, radio, and radiation. In this book we will see what few may see, hear what few have heard, and sense what few have sensed. The book should prove to be extremely helpful in aiding the reader to understand and appreciate some of the unseen and unheard energies all around us, as well as to help the reader sense and monitor these energies. This book is written so that the interested reader can readily build, test, and explore the fascinating and often mysterious world of natural phenomena. We will introduce the reader to many different types of sensors, detectors, and transducers, which convert one form of energy to another.

Our hope is that *Electronic Sensors for the Evil Genius: 54 Electrifying Projects* will inspire a student to construct a science fair project or two or perhaps send the inquisitive reader on a lifelong quest to investigate the natural world through electronics sensing and detection.

Electronic Sensors for the Evil Genius: 54 Electrifying Projects provides extensive photos, schematics, tables, and diagrams. The appendix provides parts suppliers and project kits sources.

Chapter 1—Sound Energy

Sound energy is a very exciting starting point for exploring and observing natural phenomena all around us. The sound waves humans hear with our ears are but a very limited range of the audio spectrum. Our perception of sound allows us to "hear" only a narrow slice of energy between 20 Hz to 15 KHz. In fact a whole range of audio exists both above and below our range, which we cannot perceive at all but are in fact very interesting to explore.

In this chapter we will explore the interesting worlds of audible, ultrasonic, and infrasonic sounds. You will investigate how to listen to high-frequency sounds of animals and remote conversations and how to track down noise and machine faults with an electronic stethoscope. You will discover a whole new universe of underwater sounds after building a hydrophone and an audio amplifier. You will learn that a longitudinal mechanical wave whose frequency is below the audible range is called an *infrasonic* wave (*infrared* light waves are waves below red light), and one whose frequency is above the audible range is called an *ultrasonic* wave (*ultraviolet* waves are above violet light). The longest wavelength sound waves that can affect the normal human ear (20 Hz) are a thousand times as long as the shortest waves to which the ear is sensitive (20,000 Hz).

In this chapter you will construct an ultrasonic listener, which you can use to listen to insect and other sounds that are above the human hearing range. We will also explore infrasonic waves, which are usually generated by large sources, such as barometric or weather fronts or by earthquakes. You learn how to construct your own microbarograph, which will allow you to detect these very long sound pressure waves produced by barometric changes and approaching storms.

Chapter 2—Light Detection and Measurement

Although about ten million shades of color can be identified by the human eye in the visible spectrum of light, the light that produces those colors spans only a narrow spread of wavelengths. This frequency density is comparable to crowding all the world's human-made radio frequencies into a narrow frequency range from 550 KHz to 880 KHz in the standard AM radio broadcast band. The eye is indeed an amazing electromagnetic receiver; consider that if you glance at a yellow dress for just one second, the electrons in the retinas of your eyes must vibrate about 5×10^{15} times during the interval to receive the yellow. If you were to count all the waves that beat upon all the shores on Earth, you would have to count for ten million years in order to count the same number as that of the oscillations in one second of yellow light.

In this chapter we will take a closer look at light sensors such as photocells and solar cells and how we can use them to detect light or the absence of light. You will learn how to measure the solar constant, how to measure ultraviolet light, and how to detect ozone in the atmosphere. You will also use light sensors for optical listening, that is, listen to the sound that light makes when modulated by movement. You can do this by listening with the ear, a transducer, and an audio amplifier in the amplitude domain of light, rather than by looking through the frequency domain with light received by the eye. After building the opto-listener, you will be able to listen to electronic displays, "singing" automotive headlamps, burning flames, and lightning. You will be able to listen to just about any light source to "see" what it sounds like. In addition, you will learn how to measure the speed of objects using light, by constructing your own optical tachometer. Finally, you will look at how you can

detect and measure water pollution in water using the optical turbidity meter.

Chapter 3—Heat Detection

Heat is transferred from one place to another in three ways: via conduction, convection, or radiation. *Conduction* is the process of transferring heat from molecule to molecule in a substance. When one end of an iron rod is placed in a fire, the other end soon gets hot because the heat is transferred from one end of the iron to the other end by conduction (from molecule to molecule). *Convection* is the process of transmitting heat by means of the movement of heated matter from one place to another. Convection thus takes place in liquids and gases. A room is heated by means of convection by circulating warm air through the room. This brings us to *radiation*. In both conduction and convection, heat is transmitted, or transported, by moving particles (e.g., molecules or air). However heat can also travel where matter does not exist. For example, the heat from the sun reaches the earth across the 93 millions of miles of space. When a cloud passes between the sun and a point on Earth, the heat at that point is diminished or cut off. This is due to the fact that heat is transmitted or radiated by waves.

Heat waves and light waves are of the same nature; they are both electromagnetic radiations that differ only in wavelength, heat waves being longer than light. Heat waves near the radio portion of the spectrum are called infrared.

In this interesting chapter, you will construct an infrared flame detector, which can sense a match or flame up to 3 feet away. The reader will also learn how to construct a freeze alarm, which could be used to alert you of icy driving conditions. You will construct an overtemperature monitor, which you could utilize to warn you of an overheating condition in a machine or in your refrigerator. You will also read about an analog data-logger for sending temperature data remotely via radio link; or you could use the system to record readings in the field and transport the data back to your laboratory. More advanced projects include an LCD thermometer, a night vision viewer, and an infrared motion detector, which can sense the body heat of an intruder up to 50 feet away. The infrared motion detector could be used to create your own home alarm system.

Chapter 4—Fluid Sensing

In this chapter you will explore liquid sensors, another very interesting and important aspect of sensing. Your first project in this chapter is a simple yet useful rain detector, which can detect the earliest signs of rain drops and will allow you a few precious moments to roll up your car windows or bring in your laundry. When used as part of a weather data collection system, the exact time of a shower can be recorded. In this chapter, you will also learn how to build a fluid or liquid sensor as well as fluid level indicator, which can be used to indicate how much fuel or water is left in a container or tank. Weather fans will also learn how to construct a humidity monitor to measure humidity around your home or shop. Junior scientists will learn about pH and how to build and use a pH meter using easy-to-locate, low-cost components. Nature- and ecology-minded readers will learn about how to build and utilize a stream-gauge water level monitor for studying river and stream flow and runoff.

Chapter 5—Gas Sensing

Air and gas sensing always seems so ephemeral, because air and many gases cannot be seen and often cannot be smelled. But once again modern electronics comes to our assistance in helping us to sense both air and various gases in the air and atmosphere. In this chapter you will learn how to build an air pressure sensing switch, which can be utilized to detect doors moving or vehicles approaching. Your next project is an electronic sniffer, which can detect a number of different gasses and sound an alarm. With the bargraph pressure sensor, you can monitor water or air pressure changes and display the level on an LED bargraph display. This chapter also introduces the new Pellistor combustible gas sensors. You will learn how to operate and construct a toxic gas sensor.

And weather enthusiasts will be eager to construct the electronic barometer project, which can be used as the basis of a weather station.

Chapter 6—Vibration Sensing

Vibration sensing is an exciting aspect of sensing technology. Seismology is the study of vibration and is primarily used for detecting and monitoring ground vibrations or earthquakes. Seismology is also used to study bomb blasts to determine signatures and locations, to verify nuclear test ban treaties (in other words, for keeping tabs on other countries and their bomb detonations).

Vibration sensing can be used also to solve industrial problems and to detect earthquakes deep in the earth. Did you know that vibration sensing can be used to detect engine or motor vibration problems, which can be monitored over time in studying developing machine problems? In the first project in this chapter, you will build a vibration hour meter. This simple yet unique circuit will permit recording the length of time a machine or natural event occurs, using an off-the-shelf hour meter. As long as the vibration persists the hour meter will record the event.

Your next project is a vibration alarm, which can notify you of intruding persons or animals. This project uses a commonly available audio speaker as a sensor to detect vibration. The vibration alarm project could also be used to scare away pesky animals from your garden or to warn you of approaching intruders. Our next project, the piezo seismic alarm sensor, uses a gas barbeque piezo sparker mechanism as the seismic sensor. The vibration alarm, or the piezo seismic alarm, might be an interesting science fair project for the budding scientist. Our advanced project, the AS-1 seismograph, is capable of detecting earthquakes all around the world. This seismograph can be used for serious amateur research and observations. Many schools across the country are currently using the AS-1 for seismic research and display. Qualified schools can actually receive an AS-1 for free.

Chapter 7—Detecting Fields

Magnetic fields are all around us, but our normal senses cannot detect the presence of these fields. Our senses must be extended or enhanced in order to sense or detect these magnetic fields. A transducer, or sensor, usually some type of coil, can be used with an amplifier to detect the presence of a magnetic field. You will learn that a magnetic field radiates on its own, a limited distance, but can also join with an electric field to form an electromagnetic field. A magnetic field becomes a complement to an electrical field when the two are at 90 degrees, to form a complex time-oriented wave called an electromagnetic wave. Electromagnetic waves radiate out into space as radio waves.

In this chapter you will read about different types of magnetic sensors, from the small induction pickup coil, which can be used in conjunction with the mag-ear project to listen to telephone conversations or to locate hidden electrical conduits and hidden metal, to larger coil detectors that can be used to detect magnetic fields produced by moving cars and trains. In this chapter you also learn how to construct an electronic compass, called an ELF radiation monitor, which can be used to survey your home electronics and appliances for potentially dangerous low-frequency magnetic fields. In this chapter you will discover, how a sudden ionospheric disturbance receiver can be used for radio propagation studies; you will also learn how to build and utilize one. Your final project is an Earth-field magnetometer, which can be used to detect solar magnetic storms originating from the sun.

Chapter 8—Sensing Electronic Fields

The earliest study of nature's wonders began with the observations relating to static electricity or electrostatics. Many early observations of static electricity involved animal skins and hair and glass or stone objects. These early observations lead to the discovery of electrical attraction principles, electrical waves, and later the basis for electronics as we know it today. In this chapter you will discover and learn about electrostatic fundamentals, electric fields, and electromagnetic fields. When a magnetic field is enjoined at 90 degrees with an electric field an electromagnetic field is produced.

In this chapter you will learn that all electromagnetic energy, regardless of its frequency, has certain common properties. Two things magically bind electric and magnetic energy together into electromagnetic radiation: the traveling electric and magnetic fields are (1) always at right angles to each other and always at right angles to the direction of propagation, and (2) they are always becoming weaker with the distance they travel.

You will discover applications for the classical electroscope, the Leyden jar, the static tube, as well as how to build and use a cloud chamber to detect alpha particles. Practical projects include an ion detector, an electronic electroscope, and an atmospheric electricity monitor. Advanced projects presented for junior scientists include an advanced electronic electroscope and a cloud charge monitor, which can detect and display the charges from clouds traveling overhead. The final project of this chapter is an electric field-disturbance monitor, which can be used to detect human bodies in an electrical field and sound an alarm. The electric field-disturbance monitor could be used for serious electric field research or could form the basis for a home or camping alarm system.

Chapter 9—Radio Projects

Electromagnetic energy encompasses an extremely wide frequency range. A classic example of a natural broadband transmitter of electromagnetic energy is a lightning flash or streak. A lightning flash exhibits a fantastic amount of energy being radiated into space as it generates an electromagnetic signal. This signal typically covers frequencies from a few hertz up through the broadcast band, where we can hear static on the AM radio. Natural wideband radio frequency storms, on Jupiter for example, can be detected and observed in the shortwave frequency bands.

In this chapter you look at radio frequency energy, both natural radio energy created by lightning and planetary storms as well as radio frequencies generated by humans for television, communications, and radar. Radio frequency energy covers a broad range from the low end of the radio spectrum (10 to 25 KHz) used by high-power navy stations that communicate with submerged nuclear submarines, to the familiar AM broadcast band from 550 to 1,600 KHz, to the shortwave bands from 2,000 KHz to 30,000 KHz, on up to the very high-frequency television channels covering 54 to 216 MHz, through the very popular frequency modulation FM band from 88 to 108 MHz, on through the radar frequency band of 1,000 to 15,000 MHz, and extending through approximately 300 GHz. The radio-frequency spectrum actually extends almost up to the lower limit of visible light frequencies.

You will explore some different types of radio receivers that you can construct and use. Your first project is an electronic lightning detector, which can be used to warn of oncoming electrical storms, a great project for weather enthusiasts. Your next project is the ELF natural radio, which can be used to listen to those mysterious low-frequency sounds produced by Mother Nature, such as tweeks, pops, the dawn chorus, as well as whistlers. These ascending and descending frequency sweeps are caused by electrical storms on the other side of the earth. Why not build your own shortwave receiver and explore the world of radio from foreign broadcasters on the other side of the globe. Listen to exciting music and news from European and African radio stations. The advanced project, the Jupiter Radio telescope project, will permit you to listen to the strange sounds of planetary storms on Jupiter. This radio receiver project is a great starting point for amateur research projects in radio astronomy.

Chapter 10—Radiation Sensing

The radiation spectrum is usually broken down into electromagnetic radiation and ionizing radiation. Electromagnetic radiation is visible light and longer wave lengths like heat and radio. The bulk of the energy from the sun is in this range. We use this light to see, grow plants, and to power solar cells.

The second type of radiation is called ionization radiation. Ionizing radiation is usually thought of as high-energy and high-speed particles, though sometimes as very short wavelength waves. The particles can be photons, electrons, protons, and ionized elements, such as helium and iron. The ionized elements have been stripped of their electrons.

When these high-speed particles pass through matter, they can do damage, possibly deep inside the matter. As they pass through, they leave a trail of ionized (missing some electrons) particles behind them. The number of ionized particles per centimeter of path depends on the type of particle and its velocity. A bigger and higher speed particle will do more damage.

On Earth, radiation can come from rocks and minerals and even from the soil as radon gas. Natural ionizing radiation also comes from our sun and also distant parts of the universe. In this chapter you will learn how to construct and utilize a cloud chamber for detecting low ionizing alpha particles. You will learn how to detect ionizing radiation using the low-cost electronic ion chamber, which can be constructed using four commonly available transistors. A more advanced electronic ion chamber will allow you to conduct more serious radiation studies. Finally you will learn how to build your own battery-powered Geiger counter, which can be utilized for detecting radioactive rock formations, such as uranium, and for radiation field studies.

Electronic Sensors for the Evil Genius

Chapter One

Sound Energy

Sound energy is a very exciting starting point for exploring and observing natural phenomena all around us. The sound waves we hear with our ears are but a very limited range of the total audio spectrum. Our perception of sound allows us to "hear" only the narrow slice of energy between 20 Hz and 15 KHz on this spectrum. In fact a whole range of audio exists both above and below the range of what we can perceive, which is in fact very interesting to explore.

In this chapter we will explore the interesting world of sound. We will investigate audio sounds in the range of human perception as well as those in the infrasonic and ultrasonic range, which are all around us but we are rarely aware of. You will discover how to listen to the high-frequency sounds of animals and remote conversations and how to track down noise with an electronic stethoscope. We will also learn how to explore a whole new universe of underwater sounds using a hydrophone and an audio amplifier.

The keys to our explorations will be centered around various types of microphones and audio amplifiers and how you might be able to use them to listen to both natural and man-made sounds.

An audio amplifier is an electronic circuit that is used to increase the level of sound. The input of the amplifier takes a small or low-level signal and amplifies it so we can comfortably hear it. We might use an audio amplifier to increase audio levels to help us hear nearby sounds as well as distant sounds more clearly, or simply to increase the audio level for listening to music or a telephone conversation. We will use the principle of amplification in most of the projects in this chapter, but let's start with the basics of sound energy.

Sound Energy

Sound waves are basically longitudinal mechanical waves that can be propagated in solids, liquids, and gases. However unlike electromagnetic waves, they cannot travel in a vacuum.

Sound in Air

Figure 1-1 illustrates a sound wave traveling to the right (long horizontal arrow dissecting vertical waves). The air particles move back and forth (shorter arrows above vertical waves) to alternately compress the surrounding air on a forward movement and rarify the air on a backward movement. The air transmits these disturbances outward from the sound as a wave.

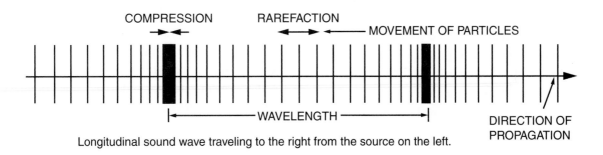

COMPRESSION   RAREFACTION   MOVEMENT OF PARTICLES

|← WAVELENGTH →|

DIRECTION OF PROPAGATION

Longitudinal sound wave traveling to the right from the source on the left.

Figure 1-1   *Longitudinal sound wave traveling to the right from a source on the left*

Sound waves are confined to the frequency range that can stimulate the human ear and brain to the sensation of hearing. We define the audible hearing range as the range from 20 to about 20,000 Hz, although humans can seldom hear sounds as high as 20,000 Hz. As one becomes older, one can hear less and less at the high end of the range. This range is also the exact range to which the average high-fidelity or stereo amplifier is tailored.

Ultrasonic Waves

A very interesting world of sound waves exists above the human hearing range, in the range between 20 KHz and 50 KHz. Insects and animals such as bats create many unique sounds that cannot be heard by our ears. Gas and chemical leaks as well as many machine squeaks and grinds are all silent to humans, but they go on all around us continuously, without us knowing it. Ultrasonic waves are also created by intrusion detectors. These devices flood a room or area with "silent sound" in order to sense an intruder by detecting "beats" of the moving object and the original signal. They normally operate at 40 KHz and cannot be heard by the human ear. Most people are familiar with these *ultrasonic waves* that exist just above the human hearing range, but many people are not familiar with sound waves far below the human hearing range called *infrasonics*.

Infrasonic Waves

A longitudinal mechanical wave whose frequency is below the audible range is called an *infrasonic* wave (*infrared* light waves are waves below red light). The wave whose frequency is above the audible range is called an *ultrasonic* wave (*ultraviolet* waves are above violet light).

The longest-wavelength sound waves that can affect the normal human ear (20 Hz) are a thousand times as long as the shortest waves audible to the human ear (20,000 Hz). As we have observed, the longest light waves (red) that can affect the eye are

not even twice as long as the shortest light waves visible to the eye (violet). The ear, however, has a range of 10 to 12 octaves; the eye range is but one octave. (The interval between two frequencies, one of which has twice the frequency of the other, is an *octave*—in this case, 400 Hz to 800 Hz.)

Infrasonic waves of special interest are usually generated by large sources, such as an earthquake. Without resorting to such upheaval as an earthquake, you can sense these kinds of waves while driving behind a large trailer truck on a highway. The large frontal surface area of the truck (which often is flat) buffets the wind and sets up infrasonic waves, which are impressed on the ears as a "feeling" sensation rather than a sound sensation. This buffeting also makes steering the car more difficult. Atmospheric wave fronts and meteors traveling though the atmosphere also create infrasonic waves that can be detected.

If you want to "hear" an extremely low-frequency wave, take a trip in a fast elevator in a tall building. You will be going from a position of high air pressure (on the ground floor) to a position of low air pressure (at the highest floor reached) in perhaps 30 to 60 seconds. Doubling this time to get a full wavelength of a complete cycle, or period, of the wave is 60 to 120 seconds. Frequency is related to period by the expression:

$$f = 1/T$$

where f is the frequency in hertz, and T is the period in seconds.

We find that the frequency is $1/60$ second, or 0.0166 cycles per second. This is the same as one cycle every 1 to 2 minutes, which is way below the threshold of hearing. When you are rising rapidly in the elevator (or in an aircraft), you should swallow or clear your throat every so often to equalize the pressure inside with that outside your ears. When a weather front passes through your area, you may have "heard" an infrasonic signal whose period is 3 hours! Watch a barometer to note this "rapid" change of atmospheric pressure. The physical process is the same for weather pressure changes and sound. It is just much slower for weather.

Temperature and Sound

At a temperature of 70°F, the speed of sound in air at sea level is 1,130 feet per second. Because temperature affects the speed of sound, at high temperatures the molecules transmitting the sound (for example, air or metal) move faster, and the speed of sound is increased. The speed of sound is 1,088 feet per second at 32°F (the freezing point of water) because the molecules are slowed down.

Sound Pressure

Sounds are pressure waves that vary around the normal air pressure of 15 pounds per square inch. Therefore, the amplitude measured as an *average* pressure does not convey any meaning. Instead, the *root-mean-square* (RMS) pressure is usually used to describe the magnitude of a sound. Because we can hear sounds over an extremely wide range of sound pressures (from about 0.0001 to 1,000 microbars), it is customary to work with *sound pressure level* (SPL) instead of using the value of sound directly. SPL is defined by the expression:

$$L = 20 \log (P_1/P_0)$$

where L is the sound pressure level of sound pressure P_1, and p_0 is a reference pressure. SPL is often used in defining the performance of a high-fidelity, high-power speaker.

The Decibel

The unit of sound pressure level is the *decibel* (dB). The dB is a relative unit and refers to a ratio of sound pressure (for example, P_1 to P_2). Thus, we say that the difference between two sound pressure levels L_1 and L_2 is:

$$L_2 - L_1 = \tfrac{1}{2} \log (P_2/P_0) - 20 \log (P_1/P_0)$$
$$= 20 \log (P_2/P_1)$$

Here we have assumed that the sound level P_2 is greater, or stronger, than sound level P_1. Note that the reference pressure, P_0, does not affect the difference so long as the same reference pressure is used in both expressions of the two levels.

The pressure level of $P_0 = 0.0002$ microbar has been adopted as a standard reference pressure because it is close to the lowest level the human observer can hear when the sound is a 1,000 Hz tone. Sound at a level of 140 dB above the 0 dB level is threshold of pain, as we might observe if we were coming from a nearby jet aircraft without having on ear protectors. The 0 dB level is also considered to be the level of a whisper or quiet footsteps. A sound that is 140 dB above threshold is 10 million times stronger than the weakest sound that can be heard. From this, we can see that the normal ear has a fantastic signal-level handling capability. However, in order to retain this hearing capability as long as possible as we age, very strong sound levels should be avoided unless hearing protectors or earplugs are used.

Types of Microphones

Microphones are conversion devices. Microphones are also known as *transducers*, as they convert mechanical waves to electrical waves. Once converted to electrical waves, these waves can be amplified. Microphones are available in many different types and styles:

- **Carbon microphone**   A microphone using a flexible diaphragm, which moves in response to sound waves and applies a varying pressure to a container filled with carbon granules, causing the resistance of the microphone to vary correspondingly.

- **Piezoelectric microphone**   A microphone in which deformation of a piezoelectric bar by the action of sound waves generates an output voltage between the faces of the bar; also known as *crystal microphone*.

- **Magnetic microphone**   A microphone employing a diaphragm acted upon by sound waves and connected to an armature, which varies the reluctance in a magnetic field surrounded by a coil. Applications include miniature microphones for hearing aids and guitar pickups.

- **Dynamic microphone**   A conductor (usually a coil attached to diaphragm or ribbon) flexibly suspended in the field of a fixed magnet is vibrated by sound waves. This induces in the conductor an AC voltage that varies in step with the sound waves.

- **Electrostatic microphone**   A flexible diaphragm and a fixed electrode together form a two-plate air capacitor whose capacitance varies in step with the sound waves that vibrate the diaphragm. This is also known as a *capacitor microphone* or *condenser microphone*. With electret microphones, one of the electrodes carries a permanent charge.

Microphones can also have *directional* or *nondirectional* characteristics. Most microphones are simple, nondirectional types, which pick up sounds from all directions equally. These microphones are called *omnidirectional*. Directional microphones are available in a few major classifications. *Cardiod microphones* are one type of directional microphone. This type of microphone has a heart-shaped sound acceptance pattern. These microphones are used by entertainers to pick up their own voice, but to reduce sounds from other directions. Shot-gun microphones are highly directional microphones, which are used by bird enthusiasts to listen to distant bird sounds and by private investigators to hear conversations from a distance. Movie studios often use shot-gun microphones with long booms to pick up conversations of two people talking on a movie set. Highly directional microphones are also known as *parabolic microphones*. Parabolic microphones are very sensitive and highly directional. They are designed to pick up sounds very far away and amplify them through an electronic amplifier. Parabolic microphones are

very selective and the sound pickup pattern is very narrow (only a few degrees). Often when these microphones are used they must scan an area until the desired sounds are heard.

Simple Microphone Conversion

In order to improve the directional characteristics of an omnidirectional microphone, a simple adaptation can be made that greatly improves the direction abilities of the microphone. This adaptation can be made by doing the following. Cut out the small end of a disposable paper coffee cup (the kind that fits in a plastic holder). The small end of the cup is just the right size to fit over the small microphone. Another alternative is to find a large funnel and place it in front of a small sensitive electret microphone. Next, use black electrical tape to secure the cup to the microphone and also to prevent sound from getting into the microphone except through the wide funnel end. The microphone with the new shield is small enough to swing around at various sounds. The directional pickup pattern will have approximately a 60-degree width. This pattern will exclude sounds arriving from the back and sides of the microphone. It is helpful when listening for bird calls and other sounds that might be as much as 50 to 75 feet away.

The High-Gain Parabolic Microphone

Almost everyone is familiar with TV broadcasts of football games where they use parabolic reflectors to pick up crowd noises, band music, and play calls from the quarterback. A high-direction parabolic microphone is shown Figure 1-2. The parabolic reflector is made of plastic with a focal point about 6 inches in front of the center of the parabolic dish. This dish will provide greatly increased sensitivity when the microphone is placed at the focal point, facing inward toward the center of the dish. These parabolic units

Figure 1-2   *Parabolic microphone*

are generally light enough to carry into the field for listening to sport or wildlife activity, such as tracking game, or for listening to distant voice conversations. The unit is particularly effective when listening across water such as lakes or ponds because water absorbs little of the sound energy, allowing the microphone to pick it up easily. The dish has almost a pencil beam pattern, much like a flashlight, so you should scan slowly for sounds from a distance.

Amplifying Sounds—The Audio Amplifier

A microphone and an amplifier can be used outdoors to listen to the sounds of birds, autos, trains, countryside sounds, and people. An amplifier used along with a tape recorder will allow you to record unusual outdoor sounds. It can aid in hunting by amplifying the sounds of approaching animals or game. By placing several microphones many hundreds of feet away from a hunting shelter or blind, you could listen for game approaching from unobserved directions.

The diagram shown in Figure 1-3 illustrates a microphone pre-amplifier. This audio amplifier utilizes a TL084 op-amp to amplify the microphone signal. The op-amp amplifier has an overall gain of 27 dB. Potentiometer R6 in the feedback path provides a gain control. This microphone pre-amplifier was designed for an electret microphone, and a bias resistor R1 is used if you utilize an electret microphone. If you choose to use a dynamic microphone, eliminate this resistor. The microphone pre-amp circuit is powered from a 9-volt transistor radio battery through

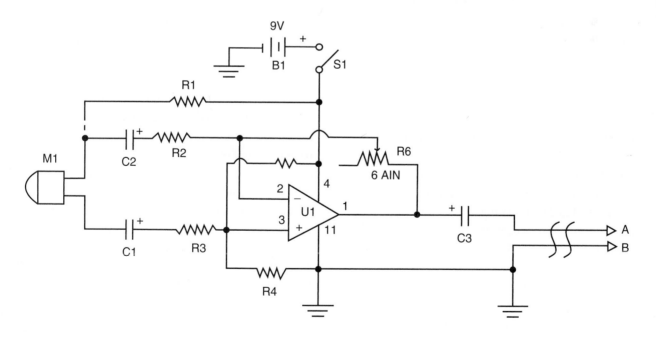

Figure 1-3   *Microphone pre-amplifier circuit*

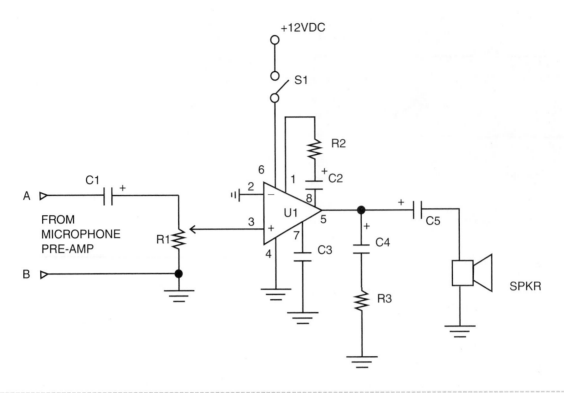

+12VDC

S1

R2

C2

C1 +

A ▷

FROM
MICROPHONE
PRE-AMP

R1

B ▷

6

2

3

U1

1

8

5

7

C3

C4

C5

+

R3

SPKR

Figure 1-4   *High-power audio amplifier circuit*

switch S1. The output capacitor at C4 is used to couple the pre-amp circuit to a high-power audio amplifier. A shielded microphone cable should be used to connect the microphone pre-amp to the power amplifier (for example, if the microphone is used in a probe configuration and the circuits are separated by any distance). If both circuits are placed on the same circuit board then no cable is required.

A high-power audio amplifier is depicted in Figure 1-4. This audio amplifier utilizes a high-output integrated circuit amplifier shown at U1. The audio input signal is fed to the input potentiometer at R1. The potentiometer is coupled to the input of the amplifier through pin 3. Pin 2 of the audio amplifier is grounded. The gain of the audio amplifier can be varied from 20 to 200 by adding the network formed by R2 and C1. A bypass capacitor can be added from ground to pin 7 on the IC if desired and is optional. Output sound shaping is performed by the network of C3 and R3. The output of the audio amplifier is coupled to an 8-ohm speaker via a 2,200 uF electrolytic capacitor. The circuit is powered by a 9- to 12-

volt battery at pin 5 of the LM386 amplifier IC. Note for high output a heatsink should be mounted on the LM386 amplifier. Microphone pre-amplifier shown previously can be combined with the audio power amplifier to form a powerful audio amplification system, which can be used with many different types of microphones to aid in listening to distant sounds.

A simple but interesting application for both preamplifier and power amplifier described previously is to construct a "rain" microphone. Locate a small, 8- to 10-inch plastic bucket. About halfway down inside the bucket mount the electret microphone on a round sheet of plywood or plastic, so that the microphone points to the closed end of the bucket. Turn the bucket upside down and cut a slot at the bottom for the microphone wire to exit. Finally, mount the upside down bucket on the ground or on your roof top and run a length of shielded cable from the microphone to the pre-amplifier inside your house. You will now be able to hear rain drops as they begin to fall on the bucket placed outside, so you will know immediately when it begins to rain, day or night.

Electret Microphone Pre-Amplifier Part List

Required Parts

R1, R2   10K ohm, 1/4-watt, 5% resistor

R3   1K ohm, 1/4-watt, 5% resistor

R4, R5   100K ohm, 1/4-watt, 5% resistor

R6   1-megohm potentiometer

C1, C3   1 uF, 35-volt tantalum capacitor

U1   TL084 op-amp (Texas Instruments)

M1   dynamic or electret microphone (see text)

S1   SPST toggle switch

B1   9-volt battery

Miscellaneous   PC circuit board, wire, shielded wire cable, etc.

High-Power Audio Amplifier Parts List

Required Parts

R1   10K potentiometer

R2   1.5K ohm, 1/4-watt, 5% resistor

R3   10 ohm, 1/4-watt, 5% resistor

C1   2.2 uF, 35-volt electrolytic capacitor

C2   10 uF, 35-volt electrolytic capacitor

C3   0.01 uF, 35-volt disc capacitor

C4   0.05 uF, 35-volt disc capacitor

C5   220 uF, 35-volt electrolytic capacitor

U1   LM386 audio amplifier IC (National)

S1   SPST toggle switch

SPKR   8-ohm speaker

Electronic Stethoscope

Stethoscopes are not useful only for doctors, but for home mechanics, exterminators, spies, and any number of other applications. Standard stethoscopes provide no amplification, which limits their use. Our first exciting project uses the ubiquitous op-amp to greatly amplify a standard stethoscope, and the circuit shown also incorporates a low-pass filter to remove background noise.

Here's an old mechanic's trick for finding the source of a funny noise in a car's engine. By using a length of garden hose, move one end of the hose around under the hood until you pinpoint the noise, which is directed through the hose so that the sound coming in travels through the hose to the ear of the mechanic. It's probably the simplest diagnostic tool used with modern cars, but often the most effective for situations such as isolating a knocking valve.

A garden hose would be too large for finding a problem with a small part, but a soda straw cut down to 3 inches in length works quite well. To design a somewhat efficient diagnostic tool or listening device, you need to amplify the sound. So we need a way to couple a soda straw or small tube to an electronic amplifier.

The *electronic stethoscope* schematic diagram is shown in Figure 1-5. The electronic stethoscope begins with the sensitive electret microphone, which is biased with resistor R1. The audio from the microphone is fed through capacitor C2 and R2 and then sent to the minus (−) input of op-amp U1. The output from the first op-amp amplifier is then coupled to the next op-amp at U2 through resistor R5 and R6. Op-amp U2 is directly coupled to the final op-amp amplifier section at U3. The output of U2 is also fed to a level meter consisting of the op-amp stage at U4, which is used to drive a bicolor *light-emitting diode* (LED) at D1. The audio output at U3 is sent to the final amplification stage at U5, through the volume control at R11. The volume control is coupled to U5 via a 0.01 uF capacitor at C5. The final amplification stage of the electronic stethoscope is made of an

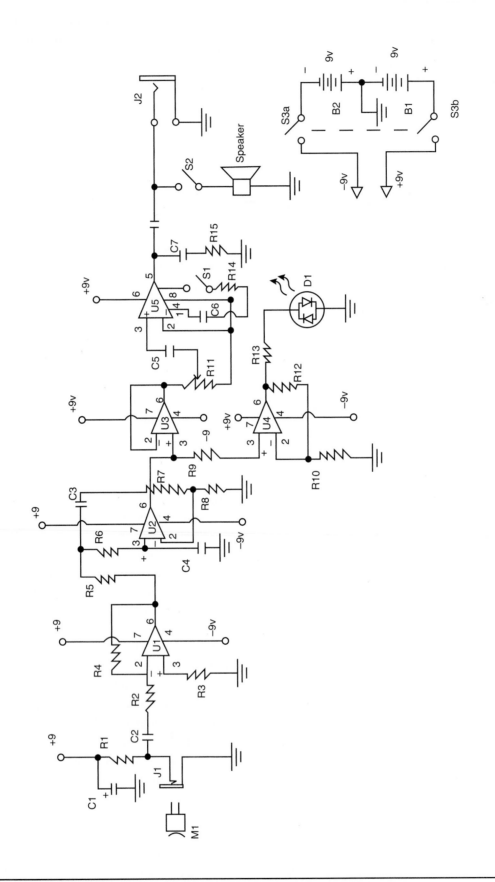

Figure 1-5   *Electronic stethoscope circuit*

LM386 power amplifier integrated circuit. The gain of the final amplifier can be adjusted from 20 to 200 by switching in the resistor capacitor network composed of R14 and C6, using switch S1. The output from the LM386 amplifier is finally sent to an 8-ohm speaker or headphone via capacitor C8. A headphone jack is provided at J2. Switch S2 is used to switch in the speaker in addition to the headphone, if desired. This added level of amplification should be used only for signals that are extremely small. Generally for troubleshooting and detective purposes, headphone operation is desirable, but in certain instances a speaker can be preferred.

Because this circuit uses op-amps, you will notice that the circuit requires both plus and minus power supplies. Two 9-volt batteries are connected together as shown and are connected to the circuit through DPST switch at S3. Two plastic battery holders were used to hold the batteries in place.

Construction of the electronic stethoscope is pretty straightforward. It can be constructed on a small perf-board or circuit as desired. The circuit utilizes the ubiquitous LM741, eight-pin op-amp. It is recommended that integrated circuit sockets be used for this project, in the event of a circuit failure at a later date. You will be able to repair the circuit much more easily if a socket is used. When installing the integrated circuits, be sure to orient them correctly in order to avoid damage at power-up. Generally integrated circuits have some sort of markings to indicate the correct orientation. The IC package will have either a small indented circle near pin 1 or a small rectangular cutout on the top center of the IC package. If you see a small indented circle or a cutout on the top of the IC, then pin 1 will be just to the left. Note that the electrolytic capacitors have polarity and must be installed with respect to the polarity. These components will have either a small minus or plus marking on them.

Once the circuit has been built, use a small metal box to house the circuit and batteries. The three switches, LED, and a small speaker were all mounted on the top of the enclosure, whereas the output jack for the headphone was mounted at the rear of the chassis panel. Some builders may elect to omit the speaker all together, if desired. Because the microphone needs to be remote from the chassis enclosure, you can elect to install a microphone jack on the front panel, as shown at J1, or simply run a shielded wire from the circuit to the microphone. Depending upon your application, you may want to construct a 4- to 8-foot shielded microphone cable leading back to the chassis enclosure.

The most difficult task in constructing the electronic stethoscope is mounting a soda straw or small length of plastic or nylon tubing to the microphone. You will need to devise a method to reduce or enlarge the diameter of the tubing to match the size of the small electret microphone.

When you build this project, you may discover that it wails when the speaker is brought close to the microphone or when you disconnect the potentiometer's connection to ground. This is caused by positive feedback in which the output from the speaker is picked up by the microphone and passed to the amplifier. Positive feedback should be avoided whenever possible. This is a secondary reason for using the soda straw with the microphone; it helps keep sound that is not directly in front of the microphone from being picked up and amplified.

The electronic stethoscope is a fun project with many applications. This project can serve many needs, including detecting noises, vibrations, and leaks by using tubing coupled to the microphone. By changing the microphone configuration from a tube to a parabola, you could use this project to eavesdrop on remote conversations, listen to distant bird or animal calls, or help you to track deer or other animals in the woods.

Required Parts

R1   10K-ohm, 1/4-watt resistor

R2, R3, R9   2.2K-ohm, 1/4-watt resistor

R4   47K-ohm, 1/4-watt resistor

R5, R6, R7   33K-ohm, 1/4-watt resistor

R8   56K-ohm, 1/4-watt resistor

R10   4.7K-ohm, 1/4-watt resistor

R11   10K potentiometer

R12   330K-ohm, 1/4-watt resistor

R13   1K-ohm, 1/4-watt resistor

R14   1.5K-ohm, 1/4-watt resistor

R15   3.9-ohm, 1/4-watt resistor

C1   470 uF electrolytic capacitor, 35 volts

C2, C3, C4   0.047 uF disc capacitor, 35 volts

C5   0.05 uF capacitor, 35 volts

C6   10 uF electrolytic capacitor, 35 volts

C7   0.01 uF capacitor, 35 volts

C8   220 uF electrolytic capacitor, 35 volts

D1   bicolor LED

S1, S2   SPST toggle switch

S3   DPST toggle switch

J1, J2   1/8-inch mini phone jack

M1   electret microphone

SPK   8-ohm speaker

B1, B2   9-volt transistor radio batteries

U1, U2, U3, U4   LM741 op-amp

U5   LM386 power amplifier

Miscellaneous PC circuit board, IC sockets, battery clips, wire, hardware, etc.

Underwater Hydrophone

How would you like to listen to underwater sounds that you have never heard before? Not just the usual sounds that you might hear while swimming, but truly unusual underwater sounds made by creatures of the aquatic world. We will explore the use of the hydrophone and the sounds you can expect to hear in many places where water is found.

The *hydrophone* is an underwater listening device, microphone, or electroacoustic receiving transducer, designed specifically for continued use in fresh or salt water. It operates in water in much the same manner that an ordinary microphone operates in air. It converts audio sound waves in water into analog electrical signals, which are then amplified by your audio amplifier to a level where you can hear them. You can use this device to listen to amplified sounds, or you can tape record underwater sounds of all types.

Water Sounds You Can Hear

You can use your hydrophone almost anywhere you go—home, on vacation at the seashore, aboard ship, at the lake, almost anyplace. You will be able to listen to the sounds of the sea, lakes, ponds, streams, pools, creeks, and rivers. A number of extremely unusual and unknown sounds can be heard when listening underwater. You'll hear the clicks, squiggles, and musical sounds made by fish, shrimp, crabs, whales, porpoises, and other marine life talking back and forth. The hydrophone will pick up sounds made by the propellers or turning screws of passing ships, motor boats, and submarines. You will be able to hear the sounds of people diving and swimming in your swimming pool. Then when you listen to your home fish aquarium, you can try to identify the minute sounds your guppies make. You might even find yourself listening to the stagnant pool of water in your back yard to see if you can hear the sounds of mosquito larva.

Hydrophone Listener

The hydrophone listening system is composed of two parts, a hydrophone or microphone pre-amplifier assembly and an electronic amplifier assembly linked together by coaxial cable. Figures 1-6 and 1-7 illustrate how the hydrophone or sending unit is assembled. An electret microphone pre-amplifier electronics circuit is shown in Figure 1-8. It is mounted in a small soft plastic film canister. The electret microphone is used as the hydrophone microphone and can be purchased in your local Radio Shack store. Power is applied to the electret microphone by the bias resistor at R1. The audio output from the electret microphone is then coupled to capacitor C1. The output of C1 is fed to a network consisting of two 27k input resistors at R2 and R3, which act as a high-pass filter. This network couples the microphone audio to the op-amp at U1, a Texas Instruments TL072. A feed-back loop at the minus input of the op-amp consists of a 27K and a 1.5K-ohm resistor, along with a 10 uF capacitor. The capacitor at C2 reduces the DC gain to avoid excessive offset problems. Capacitor C3 and resistor R4 set the high-frequency rolloff at the output. Capacitor C4 at the output blocks DC; it forms a high-pass filter when

used in conjunction with the 10K-ohm potentiometer at R8. The electret microphone electronics can be mounted on a small circuit board. Remember to observe the correct polarity when installing capacitors and the op-amp, to ensure that the circuit will operate properly when power is applied.

The transducer or pre-amp board and the amplifier board are connected via a length of coax cable. Now, you will need to determine the length of cable you will need between the microphone transducer and the amplifier board that will best suit your particular application. You might wish to start with a 15- to 20-foot cable length between the transducer and amplifier boards. Next, you will need a small plastic film canister; drill a hold to pass the coax cable through the top of the film can. Be sure to drill the hole for the coax cable undersized, so it will be a tight fit. Pass the transducer end of the coax through the

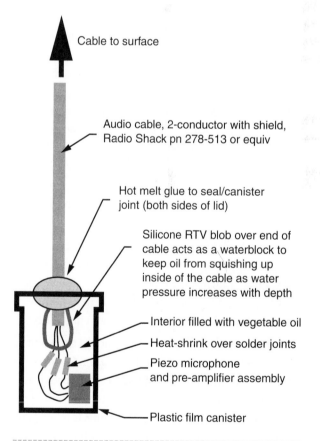

Cable to surface

Audio cable, 2-conductor with shield, Radio Shack pn 278-513 or equiv

Hot melt glue to seal/canister joint (both sides of lid)

Silicone RTV blob over end of cable acts as a waterblock to keep oil from squishing up inside of the cable as water pressure increases with depth

Interior filled with vegetable oil

Heat-shrink over solder joints

Piezo microphone and pre-amplifier assembly

Plastic film canister

Figure 1-7   *Assembling the hydrophone or sending unit*

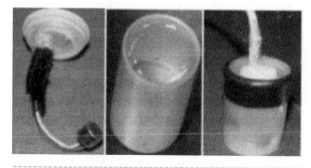

Figure 1-6   *Assembling the hydrophone or sending unit*

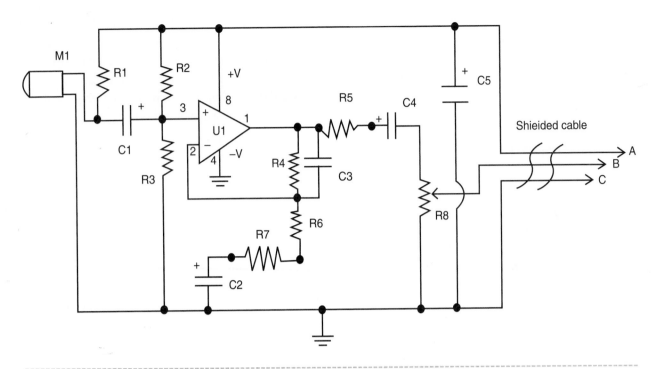

Figure 1-8 *Hydrophone pre-amplifier circuit*

top of the film canister (refer again to Figures 1-6 and 1-7). Next, solder the cable to the transducer or sender board, and then secure the electronics board in the film can so that it does not rattle around inside the can. Fill the canister with light mineral oil nearly to the top of the container. Locate some silicone sealant or *room-temperature vulcanizing* (RTV) sealant and run a bead all around the coax cable on both sides of the film can top cover. Place the top of the film can over the film can and secure it to the can. And finally, run a bead of silicone all around the top cover of the canister.

At the opposite end of the transducer cable, you will want to attach some type of connector such as a 1/8 -inch mini audio connector that will mate to a phone jack on the electronics amplifier chassis box.

The hydrophone main amplifier board shown in Figure 1-9 centers around the LM380, a 2.5-watt audio power amplifier. The shielded cable from the microphone pre-amp connects to a screw terminal strip. Actually, any three terminal connectors set may be used instead. Terminal A is the 12-volt bias to power the electret microphone. Terminal B is the actual audio output signal from the microphone pre-

amp to the main amplifier input. Terminal C is the system ground connection between the microphone pre-amp and the main amplifier. The audio output from the pre-amplifier board is fed to the main audio amplifier through capacitor C1. The output from C1 is immediately coupled to a 50K-ohm potentiometer, which is secondly coupled to the op-amp through capacitor C2, a 2.2 uF electrolytic capacitor. The output of the audio amplifier is conditioned by the resistive and capacitor network of R3 and C4. The 8-ohm speaker is coupled to the amplifier through capacitor C5.

Note, that the hydrophone is powered by the coax cable from the 12-volt battery through switch S1:a. The remote audio power amplifier is also powered by the 12-volt battery. Pins 3, 4, 5, 7, 10, 11, and 12 are all tied to ground on the circuit board. Construction of the main hydrophone amplifier is constructed on a small glass epoxy circuit board. You could also elect to construct the circuit on a perf-board if desired.

When assembling the main hydrophone amplifier, try and locate an integrated circuit socket for the LM380 amplifier. Be careful to observe the correct polarity of all of the capacitors during installation.

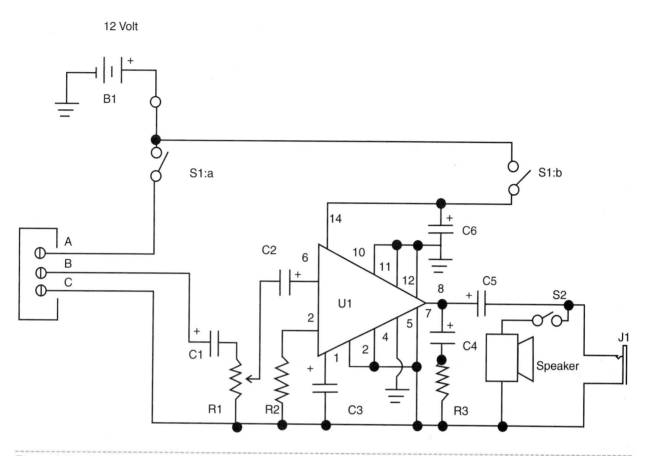

Figure 1-9 *Hydrophone amplifier circuit*

When installing the integrated circuit, you must be sure to insert the part correctly to avoid damage to the circuit. Integrated circuits are always marked with either a small indented circle to the right of pin 1 or with a rectangular cutout at the top center of the IC package. After installing all of the components on the circuit board, be sure to look for cold solder joints, shorts, and stray wire leads left on the board after trimming component leads.

The main hydrophone amplifier circuit is housed in a 5 × 7 × 2-inch aluminum chassis box. The two toggle switches S1 and S2, the volume control, and the input jack at J1 are all mounted on the front panel of the chassis box. The headphone jack at J2 is mounted on the rear panel of the enclosure. In order to supply power to the hydrophone circuits, it is advisable to use 2 four-cell AA plastic battery holders, which can be mounted to the top cover of the aluminum chassis box. The eight AA batteries will

provide a decent amount of current to power the circuit for a reasonable period of time.

The hydrophone may be used as a fish finder by mounting it on a pole or handle so you can hold it over the edge of a small boat or seawall. A piece of electrical conduit or a bamboo pole can be used. Flatten the conduit on the end to go into the water so you can mount and tape the unit to it. Be sure not to cover the flat surface of the unit with tape to assure maximum sensitivity to sound while in the water. Use a hose clamp, lacing twine, or electrical tape to attach the hydrophone to the pole. If you want to run your hydrophone down to greater depths, 50 to 300 feet or so, be sure to use shielded cable the whole length of the run to avoid picking up too much hum from any local power lines. Solder all electrical connections you make and apply sufficient silicone rubber and electrical tape to all connections to assure lasting waterproof connections and splices.

Operation

After you arrive at an area of interest, plug the hydrophone into the amplifier, turn the unit on, and adjust the volume to a point where you can hear a slight rap as you tap the surface of the hydrophone. This will tell you that the unit and hydrophone are operating properly before you attempt to use it. When you drop the unit into water, you will hear a very short and loud click as the unit contacts the surface of the water. You can also use this sound to tell if the unit is up to snuff; if you don't hear this loud click, you should begin to suspect that the battery voltage is dropping, that you have a loose connection to the hydrophone, or that the volume is not turned up high enough.

Swimming Pool Alert

You will hear this click each time the hydrophone contacts the surface of the water, no matter how many times it has been dipped in. This contact noise appears to be related to static electricity accumulation on the unit, which is rapidly discharged upon contact with the water surface. You can adapt this noise to a useful function as a *splash monitor* in swimming pools by suspending the hydrophone a few inches above the surface level of the pool. The amplifier output will be silent until the surface of the pool is disturbed by a swimmer or object falling into the pool. You will then hear a series of clicks as the water is splashed against the hydrophone. The amplifier can be placed in the house where it can be monitored by an adult. A long, shielded cable is then run to the hydrophone. The amplifier can be left on all the time as a pool safety device if the amplifier unit is powered by a 9-volt battery charger operated from 120 VAC. These chargers are available at any local audio supply house for a nominal fee. Commercial pool alarms such as this cost over $100.

Home Aquariums

The home aquarium is another interesting place to start your underwater listening. You will be surprised at the musical sounds, chirps, and beeps that your small fish make. With experience, you will be able to tell which species of fish makes the most unusual sounds, which ones don't say anything, when they are most active, and so on.

If you would like to try some experimenting with your favorite little guppy, place a radio near the aquarium so its loudspeaker is as close to the tank as possible. You want the sound from the speaker to vibrate the aquarium so the fish will be able to hear the sounds in the water. You might then move your audio amplifier into the other room while you listen to any sounds your fish make as a result of your playing music for them. With small fish as you have in your aquarium, you should be able to hear them when they are about 6 inches to 1 foot away from the hydrophone. As they get closer to the pickup unit, their sounds will get louder. You will be simply amazed at the sounds you will hear. And all along, you thought you had silent fish!

Lakes, Ponds, and Streams

The sounds you hear in the country at lakes, ponds, and streams will be different from what you might hear at home in your fish tank. A running stream will be bubbling and gurgling thereby masking some of the sounds of aquatic life. But listen anyway for this is the way you learn things. You may observe phenomena never before observed, and the only way you can tell is to listen and become an expert at what you do.

Lakes, ponds, and stagnant pools of water usually have some live activity that you should be able to hear. If you can get close enough to see some of the aquatic life from a boat or from the shore, you will be able to tell if crayfish, shrimp, or minnows make sounds, which you can begin to recognize as belonging to that species. Do the fish and crayfish make the same amount of noise when it rains, or don't they care? Do lobsters and oysters make sounds you can recognize? As you listen and become experienced, make audio tape recordings on a battery-operated cassette tape recorder, study the sounds you hear from the different fish, and you may become an expert in your own right. Your project will add yet another reason why you should go to the lake to relax over the weekend. It's all in the interest of science.

Seasides and Oceans

A pier at the seashore is a perfect place to drop your hydrophone into deep water to do your listening. At the seashore with the sound of breaking waves and the wind blowing, it may be difficult to hear the sounds of life with your amplifier. In this case, use a battery-powered tape recorder to record the sounds from your amplifier so you can listen to them later when it is not so noisy. A tape recorder amplifier will not have sufficient audio gain to record the fishy sounds directly so it will be necessary to feed your audio amplifier output into the tape recorder input.

To assure proper tone recording, use an impedance matching transformer. This circuit matches the low impedance of the loudspeaker output (4 to 8 ohms) to the high input impedance of the tape recorder (nominal 1,000 ohms).

An alternate recording method is to take an induction pickup coil and simply place it over the loudspeaker of the amplifier and then connect it to the input of a tape recorder input. This will eliminate any matching transformers, and audio sounds will be picked up as if you used a microphone for recording.

Listening from a Boat

A motorboat or cruiser is an ideal platform from which to listen. You can move to different locations, shut off the engine or motor, drop your hydrophone over the side, and listen in total quietness to the sounds of the sea. A small rowboat is also ideal for moving out away from shore so you can drift into different locations without making noises that would frighten off schools of fish. Listening from a large seagoing vessel will pose different problems because of the distance from your position down to the water and because of the speed of the ship. The hydrophone would be pulled along by the water and this would put a strain on the cable. Ship noises such as the engine and totaling screws might mask any sea noises you are hoping to hear.

Hydrophone Transducer and Pre-Amplifier Unit Parts List

Required Parts

R1 10K ohm, 1/4-watt resistor

R2, R3, R7 27K ohm, 1/4-watt resistor

R4 33K ohm, 1/4-watt resistor

R5 100 ohm, 1/4-watt resistor

R6 1.5K ohm, 1/4-watt resistor

R8 10K potentiometer (log-taper)

C1, C4 2.2 uF, 35-volt capacitor

C2 100 uF, 35-volt electrolytic capacitor

C3 3 pF, 35-volt disc capacitor

C5 50 uF, 35-volt electrolytic capacitor

U1 TL072 op-amp (Texas Instruments)

M1 electret microphone

Miscellaneous PC board, film can, RTV compound, wire, shielded wire, etc.

Hydrophone Power Amplifier Unit Parts List

Required Parts

R1   50K ohm potentiometer

R2   47K ohm, 1/4-wwatt resistor

R3   2.7K ohm, 1/4-watt resistor

C1, C2, C3   2.2 uF, 35-volt capacitor

C4   0.1 uF, 35-volt capacitor

C5, C6   470 uF, 35-volt electrolytic capacitor

U1   LM380 audio amplifier IC

J1   1/8-inch mini jacks

S1   DPST toggle switch

S2   SPST toggle switch

SPKR   8-ohm speaker

B1   12-volt battery

Miscellaneous PC circuit board, wire, battery holders, screw terminal strip

Ultrasonic Listener

Tune in to the fascinating world of ultrasound with this ultrasonic listener-receiver. You will be able to hear sounds that are too high in frequency to be heard by human ears, such as glass breaking and electric arcing (see Figure 1-10). This project will enable you to listen to a world of sound that few people even know exists. Numerous possible applications exist for the ultrasonic listener, from the detection of leaking gases and liquids, to the mechanical wear of bearings or rotational and reciprocating devices, to electrical leakage on power-line insulators. A whole world of sounds coming from living creatures is also audible. Simple events like a cat walking across wet grass, the rattling of key chains, and even a collapsing plastic bag can all be heard clearly. On a warm sum-

Figure 1-10   *Ultrasonic listener*

mer night, the sounds that can be heard are remarkable, as bats and small insects perform a cacophony of nature's own orchestra at its best. The handheld ultrasonic listener can easily detect and locate these high-frequency sounds.

The addition of a parabolic reflector further enhances the performance of this project. Because ultrasonic frequency spectrums are beyond our hearing range, they can be listened to only by indirect means, such as frequency heterodyning. *Frequency heterodyning* is a method widely used in modern radio receivers. In an ultrasonic receiver with frequency heterodyning, a *local oscillator* (LO) is used to generate a square wave. The output from the LO is from 20 to 100 KHz. The incoming signal, or *frequency input* (FI), is first picked up by an ultrasonic transducer and then amplified by a three-stage amplifier. The input signal is then mixed by the mixer section to produce a sum (LO + FI) of the frequencies and a difference of frequencies (LO − FI) at the mixer output. Because the sum of frequency (LO + FI) is too high to be heard, it is filtered out. The difference in frequency (LO − FI) is just within the audio range of frequencies; therefore after amplification, it can be heard from a loudspeaker or headphones. A system block diagram is shown in Figure 1-11.

A special *piezoelectric* ultrasonic transducer acts as a microphone and is used to detect the high-frequency ultrasound waves. Whenever sound pressure is applied to the transducer, it will produce a small voltage at its output terminal. The frequency response of this transducer peaks around 40 KHz but will work from 20 to 100 KHz. This weak signal from the transducer is passed to input pin 9 of the CD14069 *integrated circuit* (IC) at U1:a, a hex inverter IC. This digital IC operates in the linear mode by connecting feedback resistors R2, R3, and R5 from the inverters' outputs back to their inputs. The weak signal from the transducer goes through three stages of amplification at U1:d, U1:e, and U1:f. The signal is then rectified and coupled to the mixer by C7, as shown in Figure 1-12.

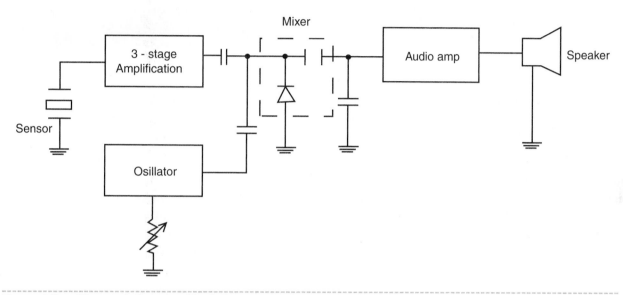

Figure 1-11   *Ultrasonic listening block diagram*

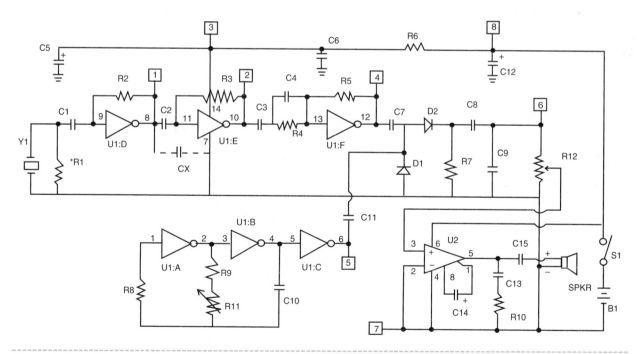

Figure 1-12   *Ultrasonic listener circuit*

Integrated circuits U1:a and U1:b are connected together to form an astable oscillator. Its frequency of oscillation is determined by R9, VR2, and C10. This frequency can be varied from 20 KHz to 100 KHz by variable resistor VR2. The square wave generated by U1:a and U1:b is buffered by U1:c and is coupled to the mixer circuit via C11.

In the mixer circuit, the amplified incoming signal (FI) from C7, and the signal of the local oscillator (LO) from C11 are mixed by diodes D1 and D2 and the sum of frequencies (LO + FI). They are filtered by C9, as shown in Figure 1-12. The difference of frequencies (LO − FI) are amplified by an audio amplifier at U2 and the LM386 power amplifier IC. The LM386 amplifier can provide up to 1 watt of output to an 8-ohm speaker. VR1 is a volume control, which is used to control the output of the sound level from the power amplifier.

Low-frequency instability is prevented by C5, R6, and C12 in the power supply section. High-frequency stability of U2 is enhanced by C13 and R10. The ultrasonic listener is powered from a standard 9-volt transistor radio battery. The ultrasonic listener is enhanced for listening to the high-frequency sounds made by bats.

The ultrasonic listener is constructed on a small glass epoxy circuit board, measuring $2\frac{1}{2} \times 1\frac{3}{4}$ inches. The compact circuit board houses all the components including the potentiometers, integrated circuits, capacitors, and diodes. The ultrasonic transducer is mounted off board: to the side of a small plastic enclosure after small sound input holes have been drilled into the plastic case (see Figure 1-13).

The original circuit was modified to aid the experimenter in listening to the high-frequency sounds of bats. These changes include eliminating resistor R1 and adding a capacitor at Cx between U1:d pin 8 and U1:c pin 7 as shown.

When installing the components on the printed circuit board, be careful to observe the polarity of the capacitors and especially the two diodes and the integrated circuits. If possible use integrated circuit sockets for the two ICs, in the event of trouble at a later date. The integrated circuits have a notch at the top of the package and usually to the left of the notch is

Figure 1-13   *Ultrasonic listener circuit board*

pin 1. Some integrated circuits have a small circle or dot placed near pin 1. Make sure you orient the IC correctly and double-check your wiring before applying power to the circuit.

A rear panel contains the on-off and volume, the tuning control potentiometer, and the headphone jack. The front of the unit contains the directional receiving transducer. The handle houses the batteries. The addition of an optional parabolic reflector greatly enhances the device's performance, providing super-high gain and directivity.

An excellent demonstration of this ultrasonic microphone involves Doppler shifts. *Doppler shift* is when an observer moving toward a source of sound experiences an increasing frequency. This is easy to visualize when you realize that sound propagates as a longitudinal wave at a relatively constant velocity. As the observer moves toward the direction of the sound source, he or she intercepts more waves in a shorter period of time, thus hearing a sound that seems to be shorter in wavelength or higher in frequency. A fun game for both children and adults is to hide a small test oscillator and have your opponent attempt to locate it in a minimal amount of time.

Applications

One of the most interesting sources of high-frequency sound is the many species of insects emitting their mating and warning calls. A whole new world of natural sound awaits the user, day and night. Many man-made devices also generate high-frequency sounds easily detected by the device. The following list represents only a small fraction of the potential sources of high-frequency sounds that you can explore with the ultrasonic listener:

- Leaking gases and rushing air

- Water from sprinklers or leaks

- High-voltage corona leakage, sparking devices, or light lightning

- Fires and chemical reactions

- Animals walking in wet grass or in the brush (an excellent aid for hunters or trackers)

- Pets moving in the darkness at night (good for finding lost pets)

- Chattering insects and bats

- High-frequency oscillators in computer monitors, TV sets

- Mechanical bearings

- Defects that are developing in automobiles (shakes, rattles, and squeaks)

Ultrasonic Listener Parts List

Required Parts

R1, R9   10K ohm, 1/4-watt resistor

R2, R3, R5   1 megohm, 1/4-watt resistor

R4   100K ohm, 1/4-watt resistor

R6   470 ohm, 1/4-watt resistor

R7, R8   470K ohm, 1/4-watt resistor

R10   10 ohm, 1/4-watt resistor

VR1   10K ohm potentiometer

VR2   200K ohm potentiometer

D1, D2   1N4148 silicon diode

C1, C2, C7   0.01 uF, 25-volt ceramic capacitor

C4, C11   20 pF, 25-volt ceramic capacitor

C5   200 uF, 15-volt electrolytic capacitor

C6, C13   0.04 uF, 25-volt ceramic capacitor

C3, C8, C9   0.022 uF, 25-volt ceramic capacitor

C10   100 pF, 25-volt ceramic capacitor

C12   100 uF, 16-volt electrolytic capacitor

C14   10 uF, 25-volt electrolytic capacitor

C15   47 uF, 16-volt electrolytic capacitor

Cx   100 pF, 25-volt capacitor

U1   CD4069 IC

U2   LM386 audio amplifier IC

SPKR   8-ohm speaker

BT   9-volt transistor radio battery

Y1   ultrasonic transducer

Miscellaneous   PC board, wire, sockets, hardware, coax cable, connectors, battery clip, plastic enclosure, etc.

Infrasonics

Most everybody has heard about ultrasound, sounds that are too high pitched to hear unless you're a dog or a bat. *Infrasound*, on the other hand, is sound that is too low to hear. Human ears can register sound down to about 20 Hz, the deepest bass note. Infrasonics are generally the domain of scientists and researchers who are interested in sounds 10 Hz and below, all the way down to 0.001 Hz. In fact, this frequency range is the same one that seismographs use for monitoring earthquakes. Some peculiar things go on down there. Earthquakes, as you might expect, shake the air at the same time they shake the ground, creating seismic infrasound that can travel far from the epicenter. Volcanoes make some pretty impressive infrasound.

Like waves from a rock thrown in a lake, infrasonic signals move around the planet in concentric circles. They are big, slow, and long lasting. The infrasound wave from the nuclear explosion from a test in China can take more than 6 hours to get to Alaska. The wave registered again 37 hours later as it completed another lap around Earth.

Big storms at sea generate waves in the air, called *microbaroms*, from the water waves beneath. Microbaroms are particularly troubling to the Department of Defense and they befuddle electronic equipment because their frequency is close to that of nuclear explosions. Microbaroms are so prevalent in winter that scientists have been able to track the movement of a storm in the Gulf of Alaska. The atmosphere itself makes infrasound as bodies of air move over mountain ranges. This is like the sound you make by blowing over your teeth, only about 10,000 times lower in pitch. Tornadoes and turbulence in the upper air produce infrasound. And so might *sprites*, those mysterious upside-down lightning strikes high above large thunderstorms. Other infrasound comes from space. Auroras make sounds in the 0.1 to 0.01 Hz range that can travel 1,000 kilometers. Meteors make infrasonic booms, too, and the listening stations hear them.

Researchers have compelling practical reasons for studying infrasonics. The U.S. Air Force has used infrasound monitors to detect other nations' above-ground nuclear blasts as well as rocket launches and supersonic jets. With the recently signed Comprehensive Test Ban Treaty, a new worldwide network of infrasound stations is being built.

Infrasonic Microbarograph

The world's greatest ocean is the atmosphere, which encompasses all human life and contains extraordinarily powerful waves. Like their oceangoing counterparts, atmospheric waves are normally generated by energetic storms. But they can also arise and spread, like ripples on a quiet pond, when a meteor or volcanic explosion violently shocks the air. Yet even the largest atmospheric tsunamis are quite difficult to detect. The pressure excursions that betray their passage are typically just a few millibars (thousandths of one atmosphere), and these tiny undulations often take tens of minutes, and sometimes even hours, to pass.

Instruments that can monitor such subtle signals are called *microbarographs* and can cost thousands of dollars. Any amateur scientist or experimenter can now build a microbarograph and observe these ephemeral infrasonic waves for about $50.

This clever microbarograph project was designed by Paul Neher, a gifted amateur scientist. The key to this microbarograph is the manometer (in this case, a U-shaped tube and sensor), which is used to balance the barometric pressure against the air pressure trapped inside an aluminum bottle that is kept a bit warmer than the surrounding air (see Figure 1-14). Because the pressure of an isolated gas varies with temperature, any shift in external pressure can be matched by changing the temperature of the air inside the bottle. Neher's instrument senses external pressure fluctuations by monitoring the height of a liquid in the manometer. A rise in external pressure presses the liquid down into the manometer and triggers a heating coil that warms the bottle. If the pressure drops, the level of the liquid rises, and the heater is kept off, thus allowing the bottle to cool a little to compensate. Monitoring the temperature changes with the circuit shown in Figure 1-14 reveals minuscule shifts in air pressure. The aluminum bottle or container is made from a completely drained, 32-ounce (about 1 liter) tire inflator bottle that has had its crimped-on cap replaced with a single-hole rubber stopper. The manometer is fashioned from a piece of

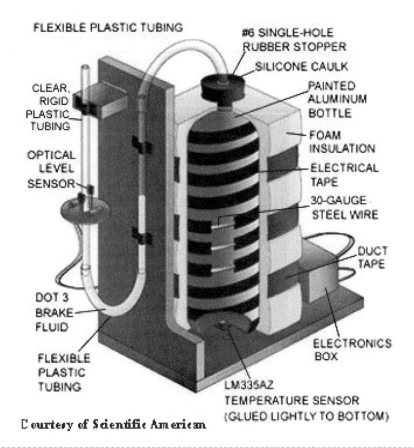

FLEXIBLE PLASTIC TUBING

#6 SINGLE-HOLE RUBBER STOPPER

SILICONE CAULK

CLEAR, RIGID PLASTIC TUBING

PAINTED ALUMINUM BOTTLE

OPTICAL LEVEL SENSOR

FOAM INSULATION

ELECTRICAL TAPE

30-GAUGE STEEL WIRE

DUCT TAPE

DOT 3 BRAKE FLUID

FLEXIBLE PLASTIC TUBING

ELECTRONICS BOX

LM335AZ TEMPERATURE SENSOR (GLUED LIGHTLY TO BOTTOM)

Courtesy of Scientific American

Figure 1-14   *Microbarograph with manometer*

glass tubing, which is bent after heating with a propane torch. But you can link two clear, rigid, plastic tubes with a short, flexible plastic hose. You'll need to fill this assembly about one-third full with a liquid that has a low viscosity and does not evaporate. Try to locate some DOT 3 clear brake fluid, which works quite well for the manometer.

The instrument senses changes in the fluid level by using the transparent liquid to focus the light from an infrared *light-emitting diode* (LED) onto a photo-transistor. When the liquid drops below the set point, the defocused light becomes too diffuse to detect. This change causes the circuit attached to the photo-transistor to send an electric current through the heater. Neher employed "beading wire," obtained from a craft store, as the heating filament.

Hobbyists often use this beading wire, a 30-gauge ($1/4$-millimeter) steel wire, to make necklaces. It has a resistance of about 1 ohm per foot (about 3 ohms per meter), which is ideal for this application (see Figure

1-15). After electrically insulating the bottle with a layer of enamel paint, wrap 10 evenly spaced turns along the length of the bottle and secure them with electrical tape. Surround the wrapped bottle with a few inches (about 10 centimeters) of an insulating material such as foam rubber or a spray insulation (which you can purchase at a hardware store). When operating, the circuit heats the bottle slightly every 10 seconds or so, replacing the tiny amount of heat that leaches through the insulation, thereby keeping the level of the fluid stable. The LM335AZ chip is a sensitive solid-state thermometer that varies its output voltage by 10 millivolts for each degree Celsius change in temperature. This marvel can measure temperatures to about 0.01°C, which corresponds to an ultimate pressure resolution approaching 20 micro-bars. That's a scant 20 millionths of one atmosphere.

To begin your observations, disconnect the tube that links the manometer to the aluminum bottle, then heat the bottle about 10°C (18°F) above room temperature by blocking the light from reaching the

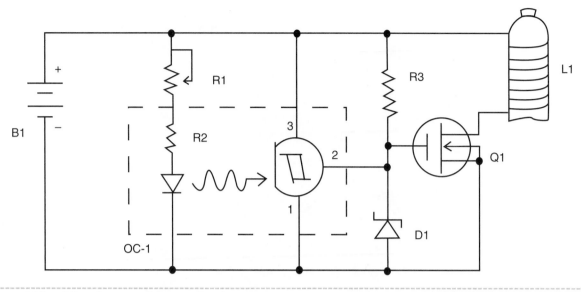

Figure 1-15   *Microbarograph heater circuit*

phototransistor. Reconnect the tube and allow the circuit to stabilize the bottle at this elevated temperature. It's easy to know if your instrument is working: just lift it. It should register the 100-microbar drop in pressure that results when you raise it about 1 yard (1 meter).

You will want to calibrate the microbarograph over a larger range than you can easily generate by shifting its height. The solution is quite simple: move the LED-phototransistor pair up or down a bit on the manometer column. The circuit will then adjust the temperature inside the bottle to raise or lower the liquid between the LED and phototransistor to match. This manipulation causes the fluid level to become uneven, with the weight of the unbalanced liquid being supported by the pressure difference between the atmosphere and the air inside the bottle. The specific gravity of DOT 3 brake fluid, which is widely available in the U.S., is 1.05. For this value, each inch (centimeter) difference in the level of the liquid corresponds to a pressure difference of 2.62 millibars (1.03 millibars). This fact allows you to set the device to a number of known pressure differences while measuring the corresponding output voltages. To make the LED phototransistor pair easy

to move, mount the assembly on a small piece of perforated circuit board with a hole in the middle that is large enough to let the manometer tube slip through; a small cable from the phototransistor assembly is wired to the amplifier circuit shown in Figure 1-16.

You can read the output of the microbarograph circuit with a digital voltmeter or record the data continuously using your computer and an analog-to-digital converter. These versatile devices, once too pricey for amateur budgets, are now quite affordable. Both Macintosh and PC aficionados should check out the Serial Box Interface, which is available for $99 from Vernier Software. PC users might also consider buying a similar unit for $100 from Radio Shack (part number 11910486). Another alternative is to check with DATAQ Instruments for their DI-194RS $24 chart recorder hardware/software combination at their Web site www.dataq.com. Any of these combinations will turn your home computer into a sophisticated data-collection station. This sensitive microbarograph will allow you to detect the infrasonic waves created by atmospheric storms that constantly blow through our lives daily.

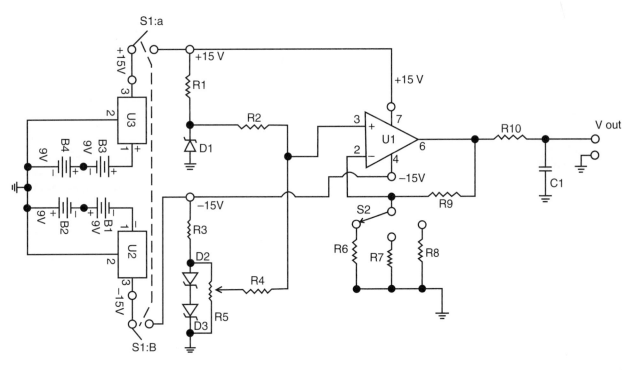

Figure 1-16   *Microbaroraph output circuit*

Microbarograph Heater Circuit Parts List

Required Parts

R1   1K potentiometer

R2   750-ohm resistor

R3   39K ohm resistor

D1   12-volt Zener diode

Q1   IRF-520 Power MOSFET

OC-1 H23LOI optocoupler

L1   heater coil (see text)

B1   12-volt auto battery or 12-volt AC adapter

Microbarograph Output Circuit Parts List

Required Parts

R1, R3 10K ohm, 1/4-watt, 5% resistor

R2, R4, R9 1 megohm, 1/4-watt, 1% resistor

R5   10K potentiometer (10 turn)

R6   100K ohm, 1/4-watt, 1% resistor

R7   10K ohm, 1/4-watt, 1% resistor

R8   1K ohm, 1/4-watt, 1% resistor

R10   15K ohm, 1/4-watt, 5% resistor

C1   10 uF, 25-volt electrolytic capacitor

D1   LM335AZ temperature sensor

D2, D3 LM336Z voltage reference

U1   OP-07CN precision op-amp

U2   LM7915 15-volt minus voltage regulator

U3   LM7815 15-volt plus voltage regulator

B1, B2, B3, B4 9-volt batteries (or replace batteries with power supply)

J1   2 RCA output jacks

S1   DPDT toggle switch (power)

S2   three-position rotary switch (gain)

Miscellaneous wire, terminals, chassis box

Additional Microbarograph Parts List

Required Parts

aluminum tire inflator bottle, painted with enamel paint

#6 single-hole rubber stopper

30-gallon wire (beading wire), 1 ohm/per foot resistance

2 wood boards used for component mounting

wood block to hold plastic tubing (manometer assembly)

2 pieces of flexible plastic tubing

2 pieces of clear rigid plastic tubing

foam insulation

electrical tape

silicone calk

duct tape

small circuit board used to house optical sensor (OC-1 assembly)

DOT 3 brake fluid (see text)

Chapter Two

Light Detection and Measurement

Early humans knew about the *light* form we now call *electromagnetic energy* many tens of thousands of years before they knew about the *radio* form of electromagnetic energy. Although our eyes are sensitive to light, we have no way to detect radio energy; we need a receiver. Eons ago, people could see the stars and make conjectures as to what they were. On the other hand, radio, as we know it, traces its infancy only to the late 1800s.

Light is defined as those wavelengths of electromagnetic energy that are visible to the human eye. The response of the human eye thus defines the frequency limits for light, as the human ear does for sound. Light covers a very narrow band of frequencies, as we can see from Figure 2-1.

When we look at light, or "see light" and "look at a scene," we are seeing or observing light in the frequency domain, that is, in different colors. Although the eye has the unique capability of seeing the differ-ence between about ten million shades of color, light has a rather narrow spread of wavelengths. This portion that is visible by humans is packed into 1.6 percent of the entire electromagnetic spectrum. This frequency density is comparable to crowding all the world's human-made radio frequencies into a narrow frequency range from 550 KHz to 880 KHz in the standard AM radio broadcast band. The eye is indeed an amazing electromagnetic receiver. Consider that if you glance at a yellow dress for just one second, the electrons in the retinas of your eyes vibrate about 5×10^{15} times during the interval. Next consider that if you were to count all the waves that beat upon all the shores on Earth, you would have to count for ten million years in order to reach the same number of oscillations of yellow light in 1 second.

The frequency units commonly used for describing audio waves would be astronomically large if they were used to describe light waves, so a different set of

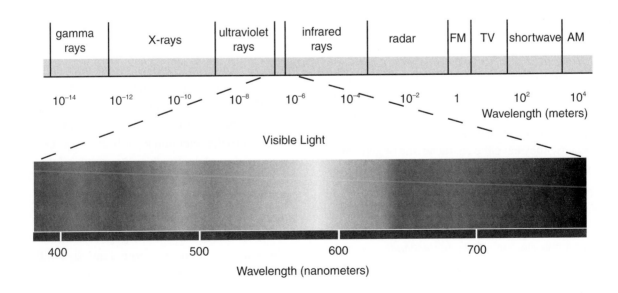

Figure 2-1 *Visible light wavelengths*

units has been defined. If we define the limits of the wavelengths of light as that where the sensitivity of the eye has dropped to 1 percent of that at its maximum sensitivity, then light covers the wavelengths from 430 to 690 nanometers (nm). And a nanometer is equal to 10^{-9} meter.

In this chapter, we will take a closer look at light sensors such as photocells and solar cells and how you can use them to detect light or the absence of light, how you can measure the solar constant, as well as how you can measure *ultraviolet* (UV) light and detect ozone in the atmosphere. We will also utilize light sensors for optical listening—that is, we will listen to the sound that light makes. We can do this by listening with the ear in the amplitude domain of light rather than by looking through the frequency domain where light is received by the eye. After building the opto-listener, you will be able to listen to electronic displays, to "singing" automotive headlamps, to burning flames, and to lightning. You will be able to listen to just about any light source to "see" what it sounds like. In addition, you will learn how to measure the speed of objects using light by constructing your own optical tachometer. Finally, we will look at how you can detect and measure pollution in water using the optical turbidity meter.

Light Detection Devices

A number of names are used to describe the *photoelectric cell*, popularly known as the *solar cell*. Some of these names are *photocell*, *electric eye*, and *photoelement*. However, the device is more properly called a *photovoltaic cell*. Two types of photovoltaic cells are available: the selenium photocell and the silicon photocell. Photovoltaic cells are different from other photosensitive (light sensitive) or photoelectric cells because they have the characteristic of generating a voltage when light strikes their sensitive surface. This self-generated voltage (up to 0.58 volts) will cause a current to flow into an externally connected circuit such as a motor or a battery.

Other photoelectric cells are *photoconductive* and *photoemissive* cells. A photoconductive cell is a semi-conductor device whose electrical resistance, or conductivity, is a function of the amount of light striking its surface. The photoconductive cell is also called a photoresistor. The photoemissive cell has a high impedance, and it readily couples into high-impedance circuit.

The selenium solar cell is a very sensitive and reliable solid-state device that can readily be used in circuit designs requiring detection of electromagnetic energy in the infrared, visible, and UV frequency ranges. The selenium solar cell can be purchased from any neighborhood radio supply store for less than $2. With it, you can easily perform all the experiments described in this chapter. The silicon solar cell will also function extremely well in these tests and demonstrations.

The solar cell is very simple in construction and fairly simple in operation. It is a fairly rugged device and consists of a metal base plate on which multiple coats of selenium compounds and precious metals are deposited. The metal base plate can be steel or aluminum. The selenium layer on the base plate is covered with a barrier layer and a transparent front electrode. The selenium layer and barrier layer are very thin, being of molecular thickness. The selenium layer is the thicker of the two, being only 0.002 to 0.003 inches thick. The overall cell is coated with a thermosetting protective resin, which provides a rugged, shatterproof case that is immune to shock and vibration. Like most solid-state devices that are rugged, a solar cell can survive a fall to the sidewalk and still perform. Such a fall would quickly dispatch a vacuum tube of years past.

When light falls on the cell, it penetrates the transparent front electrode and causes the selenium layer to release electrons, which travel across the barrier layer. The electrons are trapped, or collected, on the front electrode to form a negative charge. They cannot return to the selenium layer because of the unilateral (one-way) conductivity properties of the barrier layer. The collector strip, or ring, which collects the electrons, becomes the negative terminal of the cell and the metal base plate becomes the positive terminal of the cell. The negative lead, connected to the white strip that runs around the edge of the cell, is black, and the positive lead from the base plate is red. The solar cell is the voltage source, and the load is often a battery, which is to be charged.

A silicon solar cell, one of the most common types of photovoltaic cells, is a device that uses the photoelectric effect to generate electricity from light, thus generating solar power, or energy.

The main component of a solar cell is silicon "doped" with trace amounts of impurities. In pure silicon, each atom is fixed in a crystal lattice and bonded to other silicon atoms covalently, sharing the four valence electrons in their outer shells with them. Therefore few free electrons or positive charge carriers are available to carry a charge, and pure silicon is thus a bad conductor.

In doped silicon, atoms with three or five valence electrons are introduced to the lattice. Arsenic or phosphorus, for example, have five valence electrons. Because silicon atoms require only four of those electrons to form stable bonds, there will be one free electron that can move and thus carry charge. And because so many free electrons in silicon are doped with arsenic or phosphorus (compared to pure silicon), this sort of silicon is called *n-type silicon*.

If the silicon is doped with boron, which has three valence electrons, it will be short one electron when it bonds with silicon. This "hole" is also free to move. As so many positively-charged holes in silicon are doped with boron, this sort of silicon is called *p-type silicon*.

In a solar cell, a plate of p-type silicon is placed next to a plate of n-type silicon. At the junction between the two, electrons in the n-type plate will migrate to the p-type plate, and vice-versa for the holes in the p-type plate. After a while, enough holes and electrons would have combined to form a barrier at the junction, preventing further flow of holes and electrons.

An electric field has now formed across the p–n junction. The barrier is positive on the n-type silicon plate because the electrons have crossed over and created an excess of protons, which do not have corresponding electrons, and the barrier is negative on the p-type silicon plate, because the reverse occurs.

The electric field and the barrier act as a diode; electrons can move from the n-type plate to the p-type plate easily due to the electric field, but the reverse is difficult.

The simplest type of solar cell is a silicon diode, but research is continuing into more exotic materials with greater efficiencies. Modern solar cells are encapsulated in glass-fronted plastic sheets. They are designed with lifetimes that exceed 40 years. Sunlight provides about 1 kilowatt per square meter at the earth's surface, and most solar cells are between 8 and 12 percent efficient. In desert areas, they can operate for an average of 6 hours per day when mounted in nonrotating brackets.

Solar panels come in four varieties. Most common are rigid monocrystalline and polycrystalline silicon sheets. *Monocrystalline* silicon provides the highest efficiency of all four types but is also the most expensive. *Polycrystalline* silicon is less expensive but lower in efficiency. Also available are *amorphous* silicon solar cells, which can be applied to a variety of substrates including flexible ones, like metal foil or plastic foil. The main advantage of amorphous silicon solar cells is that they should eventually be able to be manufactured at a much lower cost than crystalline solar cells. *Nanocrystalline* silicon has also been used in the same processing systems as for amorphous silicon, and it shows slightly higher efficiency due to the increased absorption in the longer wavelengths. Experimental nonsilicon solar panels are made of *carbon nanotubes* or *quantum dots* embedded in a special plastic. These have only one-tenth the efficiency of silicon panels but could be manufactured in ordinary factories, not clean rooms, which should lower the cost. Solar cells can be used to power many kinds of equipment, including satellites, calculators, remote radiotelephones, and advertising signs. Most often, many cells are linked together to form a solar panel with increased voltage and/or current. Solar cells produce *direct current* (DC), which can be used directly, stored in a battery, or converted from DC to AC to directly power common household devices or fed into the utility grid. This DC to AC conversion is done by means of an inverter.

Historical Review

The first photoelectric effect was noticed by E. Bequerel, a French scientist, in 1838. Then, in 1873, K. Smith discovered the change in electrical resistance of selenium when it was exposed to light: the *photoconductive effect*. H. Hertz observed the photoemissive effect during his experiments with electromagnetic waves in the 1880s. Early phototubes were built with alkali metal cathodes and were a result of Hertz's work. In 1880 Alexander Graham Bell was the first to communicate by means of a beam of light when he sent his voice over a sunbeam, using homemade selenium cells as detectors.

The man credited with the invention of the solar cell—Anthony H. Lamb in 1931—was threatened with dismissal from work if he persisted in promoting it. So he turned it into the well-known Weston light meter. A pioneer in night flying and guided missiles, Lamb, at the age of 72, took out his 200th patent! The invention of the solar cell may well be one of the more important inventions of the last 100 years because of its capability to extract light energy from the sun and convert it directly to electricity. This is of utmost importance at the present time as we seek to develop alternate energy sources to supplant oil products.

The last photoelectric device to be developed was the silicon solar cell in 1914. This work was done by Bell Telephone Laboratories. Bell researchers are also working on the development of a liquid-junction photovoltaic solar cell, which may prove to be less expensive and easier to produce than an all solid-state cell. In a liquid cell, electricity is formed at the junction between a solid electrode and a water-based solution. In an all-solid solar cell, production must ensure the alignment of the different crystal layers for the cells to operate properly. However, because the liquid conforms easily to the electrodes, the alignment cost is avoided with liquid-junction cells. Present research is being conducted with one electrode made of a semiconductor and another made of carbon or a range of common metals. Both electrodes are then immersed in a solution of polysulfides and water. When light strikes the semiconductor electrode, current flows from one electrode to the other

through the solution. One of the coinventors of the transistor, Walter H. Brattain, published a description of a semiconductor–liquid junction in 1955, following the development of the silicon solar cell.

Listening to Light—Using an Opto-Listener

Through the use of the solar cell and the opto-listener, we are able to observe light in the amplitude domain. We will build and experiment with a sensitive light detector combined with a high-gain audio amplifier called an opto-listener.

The opto-listener can be used to listen to electronic displays such as "singing" automotive headlamps. Point your opto-listener at an IR remote control or at a camera electronic flash and you can "hear" what these lights sources sound like. You can listen to just about any light source to "see" what it sounds like. Let's get started.

The opto-listener is a high-gain audio amplifier circuit, which can be used to investigate pulsating, flickering, or modulating light. The circuit is easy to build and is shown in Figure 2-2. The opto-listener circuit centers on its light sensor. The light sensor for this circuit is a photovoltaic solar cell at D1. The signal from the sensor is coupled to the first high-gain amplifier at U1, via capacitor C1. The op-amp at U1 is an LM741, but it can be replaced with another newer low-noise op-amp if desired. A gain resistor is connected across the input and output pins of the op-amp at R1. A coupling capacitor at C2 is used to connect the output of the first amplifier to the audio amplifier at U2. The integrated circuit at U2 is an LM386 audio amplifier. The potentiometer at R2 is used as audio gain or volume control for the circuit, at the input of U2. The output of the audio amplifier at pin 5 is coupled to the 8-ohm speaker via capacitor C4, a 100 uF capacitor. The opto-listener can be powered from a single 9-volt transistor radio battery so that the circuit can be made very portable.

The opto-listener can be fabricated on a perfboard or on a printed circuit board. Because the circuit is in the audio frequency range, the circuit wiring

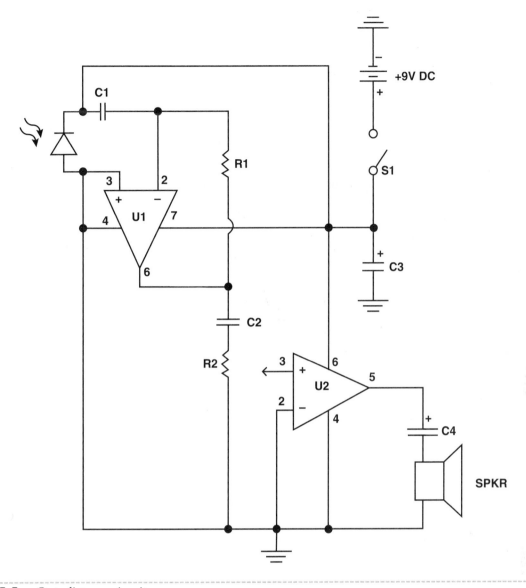

Figure 2-2   *Opto-listener circuit*

is not very critical. However wire leads between components should be kept short if possible. You can build this circuit in less than an hour, and it will provide many hours of interesting investigation into the mysteries of light and sound.

As you build the opto-listener be sure to take extra time to observe the correct polarity of the capacitors and the solar cell if used. The use of integrated circuit sockets is highly recommended in the event that one of the op-amp fails at a later date. Note that the integrated circuits will have a locator or positioning mark to denote the number one pin, which is usually at the top left of the IC package.

Generally a small square cutout can be found at the top of the IC or a small indented circle can be found near pin one of the IC; these markings help you to locate which pin is number one. Connect up a 9-volt battery power clip's positive lead in series with one of the power switch leads and connect the remaining switch lead to the plus terminal of the circuit at C3. Once the circuit has been built, you will need to double-check your wiring before applying power to the circuit to avoid burning up the circuit upon power-up. Inspect the circuit board for stray cut component leads and bare wires that may have adhered to the circuit board during construction.

Locate a small metal chassis box to house the opto-listener circuit. A $4 \times 5 \times 1\frac{1}{2}$ inch box is used for the prototype. You can elect to modify the basic circuit by adding a headphone jack and switching out the speaker if you desire. You will need to decide if you wish to have the sensor mounted directly on top of the chassis box or if you want to have the light sensor as a probe that can be remote from the chassis box. If you elect the latter light probe approach, you might want to consider using a mini jack at the input of the circuit to allow a remote sensor probe. A length of small RG-74U coax could be used from the solar cell probe back to the opto-listener circuit. The power switch and headphone jack are mounted on the front side of the chassis box, and the circuit board is mounted inside the chassis box. Once the wiring hand mounting has been completed, you can apply power to the circuit and begin your investigations into listening to light. You can further experiment by placing a magnifying lens in front of the solar cell or phototransistor to increase the opto-listener's detection range.

The human eye has a persistence of vision of about .02 seconds. Therefore a light that flashes on and off more than about 50 Hz appears to be continuously on. The human ear is much faster and responds to sound with a frequency from about 20 to 20,000 Hz. The opto-listener can transform the pulsating and flickering light that the eye cannot discern into sounds that the ear can easily hear.

Opto-Listener Parts List

Required Parts

R1   1 megohm 1/4 watt resistor

R2   10k potentiometer (chassis mount)

C1, C2, C3   0.1 uF, 35-volt disc capacitor

C4   100 uF, 35-volt electrolytic capacitor

Q1   FPT-100 phototransistor or Radio Shack (276-130)

D1   silicon solar cell

U1   LM741 op-amp

U2   LM386 audio amplifier IC

S1   SPST toggle switch

SPK   8-ohm minispeaker

Miscellaneous   PC board, chassis box, wire, jacks, sockets, etc.

Once your opto-listener has been completed we can begin exploring the "sounds" of light!

Listening to Incandescent Lamps

Plug the solar cell probe into the opto-listener and turn up the audio gain until a strong hum is heard. With the lights on at night, the hum is coming from any nearby lamp. The frequency you will hear is 120 Hz, or twice the 60 Hz power-line frequency. You should not have to turn the gain of the opto-listener up very high, as the solar cell will put out up to 0.5 volt under strong sunlight or under a bright lamp. When you turn off all the lamps, the opto-listener should become quiet as there is no longer any light striking the solar cell. With the lamp on, the 120 Hz hum should decrease considerably when you cover up the solar cell with your hand.

Listening to Fluorescent Lamps

A fluorescent lamp will also cause a 120 Hz hum in the opto-listener from the 60 Hz power line, but it will not be as soft as from the incandescent lamp. Note that both lamps put out light on either half cycle of the 60 Hz supply voltage, producing the 120 Hz hum you hear. Because the incandescent lamp is hot and cannot completely turn off (cool off between each cycle of the input voltage), its light output is more constant than the fluorescent lamp, which completely flashes on and off each half cycle.

Listening to Television Set Cathode-Ray Tubes

Your TV screen is a good subject for listening to changing light conditions that you can't see. The sound you will hear primarily is a 30 Hz buzz, as this is the U.S. picture frequency (or frame rate). The field frequency is 60 Hz, and it takes two fields to make up a picture (frame). Therefore, the horizontal line frequency is 15.75 KHz (525 lines × 30 pictures). The sound at 15.75 KHz is not as strong as the sound at 30 Hz, as the response of the opto-listener is down considerably. In addition, our hearing ability is lowered considerably at 15.75 KHz. As you place the solar cell over or near different portions of the TV screen, you will hear the buzz change slightly as the picture scene changes. Numbers and letters cause a particularly strong signal buzz.

Listening to Burning Objects

There are some interesting sounds to be heard from burning objects. We will discuss matches, sparks, and candles. Listening to the sound that the light of a burning match makes is rather impressive. Darken the room as much as possible and strike the match 4 to 6 inches away from the solar cell. You will hear a rapid, popping signal, which crackles a bit and then becomes quiet except for a slight hiss, which can be heard in the background. The popping sound is the rapid burning of the match-striking material before the wood or paper of the match starts burning. Once a steady light is output from the match, the main sound produced is the hiss of the light. The lighting sound lasts for 1 to 2 seconds, but the hiss from the light energy lasts for as long as the match emits light.

A lighted candle is another interesting subject to study with the opto-listener. Place the cell about one foot away from the candle with the audio gain set to medium on the opto-listener. After the candle is lit, it will cause a steady hiss due to the steady light output from the candle. Because the hiss (or white noise) from the light is the same as amplifier hiss, it is diffi-

cult to notice the difference. But when you cover the solar cell with your hand, the hiss will decrease or disappear entirely because it is due to the candle.

When you blow softly on the candle or clap your hands to cause an air wave, the rapidly moving flame will cause a rustle of sound heard through the opto-listener. The air movement is, in effect, modulating the candlelight beam and is detected by the solar cell. This is a form of *wind modulation* or *air-to-light transmissions*!

Listening to Lightning

When a thunderstorm with a fair amount of lightening is in the local area, your solar cell will pick up many lightning flashes, even when the cell is not aimed in the direction of the flashes. This is because the lightning illuminates the whole area, including the background, ground, sidewalks, and so on. Because lighting flashes last for only a hundred microseconds, the flashes you hear will sound like sharp clicks. At night in a place where there are no man-made lights, the effect will be most vivid as this is when the background is totally dark except for the flashes of light.

As the storm center moves off and local flashes decrease in intensity and frequency, you will still notice occasional clicks and yet your eye may not observe any lightning. This is because lightning, invisible to the eye, can be observed by the solar cell. Clicks can be heard even when the storm is 5 to 10 miles away. These clicks can be heard even when lightning is behind clouds and cannot be seen visually.

Listening to Moisture

A selenium solar cell can be used as a water moisture detector when it is connected directly across the input of a high-gain audio amplifier. The solar cell will produce broadband white noise at the output of the amplifier when the cell becomes wet with moisture. The hissing noise tapers off gradually as the moisture evaporates from the selenium surface. The noise decay rate will taper off in less than 10 seconds when the humidity is low. The decay rate will of course be slower when the humidity is high.

The wet solar cell produces most noise when it is in darkness. If you are operating in light, part of the noise output will be due to the light generating a voltage output from the cell, which will mask the noise that is due to the presence of moisture. Individual water drops hitting the solar cell can also be heard with the audio amplifier. The area that produces the greatest amount of noise is along the negative voltage pickup lead of the solar cell surface.

Measuring the Solar Constant—Using a Radiometer

The atmosphere is very transparent to certain wavelengths of sunlight. If you measure the intensity of the sunlight at one of these wavelengths over 12 hours, you can determine the solar constant. The *solar constant* is the intensity of sunlight at the top of the atmosphere.

The intensity of the light from the sun varies only slightly over the course of a *sunspot cycle*. An entire sunspot cycle is 22 years, with a number of sunspots reaching a maximum every 11 years. You can measure the approximate intensity of sunlight outside the atmosphere by using a light meter known as a radiometer to make a series of measurements over 12 hours.

Though this method is simple, it can yield surprisingly good results. This same method is used by both amateur and professional astronomers to measure the intensity of light from planets and stars. It is also used to calibrate instruments that measure amounts of ozone, water vapor, and oxygen present in a vertical column through the atmosphere.

The basic principle of this measurement technique is disarmingly simple. After you make a series of measurements of the intensity of sunlight over half a day, you then plot the results on a graph. One axis of the graph represents the logarithm of the sunlight's intensity; the other represents the thickness of the atmosphere through which the measurements were made. If the atmosphere remained stable over the period, the points on the graph should fall along a straight line. Extending this line to an atmospheric thickness of zero gives the intensity of the sunlight outside the atmosphere (see Figure 2-3).

For best results, the detector in your radiometer should have a linear response to the changing intensities of light. Also, the detector should be fitted with a filter that transmits a narrow band of sunlight. These requirements can be met by many different kinds of radiometers that you can construct.

Bouguers's Law

The principle behind the method we will use to measure the sun's radiation outside the atmosphere is known as the *exponential law of absorption*, or Bouguer's law. It was first defined in 1729 by Pierre Bouguer, a French professor, who explained how successive layers of a uniform light-absorbing medium absorb an equal fraction of light passing through the entire medium.

Assume you have a red slab of gelatin desert in front of you and a light source is over your head. If

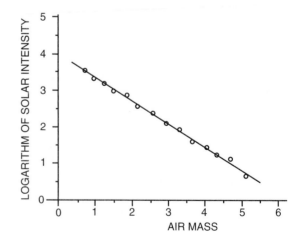

Figure 2-3   *Langley graph*

the gelatin absorbs 10 percent of a beam of white light of 10 milliwatts, the intensity of the light after passing through the gelatin is 90 percent, or 9 milliwatts. If you stack a second layer of gelatin over the first, the intensity of the light after it passes through the second layer is 90 percent of 9 milliwatts, or 8.1 m milliwatts. If you stack a third layer over the second, the intensity of the light is 90 percent of 8.1 milliwatts, or 0.73 milliwatts, and so forth.

If you plot a linear graph of the intensities of light transmitted through the various layers of the gelatin, the line that results will be curved. If you plot the intensities on a logarithmic graph, a straight line results. Such a graph is known as a *Bouguer graph*.

Beer's Law

A refinement in Bouguer's law was contributed a century later by August Beer, a German physicist. Beer's law considered the effect of the concentration of a light-absorbing substance within the medium through which the light passes. Bouguer's law and Beer's law explain the transmission of light at different angles through the same medium.

Assume there is an imaginary slab of purple gelatin in front of you. If the absorption of a beam of light is measured after it passes through the gelatin at the normal (perpendicular) angle, Bouguer's and Beer's laws can be used to calculate how much of the beam will be transmitted when the gelatin is tilted at a known angle with respect to the beam of light.

Let's agree that a slab of purple gelatin transmits 80 percent (or 0.8) of a beam of white light. Now tilt the dish so that a beam of light enters the gelatin at an angle of 30 degrees. If the gelatin doesn't slide off the dish, the path of the light through the slab forms the hypotenuse of a triangle in which the base is the bottom of the gelatin and the opposite side is a vertical line through the gelatin. The length of the hypotenuse of this triangle is the *secant* (1/sin) of 30 degrees. Because the secant of 30 degrees is 2, then the light beam passes through the equivalent of two slabs of gelatin, and 64 percent (80 percent of 80 percent) of the light is transmitted.

If the atmosphere is substituted for the gelatin and the light source is the sun or a star, the same relationship holds. If the sun is straight over a point at sea level as shown in Figure 2-4, its light passes through one thickness of atmosphere or, as it is commonly known, one *air mass*. If the sun sits at 30 degrees above the horizon as shown in Figure 2-5, then its light passes through approximately a double thickness of atmosphere or approximately two air masses. In other words, the air mass through which the sun's radiation travels is approximately the secant of the sun's angle above the horizon. Notice that the relationship is approximate and not exact. The exact computation of air mass must take into account the curvature of the earth and the nonuniformity of the atmosphere. This topic is covered in detail in a number of other resources.

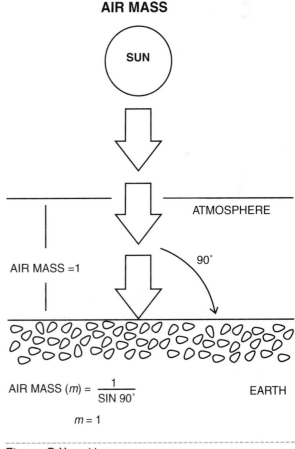

Figure 2-4   *Air mass*

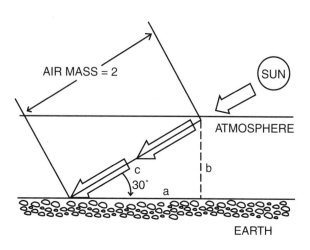

$$\text{AIR MASS } (m) = \frac{1}{\text{SIN } 30°}$$

$$m = 2$$

Figure 2-5   *Double air mass*

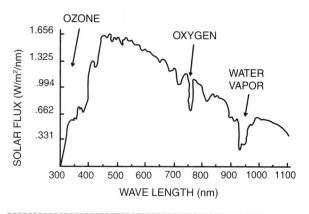

Figure 2-6   *Solar spectrum absorption bands*

Samuel Pierpont Langley of the Smithsonian Institution pioneered in the measurement of the sun's intensity. He also invented instruments for measuring solar radiation. In honor of his work, Bouguer graphs of the intensity of sunlight received through a range of air masses are often referred to as *Bouguer-Langley graphs* or sometimes simply *Langley graphs*.

Although the Bouguer method of measuring the extraterrestrial solar constant seems simple enough, it has several drawbacks. Assume for example, you use a detector that responds equally well to all the UV, visible, and infrared wavelengths in a ray of sunlight. Unfortunately, this does not mean you can measure the total intensity of all these wavelengths outside the atmosphere by making measurements from the surface of the earth. Water vapor, the most important of all the greenhouse gases, strongly absorbs many infrared wavelengths. And ozone absorbs virtually all the UV radiation having a wavelength less than approximately 295 nanometers. Variations in the total water vapor and ozone in the atmosphere from day to day and even over the course of a day can cause serious measurement errors. Carbon dioxide, oxygen, and many other gases also absorb various wavelengths of sunlight, but their affects are usually considerably less variable than are those of water vapor and ozone.

The graph in Figure 2-6 shows that there are several spectral windows where sunlight passes through the atmosphere relatively unimpeded. A very simple way to avoid the affect of water vapor and ozone on sunlight measurements is to monitor one or more wavelengths within one of these windows. While this will not give you the total radiation from the sun, it will allow you to keep track of the sun's intensity at a specific wavelength. If you do this over an extended period, you will learn much about the effect of clouds, dust, smoke, and aerosols on sunlight.

The Atmosphere versus Sunlight

During its passage through the atmosphere, the light from the sun is scattered and absorbed by molecules of gas, particles of dust, and clouds of water vapor. Some wavelengths of light are less affected (that is, absorbed less) than others are.

Certain detectors that monitor the UV radiation from the sun are designed to look at the entire sky. These detectors are said to have a global field of view. They are used because air molecules scatter UV light so well that up to half or even more of the UV that strikes your skin on a sunny day comes from the sky.

For best results when making Langley plots, however, your light detector should stare only at the sun and not at the surrounding sky. Otherwise, when smoke, haze, and dust are in the atmosphere, they will scatter sunlight toward your detector and inflate your readings. Detectors that stare only at the sun are said to have a direct field of view. The simplest way to restrict your detector's field of view is to place it at one end of a narrow tube. Such tubes are known as *collimators*. The sun subtends an angle in the sky of just over half a degree. A suitable collimator tube can easily restrict your detector field of view to a few degrees or so. Although this isn't perfect, it's much better than allowing the detector to look at the entire sky.

Filters

Many kinds of filters can be used to make Langley plots of the sun's intensity. If your budget is limited, you can start with inexpensive colored glass or plastic filters. You can make your own from colored glass or plastic available from arts and crafts shops or glass and plastic dealers. Or you can use colored camera filters. A much better approach is to use an *interference filter*. This kind of filter is made from many very thin layers of reflecting materials deposited on a glass or silica substrate. Although a colored glass filter transmits a band of light a hundred or more nanometers wide, an interference filter transmits a band typically only 10 nanometers or even less in width.

Because of the many carefully controlled steps required to make them, interference filters are much more expensive than colored glass and plastic filters. A good supplier for the novice is Edmund Scientific Company. Edmund Scientific currently stocks filters that transmit 10-nanometer bands of light at points all across the visible light and near-infrared spectrum. A 0.5-inch (12.5-millimeter) diameter visible light filter costs $38.00 plus shipping. Near-infrared and UV filters are $78.00 plus shipping.

Interference filters are also available from Micro-Coatings and Twardy Technology, Inc. All these companies supply transmission curves with their filters. If the prices for interference filters are too high for your budget, you can sometimes find surplus interference filters for bargain prices. For example, most helium–neon lasers emit a beam of red light with a wavelength of 632.8 nanometers. Surplus interference filters at this wavelength can be purchased for as little as $5 or $10. Whatever filter you select, those that are 0.5 inch (12.5 millimeters) in diameter are easiest to mount. Your budget may determine what wavelength you select.

Detectors

Many different kinds of light detectors can be used to make Langley plots. For now I recommend that you use silicon photodiodes. They are cheap, readily available and sturdy, and they generate a current that is linear with respect to the light that strikes them. Do not use a phototransistor because its output current is not linear.

Silicon photodiodes are available from various electronics suppliers. Digi-Key sells Clairex's CLD56 flat-window photodiode for around $3.18. You can also use a small silicon solar cell. You could also use Clairex's CLD71, a miniature solar cell mounted on a ceramic substrate. This diode is available from Digi-Key for around $2.00.

Another possibility is the Honeywell SD3421-002. This flat-window photodiode is available from Newark Electronics for around $3.25. Some of the other parts for the radiometer are also available from Newark Electronics (call or write for a free catalog). Because Newark has a $25 minimum order requirement, you may want to try to find the components locally.

Avoid using a detector encapsulated in a clear plastic package like those used for LEDs. The domed end of the package will act like a lens and complicate the alignment of the detector with respect to the sun. Instead, use a detector housed in a miniature metal package with a flat glass window. If you cannot find or afford a detector enclosed in a metal package, you can use a plastic encapsulated one if you first grind away the lens end of the domed package with a file or sandpaper. Polish the end of the flattened package with very fine sandpaper.

An even better approach is to purchase a detector with a built-in interference filter. EG&G Judson sells a line of such detectors designated as DF-xxxx, where xxxx is the filter's transmission wavelength in Angstrom units. (Ten Angstrom units is one manometer.) For example, a DF-5000 detector is designated to detect light with a wavelength of 500 nanometers. EG&G also sells photodiodes without filters. The DF-xxxx detectors with built-in filters cost $95. While this may seem high, you don't have to go to the trouble of mounting a filter on a detector.

EG&G will supply a calibrated detector for an additional $75.

If you cannot afford a calibrated detector, use the spectral B graph specified by the manufacturer. This graph will be within 10 percent or so of the detector's actual sensitivity. You will then need to account for the absorption of the filter. This is why it's important to request a transmission curve when you order a filter. Typical interference filters transmit from 20 to 50 percent of the light at their pass band.

In any case, having a calibrated detector is less important than having a detector that provides repeatable measurements. Although you may not know the exact value of the intensity of the light your detector receives, you will know that the trends and changes it monitors are linear with respect to one another.

A Basic Radiometer Circuit

The diagram in Figure 2-7 illustrates a simple radiometer that is well suited for making measurements of solar radiation. This radiometer uses a photovoltaic solar cell and an *operational amplifier* (or op-amp) to increase the level of the signal from the detector so that it can be read by a digital voltmeter. The op-amp is connected as a linear amplifier. In

other words, the signal from the photodiode is amplified in direct proportion to its amplitude. The level of amplification equals the resistance in ohms of R1, the feedback resistor connected between the op-amp's output and its input.

The op-amp used in this basic radiometer is a Texas Instruments TLC271. This amplifier comes in several versions, the best being the TLC271BCP; the ACP and CP versions will also work. TLC271 op-amps are available from Newark Electronics for less than $1 each. Many other op-amps with a yellow input bias current can also be used, but they may have different pin connections.

Because many different kinds of detectors will work with this circuit, you will have to determine the optimum resistance for R1 by simply experimenting. Too small a resistance will make readings difficult when the sun is low in the sky. If the resistance is too high, full sunlight will overload the amplifier. One way to find a good value for R1 is to assemble the circuit on a plastic solderless breadboard. You can then try different resistances until the circuit gives an output of several volts when the detector is exposed to the noon sun on a clear bright day.

Another way is to use a 1-megohm (1,000,000 ohms) potentiometer for the resistor. Adjust the potentiometer for the optimum reading when the sun is high in the sky at or near local noon. Measure the potentiometer's resistance with a multimeter, and

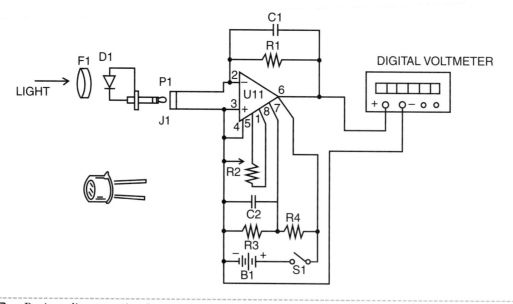

Figure 2-7   *Basic radiometer circuit*

substitute a fixed resistor with the nearest equivalent value. Or you can leave the potentiometer in the circuit permanently. Just be careful to avoid changing its setting, so your readings will always be comparable. If you choose to leave the potentiometer in the circuit, it's a good idea to mount it inside the radiometer's housing and away from curious fingers.

Radio Shack sells fixed resistors and a standard-size 1-megohm potentiometer (Radio Shack part number 271-211). They also sell a standard-size 100,000-ohm potentiometer suitable for use as R2. Miniature, multiple-turn, screw-driver-adjustable trimmer resistors are much better choices than standard-size potentiometers. Many different kinds are available from Newark Electronics and other electronics parts dealers.

The output of the op-amp U1 at pin 6 is fed directly to the digital voltmeter or digital multimeter set on the 2-volt voltage scale. The radiometer circuit is powered from a standard 9-volt transistor radio battery. The radiometer circuit can be assembled on a small piece of breadboard or circuit board with the leads kept short. When assembling the circuit, make sure that you install the integrated circuit correctly to avoid damaging the IC. IC packages will have either a small indent circle at one end or a small rectangular cutout at one edge of the plastic package. Pin 1 will always be to the left of either the indented circle or the plastic cutout. Using an integrated circuit is good practice, in the event of a later circuit failure.

The radiometer circuit can be installed in a small metal enclosure with two terminal posts for connection to a multimeter or digital voltmeter. The power switch at S1 is used to apply power to the radiometer circuit. A mini 1/8-inch phone jack is used to connect the sensor to the input of the radiometer circuit as shown in Figure 2-7.

Filter-Detector Assembly

Unless you have access to a machine shop, mating the detector with a suitable filter is probably the most difficult part of building a radiometer. After trying many approaches that don't require metal-working tools, the final result is shown in Figure 2-8. It consists of a 1/8-inch brass union coupling and some O-rings,

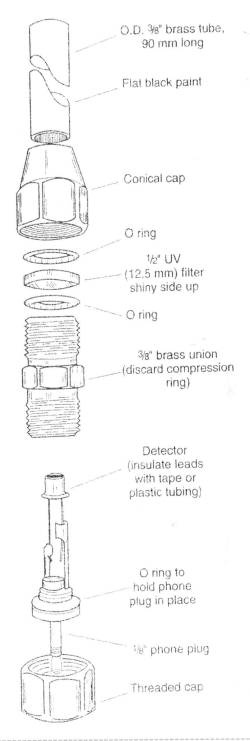

Figure 2-8 *Filter detector assembly*

O.D. 3/8" brass tube, 90 mm long
Flat black paint
Conical cap
O ring
1/2" UV (12.5 mm) filter shiny side up
O ring
3/8" brass union (discard compression ring)
Detector (insulate leads with tape or plastic tubing)
O ring to hold phone plug in place
1/8" phone plug
Threaded cap

all available from a hardware store, which hold both the detector and filter. This mounting method works only with 0.5-inch (12.5-millimeter) filters and detectors that are small enough to fit inside the hole in the union. You will need to check the dimensions of the union and detector you plan to use because the dimensions of both vary.

In the arrangement shown in Figure 2-9, the detector is soldered to a $^1/_8$-inch phone plug. I prefer this arrangement because it permits several detectors to be used with the same radiometer. I always solder the diodes cathode lead to the terminal connected to the tip of the phone plug. This is because some of my radiometers with multiple detector jacks are housed in metal enclosures, which automatically interconnect all the anode terminals of the phone jacks to ground.

Insert the plug into a union end cap and hold it in place by forcing a rubber O-ring between its threads and the threads inside the end cap. If you can't find a suitable O-ring, a thin bead of black silicone sealant might work. Just be sure not to get any of the sealant on the threads inside the end cap. After the plug is seated in the end cap, place an insulating sleeve of thin plastic (heat-shrinkable tubing or a soda straw) over the photodiode, insert it into the hole in the union, and secure it in place by rotating the end cap. The insulator is required because the metal case of some photodiodes is connected to one of the diode's terminals.

Carefully clean the filter and install it between two thin O-rings with its shiny side facing away from the detector. Some filters are thicker than others. If the O-ring and filter sandwich is so thick that the end cap cannot be screwed on, you may have to replace one or both O-rings with a ring of paper or tape. In any case, be careful not to apply too much pressure when you screw on the end cap or you may chip or crack the filter.

The filter end cap should be fitted with a collimator tube with an outside diameter of approximately $^1/_4$ inch (6.55 millimeters). A tube with a length of around $1^3/_4$ inches (about 45 millimeters) will provide a field of view of a few degrees or so. Aluminum and brass tubing available from hobby and craft stores works well. Use a cotton swab to coat the inside of the tube with flat, black paint. Attach the tube to the end cap with a fast-setting glue such as cyanoacrylate adhesive.

An Adjustable Radiometer

The radiometer can be made much more versatile by adding additional feedback resistors and a selector switch in place of the fixed resistor at R1, which is across pins 2 and 6 of the op-amp. This will permit you to easily change the radiometer's gain in fixed steps. You can increase the gain by one decade (X10) for each switch position, from a minimum of 1,000 to maximum of 1,000,000. A four-position rotary switch and four resistors (1K, 10K, 100K, and 1 megohm resistors) could be switched in place, one at a time. If you plan to assemble the instrument in a pocket-size enclosure, you will need to use a miniature rotary switch. Miniature rotary switches are available new from Newark Electronics and other electronics distributors for around $8.00 each.

Using a Solar Radiometer

Making measurements with a solar radiometer is relatively straightforward. To align your radiometer with respect to the sun, simply adjust its position until the shadow cast by the collimator tube disappears. The detector should then be staring directly at the solar disc. You can now read the voltage (or current) from the digital meter to which the radiometer is connected. If you plan to make measurements with several detectors, you might wish to use a tape recorder to record your data. You can even automate your system with the help of a chart recorder or a computer data acquisition system.

The acquisition of data for Langley graphs requires that the sun be totally open and unblocked by clouds. Always wear good-quality UV-blocking sunglasses when monitoring the sun. If clouds are present, avoid the temptation to peek at the sun to see if a cloud is in the way. If you must be sure, you can look at the sun through a number 14 welder's glass, available from a welding supply store for several dollars. Never observe the sun through a filter that is more transparent than a number 14 filter.

Performing Measurements

The object of a measuring program is to measure the sun's output on a daily basis (weather conditions permitting). It is a good idea to measure the sun's radiation at several wavelengths. The purpose of these measurements is to build up a base of data about the level of the sun's UV rays and the abundance of the total water vapor, ozone, and oxygen in a column through the atmosphere. Langley plots can be used to check the calibration of the detector. You should try to make measurements at local apparent noon, the time when the sun is at its highest position in the sky. You can convert standard time to apparent time by adding 4 minutes for each degree of longitude west of your standard meridian. In the United States, the standard meridians for Eastern, Central, Mountain, and Pacific time are, respectively 75°, 90°, 105°, and 120° west.

The equation of time explains how solar noon arrives as much as some 16 minutes early or 14 minutes late. Therefore, if you plan to make measurements at apparent noon, you will need an equation-of-time chart or diagram. For a precise knowledge of solar noon, consult any standard reference on astronomy or sundial construction available at most libraries.

The peak reading you measure will not necessarily occur precisely at apparent noon. Instead, the signal will fluctuate constantly as the radiation is attenuated and scattered by the atmosphere and its constituent gas and aerosols. For this reason I usually allow several minutes to make a measurement. Making measurements at apparent noon is not as important as the exact time recorded, so be sure to calibrate your watch every few days using WWV (from radio station WWV) or *National Institute of Standards and Technology* (NIST) time calibration on the Internet.

Equally important is knowing the angle of the sun above the horizon. You can measure the sun's angle with the help of a bubble level and a small L-bracket mounted on the enclosure in which you mount the radiometer. The bracket functions like the *gnomon*, or pointer, of a sundial. Align the bracket so that its shadow falls along a millimeter scale cemented to the enclosure. Then carefully adjust the position of the radiometer until it is perfectly level, and note the point on the scale where the shadow ends. The height of the angle bracket divided by the length of the shadow is the tangent of the sun's angle.

Besides the date, time, and sun angle, you should also note the weather and sky conditions. It's espe-cially important to describe the sky near the sun. It's also a good idea to record the barometric pressure, humidity, and temperature when you are recording your solar observations.

Basic Radiometer Parts List

Required Parts

R1	see text
R2	100K ohm potentiometer (trimpot)
R3, R4	1-megohm, 1/4-watt, 5% resistor
C1	100 pF, 35-volt Mylar capacitor
C2	0 .01 uF, 35-volt ceramic disc capacitor
D1	photodiode (see text)
B1	9-volt transistor radio battery
S1	SPST toggle switch
J1	1/8-inch mini two-circuit phone jack
P1	1/8-inch mini two-circuit phone plug
F1	filter (see text)
Miscellaneous	PC board, IC socket, enclosure, wire, terminals, collimator, voltmeter, etc.

Measuring Ultraviolet Rays—Using an Ultraviolet Radiometer

Most UV radiation that bombards the earth is barred from reaching the planet's surface because of a thick, vacuous blanket of blue toxic gas known as ozone. Were it not for this chemical shield, UV rays would strike organisms with such intensity that most organisms would perish. As human and natural activities such as volcanic eruptions alter the composition of the atmosphere, it would be wise to monitor the ozone layer and UV radiation to observe possible changes.

Thanks to many ground stations and several satellites, workers have learned much about the density and distribution of the ozone in the atmosphere. Yet investigators lack a comparable network to observe the portion of UV radiation that seeps through the ozone layer. Aside from instruments operated by the Smithsonian Institute and several other organizations, the only network monitoring stations in the United States comprises fewer than two dozen Roberston-Berger meters. (Robertson-Berger meters measure global ultraviolet-B.)

These devices are designed to detect the wavelengths of UV radiation that cause *erthema*, or reddening of the skin and eventual sunburn. Erthema develops most rapidly when the skin is exposed to UV radiation with a wavelength near 300 nanometers. This wavelength falls within the *ultraviolet-B* (UV-B) spectrum, which extends from 280 to 320 nanometers. Since 1974 the average flux of UV-B has been measured with a network of eight Robertson-Berger meters. From 1974 to 1985 the average flux fell some 0.7 percent per year. Because stratospheric ozone over the network decreased about 0.3 percent per year from 1978 to 1985, an increase in UV-B would have been expected.

With a little effort, you can construct an UV-B radiometer to record daily the flux of radiation. Comparing your observations with those of others from different regions would provide important information about how air pollution affects UV-B.

Before building a UV-B radiometer for this purpose, you will need to understand how UV-B radiation travels throughout the atmosphere. Some UV-B rays scatter because of air molecules; the remainder penetrates directly through the atmosphere. The sum of scattered and direct UV radiation is called *global radiation*.

Global radiation is of high interest in studies of the deleterious effects of UV radiation on both living systems and materials such as paints and plastics. Measurements of global radiation are also helpful in determining how clouds affect UV-B. Measurements of direct UV radiation yield valuable information about the presence and effect of absorbing and scattering agents in the earth's atmosphere. Because of the unpredictable nature of clouds and the presence of such barriers as buildings and trees, measurements of direct radiation are preferred to global ones for comparing the effects of air pollution on the relative magnitude of UV-B at two or more locations. An UV-B radiometer requires a detector and a means for selecting the wavelength to be detected. The detector signal is amplified and transmits to a digital voltmeter, an analog chart recorder, or a computerized data acquisition system.

Wavelength can be selected either with a monochromator or an optical interference filter. A *monochromator* provides a convenient but expensive way to measure UV-B across a wide range of discrete wavelengths. An *optical interference filter* offers a much cheaper and more compact method of selecting a reasonably narrow band of UV-B wavelengths. Interference filters also allow considerably more radiation to reach the detector.

UV-B interference filters, however, transmit a slightly wider band of wavelengths than do monochromators. Moreover, interference filters transmit low but detectable levels of radiation outside their specified bands, which can cause significant errors in measurement. An UV detector that eliminates a filter's secondary bands is said to be *solar blind*.

The UV-B radiometer shown in Figure 2-9 is based on *gallium phosphide diodes*. These detectors, unlike silicon photodiodes, do not respond to red light and thus provide true solar-blind operation. Gallium phosphide diodes are made by Hamamatsu Corporation. The G1961 (about $28 plus shipping) is housed in a TO-18 package and has an effective surface area of 1.0 square millimeter. The G1962 ($35) is housed in a larger TO-5 package and has a surface area of 5.2 square millimeters. Hamamtsu also offers a G1962 calibrated at 300 nanometers for an additional cost.

The photodiode is coupled directly to a Texas Instruments TLC271CP operational amplifier, which can be purchased from major electronics distributors such as Mouser or Digi-Key electronics. The TLC271CP can be damaged by static electricity, so you will need to be careful when handling the device and use antistatic handling procedures.

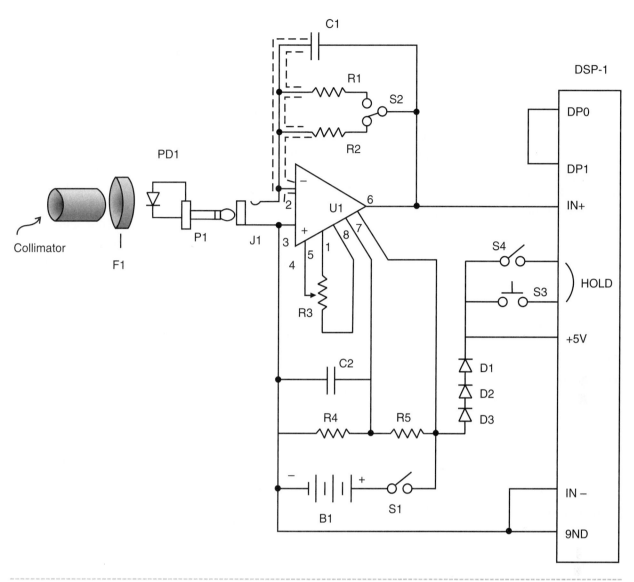

Figure 2-9 *Ultraviolet radiometer circuit*

The UV-B radiometer has two gain settings. The gain path is set up by using two resistors at R1 and R2, which is elected via switch S2. R1 and R2 are two very-high-resistance feedback resistors that are not widely available and whose values can be specified only approximately because of variations in the signals transmitted by different filters at various wavelengths. You can use two resistors to give two gain levels. Only one is needed if you plan to make measurements near noon throughout the year. The optimum resistance will probably be between 30 and 100 megohms. The exact value is not critical if 44

megohms works; then 32 megohms will work but with reduced gain.

The best values can be selected by inserting each of several resistors into the circuit temporarily while monitoring the readout, with the filtered detector pointed at the sun. The selected resistance should cause the readout to display perhaps 80 percent or so of its full range for bright, noon sun with low ozone conditions.

Resistors of more than 22 megohms are hard to find and expensive. You can make your own by soldering together several 10- and 22-megohm resistors available from electronics parts stores. Potentiometer R3 is the zero adjust control, which biases pin 5 of the op-amp. A voltage divider is formed by resistors R3 and R4, which applies a voltage to pin 8. A 9-volt battery powers the radiometer and power is applied to the circuit via switch S1.

The DP-650 +200-millivolt digital panel meters from Acculex is available for approximately $60 each plus shipping. An important advantage of the Acculex DP-650 meters is that the data in the display can be saved simply by pressing a pushbutton switch that connects pin 11 to pin 1 (+5 volts). A new reading can be made when the switch is opened. The three diodes D1 through D3 are used to drop the voltage for the display down to 5 volts. Switches S3 and S4 are used to control the display's readout. An important advantage of the UV-B radiometer is that it is powered by a single 9-volt battery, it consumes little current, and the circuit easily fits inside a compact housing.

Construction

Building the UV radiometer is pretty straightforward. The circuit can be constructed on a small circuit board. Use an integrated circuit socket on the IC; in the event of a circuit failure at a later date, you will be able to repair the circuit much easier if a socket is used. Integrated circuits have some form of identification regarding their orientation. The IC will have either a small indented circle or a small rectangular cutout on the plastic package. Pin 1 of the IC will always be to the immediate left of the indent circle or the plastic cutout. Pay attention to the orientation of the diodes when installing them at D1, D2, and D3. Note that there are two gain resistors across IC pins 2 and 6. A gain switch switches either of these two resistors into the gain path, so leads should be kept as short as possible to avoid unstable operation or circuit oscillation. Switch S2 could be mounted on the circuit board to keep leads short. Because the UV-B solar-blind radiometer includes a separate detector, filter, and amplifier, it is more difficult to assemble than an instrument in which these components are combined in a single package.

A metal chassis box is used to house the ultraviolet radiometer circuit. You will have to punch out a large hole in the chassis front panel to accept the *liquid crystal display* (LCD). You can mark the dimensions of the panel meter onto the chassis panel and drill small holes all around the inside dimension of your marking. The object is to undersize the panel meter hole, cut the sections between each hole, and then file the large hole square to accept the panel meter. The circuit board can next be aligned with the top of the chassis and a hole can be drilled for switch S2 to protrude up through the top front of the chassis box. Standoffs can be used to secure the circuit board to the top front panel of the enclosure. Switches S1, S3, and S4 can be mounted as desired on the top front panel as layout permits. An $1/8$-inch two-circuit mini jack for the input sensor is finally mounted at the top of the enclosure. A 9-volt transistor radio battery holder is secured to the bottom of the chassis to hold the battery. Once the circuit has been installed and the chassis secured, you can attach your sensor assembly and begin your UV-B measurements.

The most expensive component of the UV-B solar-blind radiometer is the optical filter. High-quality filters are made by Barr Associates. Barr makes filters on a custom basis only. Therefore, unless you are connected with an institution that can afford to place a custom order, you will need to go elsewhere.

MicroCoatings makes a 12.5-millimeter diameter filter that transmits 300-nanometer radiation and has a band pass of 10 nanometers (catalogue number ML3-300). Twardy Technology, Inc. sells a 25-millimeter-diameter filter with the same specifications for $210.

Most important in building the ultraviolet radiometer is to install the detector and the filter in a light-tight housing. If you have access to a machine shop, you can make one. Or you can install a 12.5-millimeter filter and detector in a brass compression fitting or union coupling (see Figure 2-8). Note that the basic radiometer and ultraviolet radiometer filter assemblies are almost exactly the same, except that the filters are different for each project. The coupling and the required O-rings are available from hardware stores. A two-conductor phone plug is inserted into one of the union's caps and secured in place with a rubber O-ring. The leads of the detector are inserted into an LED socket soldered to the plug's terminals. You can, however, solder the detector directly to the terminals. The cathode lead should be soldered to the terminal that is common to the tip of the plug. In either case, some couplings will accept only detectors in miniature TO-18 packages.

The filter, protected by a pair of O-rings, is installed in the second end cap. A conical cap works best but may be hard to find. If the filter and O-rings do not leave sufficient space for the end cap's threads to engage those of the union, replace one of the O-rings with a paper spacer. Screw the end cap down so that it stays in place but does not apply pressure to the filter. If necessary, cement the end cap in place with a drop of removable glue. Be sure the filter remains clean during the installation procedure.

Depending on the detector's dimensions, a conical end cap wall will give a field of view of around 10 degrees. You should therefore attach a collimator tube to the opening in the end cap to reduce the field of view to 4 degrees or less. Brass tubing can be soldered or cemented to the opening in the end cap. Coat the inside of the tube with flat, black paint.

Because the radiometer is designed to measure the direct radiation from the sun, a collimator is required to restrict the detector's field of view. Thin-walled brass tubing from a hobby shop works well. A tube with an outside diameter of 1 centimeter should slip over the detector. Wrap a layer of tape around the detector if it fits too loosely in the tube. Coat the inside of the tube with flat, black enamel. A tube around 90 millimeters long will provide a field of view of approximately 4 degrees when the tube is pushed all the way to the detector's base.

Before installing the collimator, you should clean the surface of the detector's filter because dust and oil absorb UV-B. Remove fingerprints by swabbing the surface of the filter with ethyl alcohol and wiping away the residue with lens cleaning paper. Blow away dust with clean, compressed air.

UV-B Measurements

Regularly measuring direct solar UV-B with either a calibrated or an uncalibrated detector can yield significant data. Always try to take a measurement at solar noon. For more information about determining solar noon, consult any standard reference on astronomy or sundial construction.

Rarely does the peak UV-B reading occur precisely at solar noon. Instead the signal fluctuates as the 300-nanometer radiation is attenuated and scattered by the atmosphere and its constituent gases. For this reason, allow at least 5 minutes to make a single measurement. Virtually every day for 2 years, I have measured the direct solar flux at four UV-B wavelengths and six additional wavelengths. I have found that direct solar irradiance at 300 nanometers is significantly attenuated by fog, haze, clouds, and aircraft contrails. The passage of a cold front, which raises barometric pressure, is more often than not followed by a reduction in UV-B, even when the atmosphere is exceptionally clear and dry. Low barometric pressure is typically accompanied by a decrease in ozone.

The assembled radiometer is simple to operate. First, look down the collimator tube. If you see a reflection of the pupil of your eye, the detector is perfectly centered. If not, realign the tube. After the voltmeter is connected and the radiometer's power switch is toggled on, block the opening of the collimator tube and adjust the potentiometer until the output voltage is zero. (Repeat this procedure before each measurement session.) Then point the tube toward the sun and align the tube until its shadow disappears. The detector will now be aimed directly at the sun. Record the voltage and make another measurement. You will soon discover that even on a clear day the signal level fluctuates, sometimes considerably especially around noontime and whenever the atmosphere is obscured by clouds, smoke, or dust.

Your readings will include an error factor because the detector responds to the red light that leaks through its filter. You can eliminate the error simply by following each reading with a second one during which you block the UV rays by placing a filter over the entrance of the collimator tube. An UV filter intended for a camera works well and so does a WG-345 clear glass filter.

If you have an uncalibrated detector, subtract the second reading (B) from the first (A) to get a voltage that will be correct with respect to measurements you make at other times. If you have a calibrated detector, you can compute the absolute spectral radiance at 300 nanometers in terms of watts per square meter. UV blocking filters typically reflect about 8 percent of the incident non-UV radiation. The non-UV radiation without the filter is therefore approximately equal to the B reading divided by 92 percent. Because the active area of the DFA-3000 is about 9.9 square millimeters, the detector signal must be multiplied by 101,000 to find the signal per square meter. The formula that results is

$$\frac{A - (B/.92)}{R1 \times D_r} \times \frac{101,000}{F}$$

$$\frac{1.50 - (.116/.92)}{30,000,000 \times .04} \times \frac{101,000}{10.4} = .011 \text{ watt per square}$$

meter per nanometer

where D_r is the detector's calibrated responsivity, and F is the filter's band pass. (The band pass is the number of nanometers between the two points where the filter's transmission fails to half the maximum.) The ideal filter should have a band pass of less than a nanometer. Real filters have a wide band pass.

Readings at noon on a clear August day are typically 1.5 (A) and 0.116 volt (B). Inserting these values into the previous formula yields 0.011 watt per square meter per nanometer. Remember, this is direct UV. The diffuse contribution from radiation scattered by molecules in the atmosphere adds at least 30 percent to this value at my latitude.

Digital Ultraviolet Radiometer Parts List

Required Parts

R1   (see text)

R2   (see text)

R3   100K ohm potentiometer (trimpot)

R4, R5   1-megohm, 1/4-watt, 5% resistor

C1   100 pF, 35-volt Mylar capacitor

C2   0.01 uF, 35-volt ceramic disc capacitor

PD1   Ga P photodiode (see text)

D1, D1, D3   1N901 silicon diodes

B1   9-volt transistor radio battery

S1, S4   SPST toggle switch

S2   SPDT toggle switch (gain)

S3   normally-open push-button switch (display)

DSP-1   Acculex DP-650 digital panel meter

J1   1/8-inch mini two-circuit phone jack

P1   1/8-inch mini two-circuit phone plug

F1   filter (see text)

Miscellaneous   PC board, IC socket, enclosure, wire, terminals, collimator, etc.

Measuring Ozone—Using an Ozone Meter

The total amount of ozone in a column through the atmosphere can be determined by simultaneously measuring two wavelengths of the UV radiation emitted by the sun. Advanced experimenters will find a this project to be both fun and challenging.

Ozone strongly absorbs UV radiation from the sun with a wavelength below about 330 nanometers. This absorption is so efficient that under normal conditions practically no radiation with a wavelength below 295 nanometers reaches the ground. Ozone absorbs shorter wavelengths of UV much more efficiently than it does longer wavelengths. Therefore, the amount of ozone can be measured by a method known as *direct wavelength absorption spectroscopy*. This method is, in principle, very simple. An ozone measuring instrument, for example, is simply a pair of UV radiometers installed in a single housing. The difficult part of measuring ozone is the various formulas that transform a pair of UV measurements into the amount of ozone.

Ozone-Meter

To make an ozone observation, an ozone instrument is pointed directly at the sun. Ultraviolet radiation passing through the filters strikes the two detectors, which are two-terminal photodiodes that convert light into an electrical current. The signal from each detector is amplified and sent to a miniature digital readout. Although advanced experimenters should be able to assemble a tool ozone portable spectroradiometer (TOPS) instrument, a calibrated instrument for measuring ozone, it is important to understand that the necessary pairs of UV filters, photodiodes, and high resistance resistors are not readily available.

Selecting the Filters

The most important and expensive components of an ozone-measuring instrument are the two UV filters. Most existing filter ozone-meters respond to a pair of wavelengths separated by about 20 nanometers, but this means errors can be introduced by aerosols in the atmosphere. I minimize this problem by using wavelengths only 6 nanometers apart. To make sure there is ample difference in the ozone absorption at two wavelengths this close together, it's necessary to use wavelengths close to the point at which all UV radiation is blocked by ozone because that's where the difference in absorption is most dramatic. Ultraviolet filter wavelengths of 300 and 306 nanometers were chosen for this project. These wavelengths work well at my latitude: 29° 35′ north, but they will not work well at higher latitudes during winter and spring because of the lower angle of the sun and the increased amount of ozone (see Table 2-1).

Table 2-1

Solar Wavelengths

Wavelength (nm)	Spectrum Alignment
297	ultraviolet-B and ozone absorption
300	ultraviolet-B and ozone absorption
306	ultraviolet-B and ozone reference
312	ultraviolet-B and ozone absorption
320	ultraviolet-B and ozone reference
590	ozone absorption
600	ozone absorption
630	ozone reference
700	ozone reference
760	oxygen absorption
780	oxygen reference
850	water vapor reference
940	water vapor absorption
998	water vapor reference

A better pair of wavelength ranges for locations above 350° north would be 305 to 310 nanometers for the short wavelength and 325 to 330 nanometers for the long wavelength. The disadvantage of these wavelengths is that aerosols in the atmosphere may cause more error than with more closely spaced wavelengths.

For best results, the bandpass of filters used to detect ozone must be less than the 10 nanometers, which is standard for most. If you cannot find filters with a 5 nanometers or less bandpass, you can reduce the bandpass of a 10 nanometers filter by stacking two of them.

Filters are commonly sold in diameters of 12.5 and 25 nanometers (0.5 and 1 inch). Smaller filters are cheaper and easier to mount. Stock UV interference filters cost $100 or more each, and custom filters are considerably more expensive. Manufacturers who stock filters include Twardy Technology, Inc., Micro-Coatings, and Andover Corporation. Additional manufacturers advertise in trade magazines for the optics and laser industries.

Building the Ozone-Meter

As mentioned earlier the ozone-meter consists of two identical UV-B radiometers in a one complete package. Figure 2-9 depicts the original UV-B radiometer circuit. The diagram shown in Figure 2-10 shows how both UV-B radiometer circuits are installed in an aluminum enclosure. The picture in Figure 2-11 illustrates three research ozone-meters, packaged for a field trip. If you have previous experience building miniature electronic circuits, you should be able to

assemble your own instrument with these figures as a guide. The prototype ozone-meter was installed in LMB CR-531 Crown Royal aluminum cases available from various electronics distributors and Mouser Electronics. Squeezing all the components into the CR-531 case requires careful planning. Or you can simplify assembly by using a larger cabinet if you are not skilled at dense packaging. You must, however, make sure no light can leak between the filter and detector. Be sure also that the detector views a narrow cone with an angle of less than 2 degrees.

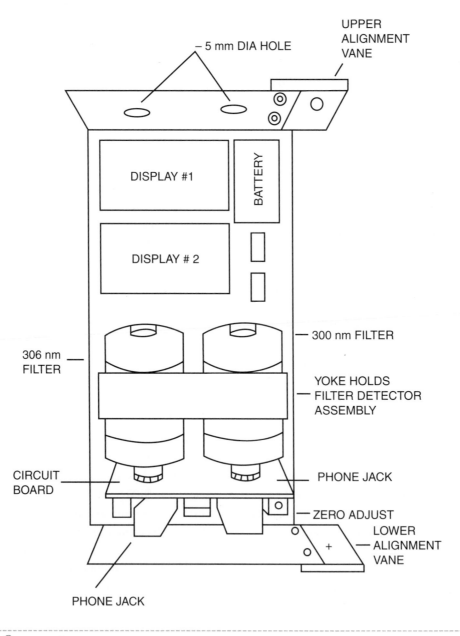

Figure 2-10   *Ozone-meter*

Figure 2-11   *Ozone-meter*

Testing and Aligning the Ozone-Meter

After you assemble the instrument, carefully check the wiring. It's especially important that the battery connections to the digital readouts and the amplifier be correct. If everything is in order, install a battery in the battery holder and switch on the instrument. Both readouts should display digits. Block both photodiodes and adjust the trimmer potentiometer R2 until the two readouts read 0 volts. A pair of aluminum vanes (see Figure 2-11) provides a means for optically aligning TOPS. Bore a small, 1 to 2 millimeter hole near the center at the upper vane. With the cabinet open, point the instrument at the sun and align it until sunlight strikes both filters. Then place a small mark where sunlight from the upper vane strikes the lower vane. If your filters are recessed, you can see when sunlight is striking them by placing a glass microscope slide over them. If you then tilt the slide at a 45 degree angle, you will see the filters reflected in the slide.

Using the Ozone-Meter

Before using the instrument, be sure the cabinet is closed and that no light leaks through any openings. If necessary, use black paper to shield the detectors from light leaks. You can also insert tubes over the detector filter assemblies. In either case, be sure that nothing blocks the sunlight reaching the detector.

To use the instrument, first make sure both amplifiers are zeroed by switching on the power in sub-

dued light or by blocking the apertures. If either readout indicates more than 0, open the case and adjust the appropriate zero potentiometer (R2). Go outdoors and point the instrument toward the sun while watching the shadow the upper alignment vane casts on the lower vane. When you see the spot of sunlight on the lower vane, align the instrument until the spot of sunlight is centered over the alignment mark. Hold the instrument securely and, assuming the instrument incorporates readouts with a hold feature, press the readout Hold button to save the data.

Noon measurements are important because that's when the level of solar UV is highest. Make at least three observations per session. Record them in a notebook or read them into a tape recorder and transcribe them at a later time. Be sure to record the standard time. Later, you can use the standard time to determine the correct local apparent time.

After a measurement session, store the instrument in a clean, dust-free spot because dust and deposits from cooking fumes will block UV. Never leave the ozone-meter instrument inside a closed vehicle. The photodiode window and both faces of both filters must be kept meticulously clean. Dust must be blown away with clean compressed air. Especially dirty filters can be cleaned with a drop of camera lens cleaner.

Several methods can be used to calibrate ozone instruments. The simplest is to compare your observations with those made by a nearby instrument. In the United States, many locations have Dobson Spectrophotometers to which you can compare your readings. If you are not near one of these instruments, another method is to compare your readings with those made by satellite. NASA operates a computer Web site from the Goddard Space Flight Center that gives worldwide measurements of ozone.

Computing the Amount of Ozone

Determining the amount of ozone in the atmosphere overhead requires several steps. First you have to find the *local mean time*, this is not noon or 12:00. To find your local mean time, first find the number of degrees between your longitude and your time meridian. Multiply the number of degrees by four to obtain the correction for your location. If you are east of the time meridian, add the correction to the standard time for your area. If you are west of the meridian, subtract the correction from the standard time. The result is what is known as your *local mean time*.

Over the course of a year, Earth's orbit causes the sun to run either ahead or behind the *local mean time* by as much as 16 minutes. The actual difference between local mean time and the actual or apparent time is called the *equation of time*.

Second you will have to determine the angle of the sun above the horizon to compute the air mass and thus the ozone amount overhead. If you measure the angle manually, be sure to do it immediately after making your measurements. You can install a bubble level on your ozone-meter. Hold the unit on its side with the upper alignment vane pointed toward the sun; when the bubble is centered, measure the length of the shadow cast by the upper vane.

The tangent of the sun's angle above the horizon is the length of the upper vane divided by the length of the vane's shadow. It's important that you record the exact time of your measurements, this time information will permit you to calculate the sun's angle electronically at a later time if you elect to do so. Various computer programs are available that give the angle of the sun for any location on Earth.

The Total Ozone Equation

The total ozone equation is given as follows

$O_3 = \log(L_1^\wedge/L_2^\wedge) - \log(L_1/L_2) - (b_1 - b_2)$ times (p times $m/1{,}013$) divided by $(a_1 - a_2)$ times m

where

$L_1^\wedge$ and $L_2^\wedge$ are the intensities of the two wavelengths outside the atmosphere

L_1 and L_2 are the intensities of the two wavelengths during a measurement

a_1 and a_2 are the absorption coefficients for ozone at the two wavelengths

b_1 and b_2 are the Rayleigh scattering coefficients for air at the two wavelengths

m is the air mass (approximately $1/\sin c$, where c is the angle of the sun above the horizon)

p is the mean barometric pressure of the observation site in millibars (inches of mercury times 33.864 gives pressure in millibars)

$L_1^\wedge/L_2^\wedge$, the ratio of the signal at the two wavelengths above Earth's atmosphere, is known as the *extraterrestrial constant*. You will need to measure this value from the ground by making a Langley graph on a very clear, dry day when the ozone amount remains fairly constant.

Record L_1 and L_2 and the time as often as possible for a few hours, ending or beginning at solar noon. Plot the log of the ratio L_1/L_2 against air mass (m, the reciprocal of the sine of the sun's angle above the horizon) on a graph. If you extend the plot to 0 air mass, you will find the approximate extraterrestrial constant. L_1/L_2 can be the ratio of the UV measurements in watts per square meter or simply the numbers read from the readouts. Because the ratio of the two signals is being measured, it's not necessary to know the calibration of the photodiodes. The ozone absorption and Raleigh scattering coefficients can be found in published tables.

Sensitive Optical Tachometer

A optical tachometer will permit you to measure the speed of rotating objects such as wheels and motors, disks, and flywheels. By attaching a small mirror to a rotating object and shining a light on the mirror, you can use a phototachometer to measure the speed of that rotating object as shown in Figure 2-12. You can build a sensitive phototachometer quite easily. The phototachometer shown in Figure 2-13 uses a phototransistor sensor and two op-amps and *field effect transistor* (FET) along with an analog meter to measure rotating objects at speeds up to 50,000 rpm.

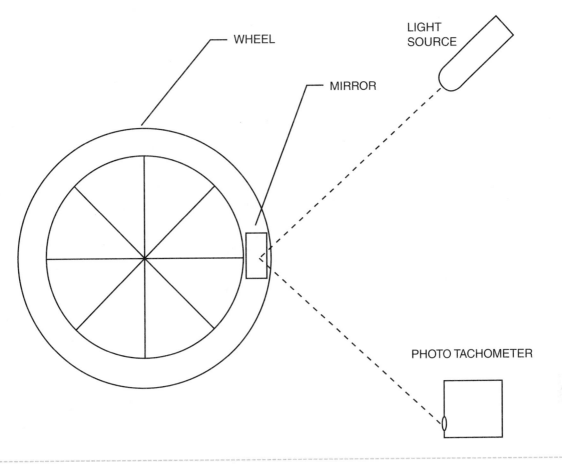

Figure 2-12   *Measuring speed with a phototachometer*

Light pulses striking the phototransistor Q1 produce voltage pulses at the input of the op-amp U1, which is connected as a schmitt trigger. (A schmitt trigger is a logic gate that reduces the problem of unwanted state changes near the voltage threshold. It prevents noisy signals from triggering a circuit.) The output pulses from U1 are then differentiated by C4/R7, giving voltage spikes that are then applied to the timer's (U2) trigger input. The output from the one-shot logic circuit passes through diode D1 and energizes the FET/R15 constant-current source to produce pulses with constant amplitude across R16, which are averaged by the meter at M1. Capacitor C11 is added to dampen the meter pointer vibration at low rpm ranges. The phototachometer can be operated using a 9-volt transistor radio battery at B1. Power is applied to the phototachometer via the power switch at S2.

The optical tachometer can be constructed on a $3\frac{1}{2} \times 6$ inch circuit board. When designing the circuit board, it is recommended that the phototransistor be mounted at one end of the circuit board so that when the circuit board is installed in an enclosure, the phototransistor will "look" outside the edge of the box. Integrated circuit sockets are highly recommended for the two integrated circuits, in the event of a circuit failure at a later date. It is much easier to replace components if sockets are used. When installing integrated circuits, you must observe the correct orientation of the IC before installing it in its socket. Integrated circuit packages will generally have either a small indented circle on one side of the package or a small cutout on the top of the IC package. Pin 1 of the IC will always be to the left of either the cutout or the small indented circle. When installing components on the circuit be careful to observe the polarity of the capacitors and diodes. There are five elec-

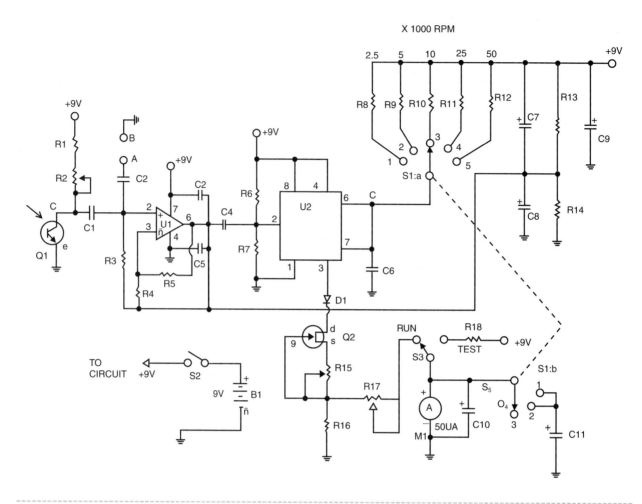

Figure 2-13 *Phototachometer circuit*

trolytic capacitors (C7, C8, C9, C10, and C11). Note the plus marking on each capacitor and its orientation with respect to the schematic diagram. A single diode is used in the circuit at D1; the band at one end of the diode is the cathode, and it should be facing the drain pin on Q2. The phototransistor at Q1 will have its collector connected to C1, while its emitter is connected to ground. The FET at Q2 has its gate connected to the junction of R15/R16, while its drain lead is connected to diode D1. The source lead of the FET is connected to one end of potentiometer R15. Also note that the meter must be installed with its plus or positive lead connected toward the rotary switch at S1.

Locate a 6 × 8 × 2½ metal chassis box in which to house the phototachometer. You will need to drill a few holes in the chassis: for the sensitivity control at R2, for the speed switch at S1, for the power switch at

S2, and for the Run/Test switch at S3. You will also have to use a chassis punch to cut out a hole for the 0 to 50 uA meter. Finally, you will have to drill a ½-inch hole so that the phototransistor can look to the outside world out the side of the chassis box.

In the initial tachometer prototype, the power switch S2, the speed switch at S1, and the Run/Test switch at S3 were all mounted on the top front of the chassis box along with the meter. The circuit board was mounted atop four ¼-inch plastic standoffs with ¾-inch 4-40 machine screws. The circuit board was oriented so the phototransistor will face the ½-inch hole that was drilled so that the phototransistor can look out the chassis box. A 9-volt battery holder was mounted on the bottom of the chassis box.

To calibrate the tachometer, R15 and R17 are first set to midposition, and the range switch is set to 2,500 rpm. A DC voltmeter is then connected across R16.

After disconnecting the wire between points C and D in the diagram, R15 is adjusted so that the voltmeter reads 1 volt. This wire is reconnected at the range switch and is set to 10,000 rpm. A 3-volt peak, 120 Hz sine wave is applied between points A and B; this is equivalent to applying 7,200 rpm.

Finally check for the rejection of low-level, 120 Hz modulation of incandescent light sources by aiming the phototransistor at a 50- to 75-watt lamp while varying the sensitivity control R2 over its range. If the meter does not remain at zero under all conditions, the input *hysteresis* is increased by increasing R4 to 10 kilo-ohms. The phototachometer is now ready for experimentation.

Sensitive Phototachometer Parts List

Required Parts

R1   3.9K ohm, 1/4-watt, 5% resistor

R2   100K ohm potentiometer (panel mount)

R3   150K ohm, 1/4-watt, 5% resistor

R4   5.1K ohm, 1/4-watt, 5% resistor

R5, R8   100k ohm, 1/4-watt, 5% resistor

R6, R7   47k ohm, 1/4-watt, 5% resistor

R9   50K ohm, 1/4-watt, 5% resistor

R10   25K ohm, 1/4-watt, 5% resistor

R11   10K ohm, 1/4-watt, 5% resistor

R12   5K ohm, 1/4-watt, 5% resistor

R13, R14   3.9K ohm, 1/4-watt, 5% resistor

R15   5K ohm calibration potentiometer (trimpot)

R16   1K ohm, 1/4-watt, 5% resistor

R17   10K ohm calibration potentiometer (trimpot)

R18   200K ohm, 1/4-watt, 5% resistor

C1   0.002 uF, 35-volt ceramic disc capacitor

C2   0.05 uF, 35-volt ceramic disc capacitor

C3, C5   0.1 uF, 35-volt ceramic disc capacitor

C4   0.001 uF, 35-volt ceramic disc capacitor

C6   0.068 uF, 35-volt ceramic disc capacitor

C7, C8, C10   20 uF, 35-volt electrolytic capacitor

C9, C11   100 uF, 35-volt electrolytic capacitor

D1   1N914 silicon diode

Q1   phototransistor ECG-3031

Q2   FET transistor ECG 312/451

U1   LM741 op-amp

U2   LM555 timer IC

M1   50 uA micorameter

S1   two-pole five-position rotary switch

S2   SPST toggle (power switch)

S3   SPDT toggle switch (run/test)

Miscellaneous   PC board, wire, IC sockets, hardware, etc.

Turbidity

Turbidity water is caused by the presence of very fine suspended matter such as clay, silt, organic and inorganic matter, soluble colored organic compounds, plankton, and other microscopic organisms. Turbidity measurements relate to the optical property of water that causes light to be scattered and absorbed rather than transmitted in straight lines through the sample. The common unit of measurement of turbidity is the *Nephelometric turbidity unit* (NTU).

In general terms, the turbidity of a fluid sample is a measure of how clear or cloudy the sample is, the degree of cloudiness being a function of the concentration of suspended solids in the liquid. In this project, a method that measures the extent to which the sample scatters light will be used.

Clarity of water is important in producing products destined for human consumption and in many manufacturing operations. Beverage producers, food processors, and potable water treatment plants drawing from a surface water source commonly rely on fluid–particle separation processes such as sedimentation and filtration to increase clarity and ensure an acceptable product. The clarity of a natural body of water is an important determinant of its condition and productivity.

Correlation of turbidity with the weight or particle number concentration of suspended matter is difficult because the size, shape, and refractive index of the particles affect the light-scattering properties of the suspension. When present, insignificant concentrations of particles consisting of light-absorbing materials such as activated carbon cause a negative interference. In low concentrations these particles tend to have a positive influence, because they contribute to turbidity. The presence of dissolved, color-causing substances that absorb light may cause a negative interference. Some commercial instruments may have the capability of either correcting for a slight color interference or optically blanking out the color effect.

Instruments that measure turbidity are called *turbidity meters*. They range in complexity and cost from battery-powered handheld units to continuous online monitoring systems. Excessive turbidity detracts from the appearance of treated water and has often been associated with unacceptable tastes and odors. Turbidity can serve as a source of nutrients for waterborne bacteria, viruses, and protozoa, which can be embedded in or adhere to particles in the raw water or become trapped within *floc* formed during water treatment; turbidity can thus interfere with the enumeration of microorganisms in finished water, as the microorganism may not be detectable or may be grossly underestimated by current detection methods. The adsorptive properties of suspended particles can also lead to a concentration of heavy metal ions and biocides in turbid waters. Turbidity can interfere with disinfection processes and the maintenance of a chlorine residual: depending on the composition of the turbidity-causing material, interference with disinfection can range from negligible to severe. Turbidity has also been related to *trihalomethane* formation in chlorinated water. Outbreaks of disease traced to chlorinated water supplies have been associated with high turbidity. The occurrence and persistence of microorganisms within distribution systems have been correlated with turbidity and other factors. The effect of turbidity on disinfection efficiency may be frequently related to the type and nature of the particulates. Surface water sources in particular may be susceptible to organic substances and undesired organisms that can impede disinfection or otherwise cause drinking water quality problems.

Appropriate technology is available to treat and monitor turbidity to low levels. Provision of treated water at or below this limit will minimize the introduction of unfavorable particulates and biological matter into the distribution system and thereby render better disinfection opportunity, effectiveness, and maintenance. Special site-specific problems may require more rigorous attention for the production of low-turbidity water. Any sudden increase in the turbidity of unfinished water indicates deteriorating quality of the raw water or loss of control in the water treatment process. Certain water supplies, such as groundwater, may contain non-organic-based turbidity, which may not seriously hinder disinfection. Therefore, a less stringent value for turbidity in water entering a distribution system may be permitted if it is demonstrated that the system has a history of acceptable microbiological quality and that a higher turbidity value will not compromise disinfection.

Electronic Turbidity Meter

The electronic turbidity meter in this project is a single-beam transmission type turbidity meter, which is ideal for a science fair project. The turbidity meter circuit shown in Figure 2-14, revolves around the clear plastic or glass flat-surfaced test cell or sample

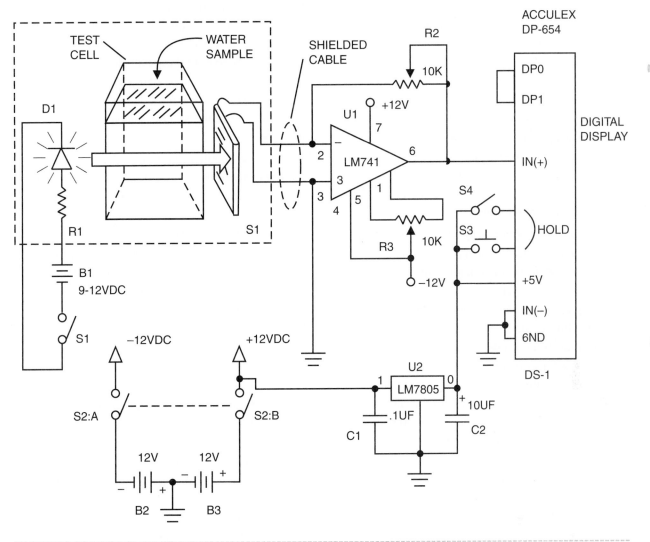

Figure 2-14   *Optical turbidity sensor*

chamber, which is housed in a "dark" light box. A light source such as a small incandescent lamp or a high-output or superbright LED is allowed to shine through the water sample in the test cell onto a silicon solar cell. The lamp is aligned with the solar cell in a straight beam path where a small light tube is placed in front of the lamp to direct the light to the solar cell.

The electronics of the circuit consists of an LM741 op-amp as the electronic amplifier. The silicon solar cell is connected directly to the op-amp as shown, with the plus (+) lead from the solar cell connected directly to the plus (+) pin of the op-map at pin 3. The minus lead (−) or the solar cell is connected to the minus input (−) of the op-amp at pin 2. The out-

put of the op-amp at pin 6 is fed directly to a Acculex DP-654 LCD display module, which will provide a good stable output reading for the turbidity meter. The output of the op-amp is coupled to the plus (+) input pin on the LCD display. The minus (−) lead of the LCD meter is connected to the display ground as well as to the circuit ground. The display hold switches S3 and S4 allow either momentary hold function or a more permanent hold feature. The hold switches are wired directly to the 5-volt power source, as is the power supply lead for the LCD meter. A 5-volt regulator at U2 is used to step down the 12-volt power source to 5 volts for the display. Note that the circuit uses two 12-volt batteries or 12-volt power supplies to power the turbidity meter. The op-amp requires both a +12-volt supply and a −12-volt power supply to operate correctly. Note also that the two batteries are wired in such a way as to connect one of the plus terminals to one of the minus terminals to form a common ground. Power switch S2 is a DPST toggle switch, which is used to separately switch in both the plus and minus power sources. The light circuit shown at the left of the diagram uses a superbright white LED as the light source placed in series with a 1K ohm resistor through a power switch to a 9- to 12-volt power source such as a 9-volt battery.

The electronic turbidity circuit can be readily constructed on perf-board or on a prototype circuit board or, if you desire, on a small printed circuit board. An integrated circuit socket is recommended for the op-amp, and will save you much grief in the event of a future failure. When installing the integrated circuit you must pay particular attention to the orientation of the op-amp to avoid damage to the IC. Most ICs have either a square cutout on the top of the IC, with pin 1 being to the left of the cutout, or to the left of the small indented circle next to pin 1 of the IC. Take your time when installing the IC to avoid damage to the IC pins. When installing the electrolytic capacitor, you will need to orient it so that the positive marking is connected to the plus (+) output of the regulator at U2. After constructing the circuit make sure to inspect the circuit board to make sure there are no cut component leads across any circuit wires or PC lands.

The prototype turbidity meter is housed in a metal chassis box measuring 6 × 6 × 4 inches. The two potentiometers used for calibration were mounted on the front of the chassis box. These two potentiometers should be chassis-mounted types so that adjustments can be made easily once the circuit is in operation. The on-off switch as well as the LCD meter switches and the LCD are all mounted on the front of the chassis box for easy control. Screw terminal strips or RCA jacks could be used to connect the circuit to a remote power supply of batteries.

The silicon solar cell is separated from the main circuit board enclosure by a length of shielded cable to the "dark chamber" that housed the lamp circuit, the lamp battery, and the test cell. A fixture made

from a block of wood and small circuit board is used to mount the LED about 4 inches from the basement of the dark chamber. A second fixture made of wood is made to hold the solar cell in perfect alignment with the lamp in the dark chamber enclosure. The rectangular plastic flat sided test cell is placed in between the lamp and the silicon solar cell assemblies.

The dark chamber that houses the lamp assembly, the test cell, and the solar cell assembly can be fabricated from wood, cardboard, or even styrofoam. It is most important that the five-sided dark chamber does not leak light and can be kept as dark as possible by taping the edges of the corners or by filling in the cracks or painting the corners. You will need to make only a small notch or hole for the silicon solar cells leads to come out of the dark chamber box. Once the circuit is assembled, the solar cell is connected, and the lamp assembly is set up in the dark chamber, you will only have to calibrate the turbidity meter before you can use it.

To calibrate the turbidity meter you will first have to fill the test cell with clear tap water or distilled water. With the source lamp off, adjust R3 to give an output on the LCD of 0.00 volts. Next turn on the source lamp with switch S1 and adjust R2 to give an output of 1 volt on the LCD voltmeter. Finally insert the water sample you have collected into the test cell and record the voltage on the LCD voltmeter. The turbidity meter is now ready to assist you in collecting and analyzing water samples.

Measurement Techniques

Determine turbidity as soon as possible after the sample is taken. Proper techniques are important in minimizing the effects of instrument variables as well as stray light and air bubbles. Regardless of the instrument used, the measurement will be more accurate, precise, and repeatable if close attention is paid to proper measurement techniques. Measure turbidity immediately to prevent temperature changes and particle flocculation and sedimentation from changing sample characteristics. If flocculation is apparent, break up aggregates by agitation (that is by swirling the water around). Avoid dilution whenever possible. Particles suspended in the original sample may dissolve or otherwise change characteristics when the temperature changes or when the sample is diluted. Remove air or other entrained gases from the sample before measurement. Preferably degas even if no bubbles are visible. Degas by applying a partial vacuum, adding a nonfoaming type surfactant, using an ultrasonic bath, or applying heat. In some cases, two or more of these techniques may be combined for more effective bubble removal.

Use sample cells or tubes of clear, colorless glass or plastic. Keep cells scrupulously clean, both inside and out, and discard if scratched or etched. Never handle them where the instrument's light beam will strike them. Use tubes with sufficient extra length or with a protective case, so that they may be handled properly. Fill cells with samples and standards that have been agitated thoroughly and allow sufficient time for bubbles to escape.

Clean sample tubes by thoroughly washing with laboratory soap inside and out followed by multiple rinses with distilled or de-ionized water. Let cells air dry. Handle sample cells only by the top to avoid dirt and fingerprints within the light path. Cells may be coated on the outside with a thin layer of silicone oil to mask minor imperfections and scratches that may contribute to stray light. Use silicone oil with the same refractive index as glass. Avoid excess oil because it may attract dirt and contaminate the sample compartment of the instrument. Using a soft, lint-free cloth, spread the oil uniformly and wipe off excess. The cell should appear to be nearly dry with little or no visible oil. Because small differences between sample cells significantly impact measurement, use either matched pairs of cells or the same cell for both standardization and sample measurement.

Gently agitate sample. Wait until air bubbles disappear and pour sample into cell. When possible, pour the well-mixed sample into the cell and immerse it in an ultrasonic bath for 1 to 2 seconds or apply the vacuum degassing technique, causing complete bubble release. Read turbidity directly from the instrument display.

Optical Turbidity Sensor Parts List

Required Parts

R1   1K ohm, 1/4-watt resistor

R2, R3   10K potentiometer (chassis mount)

D1   Superbright white LED

SI   silicon solar cell

C1   0.1 uF, 35-volt disc capacitor

C2   10 uF, 35-volt electrolytic capacitor

U1   LM741 op-amp

U2   LM7805 5-volt regulator

S1, S4   SPST toggle switch

S2   DPST toggle switch

S3   normally open push-button switch

B2, B3   12 gell cell volt battery

DS-1   Acculex DP654 digital voltmeter module

Miscellaneous test cells, PC board, IC socket, wire, connectors, dark chamber mounting assemblies, hardware, etc.

Chapter Three

Heat Detection

Heat is transferred from one place to another in one of three ways: conduction, convection, and radiation. *Conduction* is the process of transferring heat from molecule to molecule in a substance. When one end of an iron rod is placed in a fire, the other end soon gets hot because heat is transferred from one end of the iron to the other end by conduction (from molecule to molecule). *Convection* is the process of transmitting heat by means of the movement of heated matter from one place to another. Convection thus takes place in liquids and gases. A room is heated by means of convection by circulating warm air through the room. This brings us to *radiation*. In both conduction and convection, heat is transmitted, or transported, by moving particles, for example by molecules or air. However heat can also travel where matter does not exist. For example, the heat from the sun reaches the earth across the 93 millions of miles of space. When a cloud passes between the sun and a point on earth, the heat at that point is diminished or cut off. This is due to the fact that heat is transmitted or radiated by waves.

Heat waves and light waves are of the same nature; they are both electromagnetic radiations that differ only in wavelength, heat waves being longer than light waves. Heat waves near the radio portion of the spectrum are called infrared.

In this chapter we will construct an infrared flame detector, which can sense a match or flame up to 3 feet away. In this chapter, you will also learn how to construct a freeze alarm, an over-temperature monitor, and an analog data-logger for sending temperature data remotely. More advanced projects include an LCD thermometer, a night vision viewer, and an infrared motion detector, which can sense the body heat of an intruder up to 50 feet away. The infrared motion detector could be used to create your own home alarm system.

Infrared Flame Sensor Switch

The infrared detector switch is a very sensitive circuit that can be use to detect the presence of a flame, match, or a heat source, such as an iron or soldering iron, up to 3 feet away and then activate a relay. The heart of the infrared flame detector circuit is two tiny thermistors, as shown in Figure 3-1. A *thermistor* is a temperature-sensitive resistor, which changes its resistance as the temperature varies. The resistance of T1 is inversely proportional to temperature. In other words, the resistance of T1 increases as temperature decreases. Glass bead or bulb thermistors rated at 25K to 50K ohms at room temperature are recommended for this project. The heart of the infrared detector switch is two thermistors. Thermistor T1 is used as a heat sensor connected to the minus ($-$) input of an op-amp configured as a comparator. Thermistor T2 acts as a reference resistor at normal room temperature and is connected to potentiometer R2, which is used as a set-point adjustment resistor. Changes in ambient room air temperature cause equal changes in T1 and T2, but an infrared source such as a flame or hot iron affects only thermistor T1. Thermistor T1 is wired in series with a 33K ohm resistor, while T2 is wired in series with a 50K ohm potentiometer. Under normal operation, power is applied to the circuit and the circuit is allowed to stabilize for a few moments. Potentiometer R2 is adjusted until the relay just switches off; this is the switch-point threshold. The circuit is normally kept below this threshold until the thermistor detects an infrared source. Once an infrared source is detected, the op-amp changes state and activates transistor Q1, which pulls in the SPDT relay at RY-1. Because the op-amp is used in a comparator configuration, the

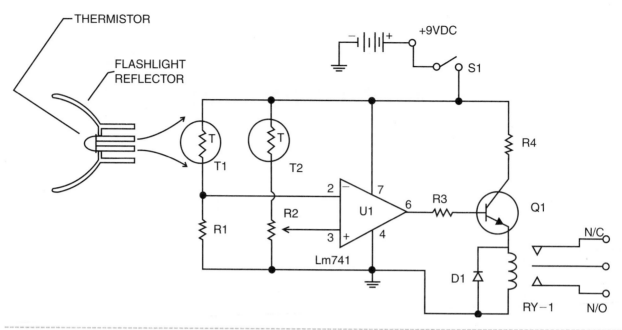

Figure 3-1   *Infrared flame sensor switch circuit*

entire circuit can be powered from a single 9-volt transistor radio battery or "wall wart" power supply. (A wall wart is a small power supply that can be plugged directly into the wall; it got its name *wart* because when plugged into a power strip, it often blocks more than the one spot it's allotted.)

The infrared switch is so sensitive that it will detect an infrared source more than 3 feet away. In order to achieve this sensitivity, thermistor T1 is mounted in the center or at the focal point of a flashlight reflector. Care must be taken to ensure that the thermistor is accurately mounted at the focal point of the reflector to ensure maximum sensitivity of the circuit.

The infrared switch circuit can be built on a perfboard or proto-board or on a small printed circuit board. The prototype circuit is constructed on a 2½ × 2 inch circuit board. The circuit layout is not critical and can be built in under an hour. Use an *integrated circuit* (IC) socket when building the circuit to enable you to repair the circuit if the need ever arises. ICs have markings on them to indicate the pinouts of the device. ICs generally will have a plastic cutout in the top center of the package. If this is the case then pin 1 will be to the left of the cutout. Some ICs may have a small indented circle on the top just to the right of

pin 1. When installing the IC be sure to align pin 1 to the socket pin corresponding to pin 1. The only other component that requires correct orientation is the diode placed at D1. The relay used in this project is an SPDT mini relay; this type of relay is used because it allows you to be able to use either a normally closed set of contacts or a normally open set of contacts depending upon your particular application. After constructing the circuit board, you will need to inspect the bottom or foil side of the circuit board to make sure that no stray or loose components leads stuck to the underside of the circuit board. You want to be sure that there are no short-circuit paths between circuit pads. Applying power to the circuit with short-circuits may destroy the circuit.

The infrared switch is placed in a metal enclosure in order to protect the circuit. A 5 × 7 inch metal chassis box is used to house the infrared switch circuit. The on-off switch at S1 and the potentiometer at R2 are mounted on the top front of the chassis box. Thermistor T2 is mounted on the circuit board, whereas thermistor T1 is at the focal point of flashlight reflector.

As mentioned earlier, the thermistor must be mounted in the exact center of the center hole of the reflector. In addition to being mounted in the center of the reflector, the thermistor must also be mounted in the focal plane of the reflector to be most effective. A small round piece of plastic or circuit board can be cut to fit over the center hole of the reflector and then glued in place. Two small holes are drilled into the small diameter piece of plastic in the center of the reflector to accept the leads from the thermistor T1. The thermistor leads can then be slid in or out of the holes to adjust them into the focal plane, and then the leads can be glued to the plastic center disk. An over-sized hole is drilled into the side of the chassis box to accept the flashlight reflector. The flashlight reflector then is mounted at the side of the chassis box, and a hole is drilled to allow the thermistor wires to be soldered to the circuit board.

The small relay is mounted to the circuit board, and the three relay contact leads are brought to a three-terminal screw terminal strip, which is mounted to the opposite end of the chassis box from the flashlight reflector. Two mounting screws are used to secure the terminal strip to the chassis box, and three holes are drilled into the box to allow the relay leads to connect to the circuit board. A 9-volt transistor radio battery holder is mounted to the bottom of the chassis box along with the circuit board.

The infrared switch unit is now ready to be tested. Connect up the thermistor T1, apply power to the infrared switch circuit, and adjust the potentiometer R2 until the relay turns off. There are many potential applications for the infrared switch circuit, from turning things off when surfaces get too hot to your next heat-seeking robot project. Your electronic infrared detector switch is now ready to "see" the world!

Infrared Flame Sensor Switch Parts List

Required Parts

T1, T2   25K to 50K ohm thermistor (Newark Electronics)

R1   33K ohm, 1/4-watt resistor

R2   50K potentiometer (chassis mount)

R3   1K ohm, 1/4-watt resistor

R4   47-ohm, 1/4-watt resistor

D1   1N4002 silicon diode

Q1   2N2222 transistor

RY-1   6-volt SPDT relay

U1   LM741 op-amp

B1   9-volt battery

S1   SPST toggle switch

Miscellaneous   PC board, wire, IC socket, terminal strip, battery holder, chassis box, flashlight reflector, etc.

Freezing Temperature Alarm

The freezing temperature alarm can be used in a variety of applications to detect when air temperature falls below zero degrees C or 32 degrees F. The freeze alarm can be used to detect freezing road conditions, informing you to slow down or change your driving habits. The freeze alarm could also be used to tell you when to sprinkle salt on your steps or to monitor experiments that require freezing.

The heart of the freeze alarm circuit as shown in Figure 3-2 is the thermistor at T1. A thermistor is a temperature-sensitive resistor, which changes its resistance as the temperature varies. The resistance of T1 is inversely proportional to temperature. In other words, the resistance of T1 increases as temperature decreases. The thermistor is a Keystone RL0503-5536K-122-MS. The thermistor produces 361K ohms at 0 degrees C and 100K ohms at 25 degrees C.

The thermistor is connected between the 5-volt power supply and the minus ($-$) input of a low-power comparator at U1. The threshold set-point of the comparator circuit is determined by the voltage divider set up by resistors R2 and R3, which bridge between the power supply and ground as shown. As the comparator threshold is reached when the thermistor reaches 32 degrees F, the comparator produces an output at pin 1. The output of U1 is

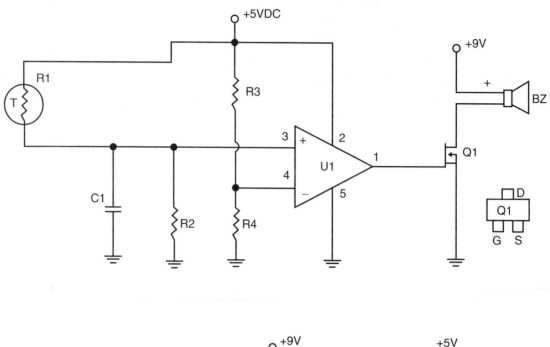

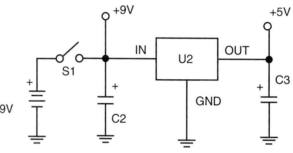

Figure 3-2 *Freezing temperature alarm circuit*

amplified by the Q1, an N-channel enhancement mode vertical DMOS FET. The three-pin N-channel FET then drives a solid-state beeper at BZ, when 32 degrees F is reached.

The freeze alarm is powered by a 9-volt transistor radio battery. The 9-volt battery is coupled to the regulator at U2 through the on-off switch at S1. Nine volts from the battery is first supplied to the buzzer. Nine volts is also fed to U2 where the voltage is reduced and regulated down to 5 volts DC to power the freeze alarm. All of the semiconductor devices are low-power small footprint devices; the comparator is a National Semiconductor device; the FET is a ZETEX product; and the regulator is from Seiko.

The freeze alarm is constructed on a small proto-board or experiments universal board, which is available from Datak (part #12-611). The circuit can be constructed in under an hour. Layout is not critical because this is not a high-frequency or RF circuit. Most components are not critical and are easily obtainable. Note, however that the resistors are all 1% resistors for accuracy of the circuit operation. The only caveat in building the circuit is to pay particular attention to the orientation of the semiconductor devices. Study the pinouts of all the semiconductor devices before soldering them into place on the circuit board. After constructing the circuit board, be sure to carefully look for cold solder joints and circuit bridges between circuit pads. Also look for loose component leads that may have attached themselves to the underside of the board. All components except for the thermistor are mounted on the circuit board.

The freeze alarm can be housed in a small metal enclosure if desired. A small $3^1/2 \times 5$ inch metal box was chosen to house the prototype freeze alarm circuit. The buzzer and power switch are mounted on the top front side of the enclosure. The circuit board is secured to the bottom of the enclosure along with the 9-volt battery holder. A two-circuit RCA jack is mounted on the side of the box to accommodate the remote thermistor.

Next, you will want to prepare a shielded sensor cable with an RCA plug at one end and the thermistor at the other end. Depending upon your application, you may wish to house the thermistor in a discarded pen, with provisions for the shielded cable to exit the top of the pen assembly. First, prepare the thermistor by spraying it and the leads with Crylon or a similar plastic coating to insulate the thermistor and the first $1/4$ inch of the leads. The thermistor leads are usually very small-gauge wire, so care must be taken when soldering the shielded cable to the thermistor. Heatshrink tubing should be put over the connection or junction between the thermistor and the shielded cable, and some strain relief should be provided.

Once the freeze alarm has been installed in its enclosure, and the thermistor cable attached to the circuit, you can install the 9-volt battery. Place the on-off switch in the off position and install the battery. To test the circuit, prepare a bowl with a number of ice cubes or crushed ice in a small amount of water. You want to create an ice bath or ice slurry. Wait about a minute for the ice to cool the water to near freezing, then apply power to the freeze alarm and immerse the thermistor probe into the ice slurry, and the buzzer should start to sound. Next remove the probe from the ice slurry and the buzzer should go quiet in just a moment. If the circuit is functioning correctly, you can turn off the circuit and your freeze alarm is now ready to serve you!

Freeze Alarm Parts List

Required Parts

T1	Keystone thermistor (see text)
R1	499K ohm, 1/4-watt, 1% resistor
R2	1 megohm, 1/4-watt, 1% resistor
R3	720K ohm, 1/4-watt, 1% resistor
C1	0.1uF, 35-volt ceramic capacitor
C2, C3	10 uf, 35-volt electrolytic capacitor
Q1	ZVN4106F N-channel enhancement mode DMOS FET (ZETEX)
U1	LMC7215 low-power comparator (National Semiconductor)
U2	S-812C50SGY-B SMT-type regulator (Seiko)
S1	SPST toggle switch
B1	9-volt transistor radio battery
BZ	QMB-12 electronic beeper (Star)
Miscellaneous	printed circuit board, battery holder, wire, chassis box, RCA jack, RCA plug, standoffs, screws, nuts, etc.

Overtemperature Alarm

The overtemperature alarm circuit is a great sensing tool and will alert you to an overheating condition. The prototype overtemperature alarm has a trip point around 150 degrees F, but this can be altered for other operating ranges if desired. The overtemperature alarm can be used to alert you of an overheating condition in your computer or other home appliance. You can easily substitute the electronic buzzer with a small relay to use the circuit to activate a siren or flashing beacon if desired.

The heart of the overtemperature alarm is the thermistor at T1. A thermistor is a temperature-sensitive resistor that changes its resistance as the temperature varies. The resistance of T1 is inversely proportional to temperature. In other words, the resistance of T1 increases as temperature decreases. A Keystone RL0503-55.36k-122MS thermistor is utilized in this project. The thermistor produces 17.89K at 150 degrees F and 100K at 77 degrees F. You can select other thermistor values for other temperature ranges.

The overtemperature alarm circuit is a rather unique circuit shown in Figure 3-3, which uses both sections of the CD4013B at U1. The CD4013B is a CMOS dual D-Type flip flop IC. The first section of the CD4013B acts as a monostable multivibrator or oscillator, producing pulses at a rate of 12 per minute, at the output of U1:A at pin 1. The oscillator's frequency is determined by the timing components around R1 and C2. The oscillator drives a FET transistor at Q1, which sends the oscillator pulse to the second CD4013B section at U1:B. Note that the data pulse lags the clock edge when the thermistor is cold, and the data pulse leads the clock pulse edge when T1 is hot. This produces a change in the output stage of U1:B. The flip flop's output at Q or pin 13 drives a ZVNL110A FET at Q2 when the state changes takes place and the set-point threshold is reached. Referring back to the diagram, the 1 millisecond pulses are 5 seconds apart. This unique low-power circuit is powered by a 3-volt, 120 mA/hr lithium cell, which will last many hours in the stand-by state.

The overtemperature alarm circuit prototype is constructed on a small proto-board or experiments universal board, which is available from Datak (part #12-611). The layout of the components is not critical, and no special parts are required in this circuit. Resistors are all 1/4-watt, 5% types. Capacitors are Mylar types except for the electrolytic capacitor at C1. When constructing the circuit pay particular attention to the pinouts of the transistors at Q1 and Q2. It is advisable to use an IC socket for the CD4013B in the event of a component failure at a later date. The IC socket will make parts substitution easy in the

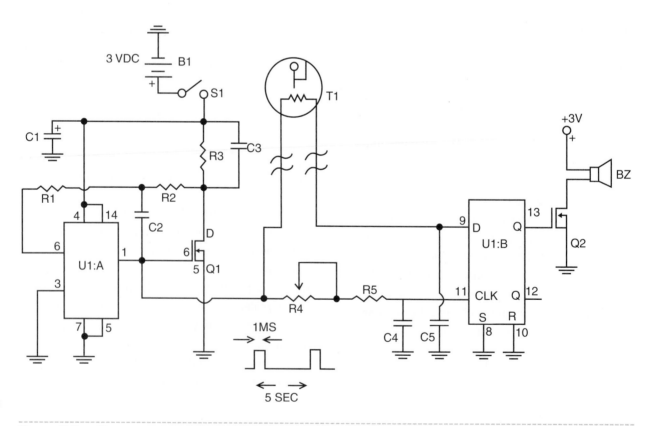

Figure 3-3   *Overtemperature alarm circuit*

event of a failure. When installing the IC be careful as to the orientation. ICs either have a square cutout at the top center of the plastic package, or a small indented circle to the right of pin 1. If the IC package has a square cutout in the top center, then pin 1 will be to the left of the cutout. All components except for the thermistor T1 will be soldered to the printed circuit board.

After constructing the circuit board, be sure to inspect the underside of the PC board for cold solder joints and bridges between circuit pads. Also look for stray component leads that may have stuck to the circuit board after cutting. Once you have completed building and inspecting the circuit board you are ready to install the circuit into an enclosure.

The prototype overtemperature alarm is installed in a $3^{1}/_{3} \times 5$ inch aluminum chassis box. The on-off switch at S1, the calibration potentiometer at R4, and the buzzer are mounted on the top front of the chassis box. The circuit board is mounted on standoffs and placed on the bottom of the chassis along with the 3-volt lithium battery holder. A two-position screw terminal strip was mounted at the rear of the chassis box to allow connection to the thermistor sensor. One screw on the terminal strip goes to pin 1 on U1, and the second screw on the terminal strip goes to pin 9 and U1. Note that a two-circuit jack cannot be used in this application because both thermistor leads are active, and neither is ground. The overtemperature alarm could be used for higher or lower temperature ranges if desired by changing the thermistor range. When selecting a different thermistor, you will want to try and roughly maintain resistor values that are not too far off the value given, to ensure that the trip point will be reached.

Next, you will want to prepare a shielded sensor cable with timed leads at one end and the thermistor at the other end. Depending upon your application, you may wish to house the thermistor in a discarded metal pen casing, with provisions for the shielded cable to exit the top of the pen assembly. First, prepare the thermistor by spraying the thermistor and the leads with Crylon or a similar plastic coating to insulate the thermistor and the first $^{1}/_{4}$ inch of the thermistor's leads. The thermistor leads are usually very small-gauge wire, so care must be taken when

soldering the shielded cable to the thermistor. Heat-shrink tubing should be put over the connection or junction between the thermistor and the shielded cable, and some sort of strain relief should be provided.

Once a thermistor housing has been decided on and built, you can move on to testing and calibrating your new overtemperature alarm. Connect the thermistor to the screw terminals on the chassis box enclosure. Make sure the on-off switch, S1, is turned off, then install the 3-volt lithium battery into its holder. You are now ready to calibrate the overtemperature alarm.

To test the circuit, you will need to heat the thermistor to 150 degrees F for the buzzer to sound. One method is to tape a calibrated glass thermometer to the thermistor housed in a metal pen housing and place the assembly next to a gas or electric stove. Bring the thermistor/thermometer assembly near the burner, to about $^{1}/_{4}$ inch above the surface. Turn on the burner, but make sure that you do not place the thermistor/thermometer on the burner surface.

Once the assembly reaches 150 degree F, the buzzer alarm should sound and your overtemperature alarm is calibrated and ready. If the circuit is functioning correctly, you can turn off the circuit and your overtemperature alarm is now ready to serve you! Feel free to experiment with different applications and different temperature ranges if desired.

Overtemperature Alarm Parts List

Required Parts

T1 RL0503-55.36k-122MS (Keystone) 17.89k@150_F/100k@77_F

R1 4.7 megohm, 1/4-watt resistor

R2 100K ohm, 1/4-watt resistor

R3 22 megohm, 1/4-watt resistor

R4 10K ohm potentiometer (panel mount)

R5 15K ohm, 1/4-watt resistor

C1 47 uF, 35-volt elec-
trolytic capacitor

C2 0.01 uF, 35-volt
ceramic disc capacitor

C3 0.22uF, 35-volt
ceramic disc capacitor

C4, C5 470 pF, 35-volt
mylar capacitor

Q1, Q2 ZVNL110A FET
transistor (ZETEX)

U1 CD4013B CMOS dual
D-type flip flop
(National
Semiconductor)

BZ MMB-01 electronic
buzzer (Star)

S1 SPST toggle switch

B1 3-volt lithium
battery

Miscellaneous circuit
board, wire, terminal
strip, chassis box,
standoffs, screws,
nuts, etc.

Analog Data Logger System

An analog data logger system is a simple way to send or record a voltage reading between two points? For example, you can take temperature measurements in one location and transmit or record the data and obtain the data and display it at a remote location in realtime or at a later date. The remote analog data logger system is shown in both Figures 3-4 and 3-5. This two-part analog data logger can be configured in two different ways. In Figure 3-4, a sensor is shown connected to the voltage-controlled oscillator circuit, which in turn is connected to a small tape recorder, which records your data at one location. In this scheme you take the recorder to a second location; you connect the tape recorder to the frequency-to-voltage circuit connected to a voltmeter and then playback the recorded data information to a digital voltmeter. In the second data logger scheme in Figure 3-5, a temperature sensor is connected to a voltage-controlled oscillator. The output of the voltage-controlled oscillator is fed directly to the audio input

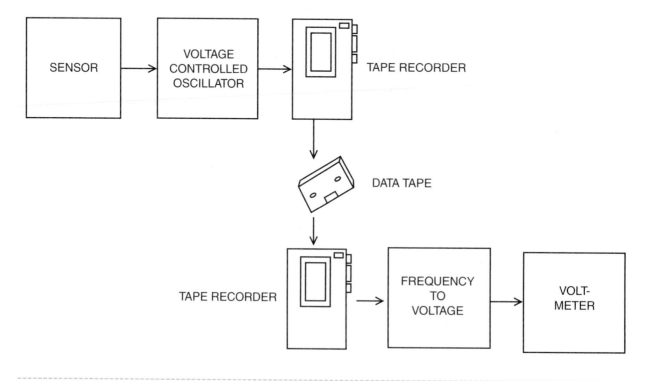

Figure 3-4   *Analog data logger system I*

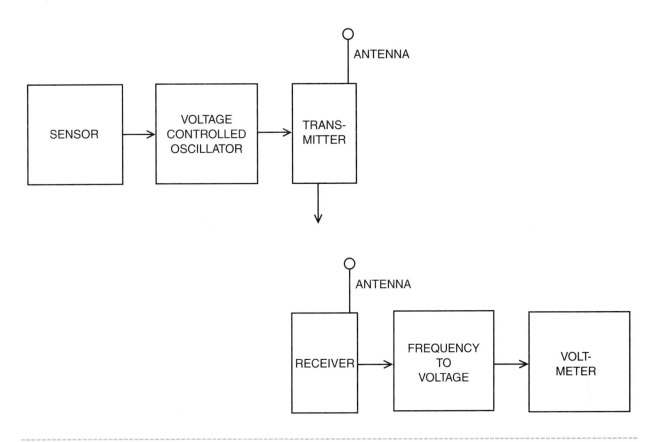

Figure 3-5 *Analog data logger system II*

of a radio transmitter, either a walkie talkie or an FM broadcast transmitter. At the receiving end of the circuit a receiver audio circuit is fed directly into a frequency-to-voltage converter. The frequency-to-voltage converter is then fed into a digital voltmeter. In this data logger scheme, you can send the data realtime from one location to another.

The sending or sensor circuit shown in Figure 3-6 shows a temperature sensor connected between pins 4 and 7 of an LM555 oscillator/timer IC. In operation, the LM555 oscillates at a frequency determined by the resistance of the temperature sensor at T1. As the temperature at T1 changes, the resistance changes, thereby changing the frequency of the oscillator. Capacitor C1 has been selected to keep the maximum frequency of the oscillator within the range of a typical tape recorder. The output of the LM555 at pin 3 of the IC is fed to a capacitor at C2, which is in turn connected to a minitransformer at L1. Transformer L1 is a 600-ohm to 600-ohm interstage type used to couple radio circuits together. (A 600-ohm to 600-

ohm is a 1:1 transformer with both windings 600 ohms each.) The transformer's output is connected to a coupling capacitor at C3. The capacitor and remaining transformer lead is connected to a mini audio plug, which couples the oscillator circuit to either a transmitter's audio input or to the audio input of a tape recorder.

The receiving or playback circuitry of the remote data logger is shown in Figure 3-7. The heart of the frequency-to-voltage converter in the display circuit is the LM331 IC. The audio input circuit to the frequency-to-voltage chip enters the circuit at capacitor C1. The audio is next coupled to a minitransformer at L1. Transformer L1 is an 8-ohm to 1K ohm interstage matching transformer used in radios. It is readily available at any Radio Shack store. The 1K output of the transformer is next coupled to a capacitor at C2, which is then fed into the frequency-to-voltage converter chip at pin 6.

In operation, one of the inputs of the LM311 is biased at a voltage determined by R2 and R3. When

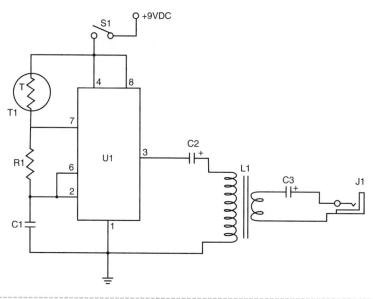

Figure 3-6   *Temperature transmitter circuit*

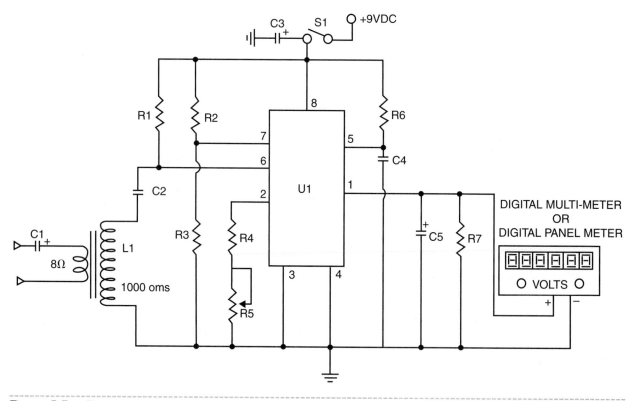

Figure 3-7   *Temperature receiver circuit*

the amplitude of the incoming exceeds that of the reference voltage provided by R2 and R3, the output of the comparator changes state until the input signal falls below the reference voltage.

The output from the comparator is connected to a monostable multivibrator (a one-shot) in the LM331.

Each time the comparator switches, the one-shot triggers and closes a current switch. This allows the output filter capacitor at C5 to charge for a time determined by the time constant of R6 and C4. Resistor R7 functions as a bleeder resistor that continually discharges the charge on capacitor C5, thereby causing the voltage on C5 at any given instant to correspond to the average charge available from the current source. Therefore, the voltage across C5 is directly proportional to the frequency of the incoming signal.

The output from the LM331 should be applied to a high-impedance voltmeter or digital panel meter. Potentiometer R5 allows you to calibrate the circuit for a particular output voltage at a specific input signal frequency. The LM331 has a linear frequency response from 1 to 10 KHz. The remote data logger receive-and-display circuit can be powered from a standard 9-volt transistor radio battery.

Constructing the remote data logger is pretty straightforward and can be done in less than 2 hours. The sending oscillator circuit is constructed on a piece of experiments proto-board from Datak (part #12-611). Consider using an IC socket for the LM555. In the event of a circuit failure, it is a simple matter to replace an IC when a socket is used. ICs must be installed correctly in order for the circuit to operate correctly. ICs usually have two different types of identifiers. On some ICs, a rectangular cutout is placed on the top of the IC. This cutout is just to the right of pin 1. Other ICs have a small indented circle next to pin 1 on the IC. All components except for the thermistor are mounted on the small circuit board. The transformer at L1 is a 600 to 600 ohm miniature matching unit. Because the transformer is a 1:1 matching transformer with the same winding on both the primary and the secondary, you can install the transformer without regard to polarity. However note the polarity of the capacitor at C3. The sending oscillator circuit is powered from a 9-volt transistor radio battery, and power is supplied through the on-off switch at S1.

The sending or oscillator circuit was housed in a small metal chassis box. The transformer was mounted to the circuit board along with all the components. The circuit board was mounted on $1/4$-inch standoff and mounted to the bottom of the chassis box. The 9-volt-battery holder was secured to the bottom of the circuit board. An RCA jack, an $1/8$-inch mono earphone jack, and S1 are mounted on the top of the chassis box. The RCA jack is installed for the thermistor connection, while the $1/8$-inch phone jack is used for the audio output connection.

Once the you have finished constructing the oscillator sending unit, you will need to carefully inspect the underside of the circuit board to make sure that there are no cold solder joints and no unwanted solder bridges between solder pads. Also look for unwanted stray component leads that may have stuck to the bottom of the circuit board.

Next, you will want to prepare a shielded sensor cable with an RCA plug at one end and the thermistor at the other end. Depending upon your application, you may wish to house the thermistor in a discarded metal pen casing, with provisions for the shielded cable to exit the top of the pen assembly. First, prepare the thermistor by spraying the thermistor and the leads with Crylon or a similar plastic coating to insulate the thermistor and the first $1/4$ inch of the thermistor's leads. The thermistor leads are usually very small-gauge wire, so care must be taken when soldering the shielded cable to the thermistor. Heatshrink tubing should be put over the connection or junction between the thermistor and the shielded cable, and some sort of strain relief should be provided. You will also want to make or purchase a 3- to 5-foot cable with a $1/8$-inch male plug on both ends to connect the oscillator audio output to the input of a recorder or transmitter. Place the on-off switch in the off position and install the battery.

Construction of the receive/display circuit is a bit more complicated than the oscillator/sending unit. The receive/display section can be built on a proto-board such as the one used in the sending unit, or you can make your own dedicated printed circuit board. All components including the transformer at L1 are mounted on the circuit board. None of the components are critical, however, particular attention should be paid to the orientation of the capacitor at C5 and the IC at U1. An IC socket for U1 is highly recommended. ICs are usually identified in one of two ways. Some ICs have a small indented circle placed on the top left side of the IC near pin 1. Other ICs have a small rectangular cutout on the top of the IC just to the left of pin 1. When installing the IC, pay particular attention to the orientation to avoid destroying the IC when power is supplied.

Calibration potentiometer R5 is a PC board trimpot. When installing the transformer at L1, be sure to identify from the packaging the two windings, which are different from each other. One winding is an 8-ohm winding and the other winding is 1K ohms. Be sure to orient the transformer correctly.

Once you have completed the receive/display board, you will need to inspect the underside of the circuit board. First check for cold solder joints, and then look over the circuit foil side to ensure that there are no unwanted solder bridges between foil traces. Finally look for stray or unwanted component leads, which may have stuck to the underside of the PC board.

The receive/display board is finally installed in a small metal chassis box, which measures $4^1/2 \times 6 \times 2$ inches. The circuit board is mounted atop $1/4$-inch plastic standoffs. A $1/8$-inch mini phone jack and the power switch S1 is installed on the top front side of the chassis box. Our prototype circuit utilizes a small terminal strip on the chassis to allow connection to an external digital multimeter. Note that you could also use a recording multimeter with RS-232 output, if desired. Alternatively you could mount an inexpensive LCD panel voltmeter on the front panel of the receiver/display unit, if desired. If you use an internal panel meter, you will need to make a two-resistor voltage divider to drop down the voltage to the panel meter, which is generally 2 volts maximum. A 9-volt transistor radio battery holder is mounted on the bottom of the chassis box. Once you have completed the circuit, turn the power switch to the off position before installing the 9-volt battery. Finally, you will need to make a cable to connect the receiver/display to the audio source that will be driving the input. Your cable will have a $1/8$-inch mini phone plug at one end and a connector of your choice on the opposite end, depending upon your application and input source.

Before using the analog data recording system, it is necessary to adjust the sensor circuit so that the frequency of the signal it generates falls within the proper range. Likewise, it's necessary to make sure the frequency-to-voltage converter is properly adjusted.

If you plan to take down the decoded voltages by hand, it's important to make sure each data sample is recorded for at least 5 seconds. This will allow the voltmeter readout to settle long enough for you to record the reading.

If you plan on using the analog data logging system with a tape recorder rather than a transmitter and receiver setup, you will need to create a calibration graph for your particular recorder for best results. In order to calibrate your recorder, you will need to store a known sequence of audio frequency tones spaced uniformly across the desired signal range. Connect an audio signal generator directly to the input of your tape recorder, and record 10- to 15-second tone bursts at 100 Hz intervals between 100 Hz and 1,000 Hz. The input volume to the tape recorder should be adjusted to $1/4$ of the range to assure that the input to the tape recorder is not over-driven or distorted. Stop the recorder and disconnect the signal generator. Connect the receive/display unit's audio input connection to the output of your tape recorder. Then connect the receiver/display unit's output connection to a digital voltmeter. Turn on the tape reorder and playback the recorded tape. As the tape is playing, take notes of the voltage readings for each of the tones recorded on the tape, and then finally make a plot on a piece of paper. At $3/8$-inch intervals along the bottom of the graph make notations of frequency. Each $3/8$-inch marking represents 0, 1, 2, 3, 4, 5 . . . Hz multiplied by 100 Hz. So the first marking is 1 KHz and so on. The vertical scale on the graph represents by voltage, starting with 0, 1, 2, 3, 4, and 5 volts at $5/8$ intervals vertically. Once you have a calibration graph, you are ready to begin testing your analog data logging system.

Connect the thermistor to the oscillator sending unit. Then connect the output of the oscillator sending unit to either a tape recorder or to the input of a transmitter, and then apply power to the circuit. If you are recording your readings, you will have to play them back into the receiver/display unit when you return to the playback location. If you are using an RF receiver in real time instead of a tape recorder, you can connect the input of the receive/display unit to audio output of the RF receiver. Now connect the output of the receiver/display unit to a digital multimeter or voltmeter and read the incoming signal from the oscillator sending unit. You should begin seeing a voltage reading on the meter display. Have a friend go over to the oscillator sending unit and place the thermistor in a bowl of crushed ice to calibrate the sensor. You should now see the voltage reading begin to change to a different value. Note the reading when the thermistor is placed in the crushed ice; you have now calibrated your analog data logging system, and it is ready for field data gathering.

Analog Data Logging System Oscillator Sending Unit Parts List

Required Parts

T1 thermistor (Radio Shack)

R1 4.7K ohm, 1/4-watt resistor

C1 0.1 uF, 35-volt ceramic disc capacitor

C2 1 uF, 35-volt electrolytic capacitor

C3 4.7 uF, 35-volt electrolytic capacitor

L1 600 to 600 ohm mini matching transformer

U1 LM555 timer/oscillator IC (National)

J1 RCA chassis jack

S1 SPST toggle power switch

B1 9-volt transistor radio battery

Miscellaneous PC board, wire, IC socket, standoffs, screws, nuts, etc.

Analog Data Logging System Receiver/Display Unit Parts List

Required Parts

R1, R2 10K ohm, 1/4-watt resistor

R3 68K ohm, 1/4-watt resistor

R4 12K ohm, 1/4-watt resistor

R5 5K ohm potentiometer (trimpot)

R6 6.8K ohm, 1/4-watt resistor

R7 100K ohm, 1/4-watt resistor

C1 4.7 uF, 35-volt electrolytic capacitor

C2 0.1 uF, 35-volt ceramic disc capacitor

C3 10 uF, 35-volt electrolytic capacitor

C4 0.01 uF, 35-volt ceramic disc capacitor

C5 1 uF, 35-volt electrolytic capacitor

U1 LM331 frequency-to-voltage converter IC (National)

L1 1K to 8-ohm mini matching transformer

Miscellaneous PC board, wire, IC socket, standoffs, screws, nuts, terminal strip, etc.

LCD Thermometer

It is often necessary to measure temperature around your home, shop, or office, and one accurate and easy method to measure temperature is to construct the digital LCD thermometer described in this section. The LCD thermometer can read and display a wide range of temperatures from −20 degrees C to +150 degrees C.

The heart of the LCD thermometer is a diode temperature sensor, the A/D converter chip, and an $3^{1}/_{2}$-digit LCD display shown in Figure 3-8. A 1S1588 silicon switching diode is used as the temperature sensor in this project. The forward voltage of the silicon diode changes in the coefficient of $-2\text{mV}/°\text{C}$ when the temperature with joint changes. Generally, the 20°C forward voltage is about 600 mV. The forward voltage becomes about 400 mV (600mV $-$ ($2\text{mV}/°\text{C} \times 100°\text{C}$)) when the temperature with joint rises by 100°C and becomes 120°C. At the thermometer this time, it is displaying the change of this voltage as the temperature. The temperature range that it can measure is within the allowable temperature of the diode. Depending on the kind of diode, it is possible from about –20°C to 150°C. Generally, a diode is enclosed in packaging of glass or plastic. So, even if the ambient temperature changes, the voltage of the diode doesn't always change immediately

The 1S1588 silicon diode is available in two different types of packages; be sure to choose the small diameter glass diode and not the larger molded plastic package diode, which is not suited for this project because it is difficult for the temperature change to spread quickly through the larger molded package. The lead wires of the diode are insulated with the heavy nylon/cloth insulated tubing to protect the diode. Shielded wire is used for the connection between the diode sensor and the A/C converter.

The analog-to-digital converter utilized in the LCD thermometer is the ubiquitous ICL7136/TC7136 analog-to-digital converter shown in Figure 3-9. The ICL7136 CMOS analog-to-digital converter accepts an analog input voltage and converts it to a digital output, through a multiplexed

Figure 3-8 *LCD thermometer*

backplane drive circuitry, and then displays on an LCD display. The 40-pin DIP A/D converter chip can be set up to accept either a $+/-200$ millivolts or $+/-2$ volts DC input voltage on pins 30 and 32.

The ICL7136's analog section divides the measurement cycle into four phases. First is the *auto-zero* (A-Z) phase, then the signal *integrate* (INT) phase, then the *deintegrate* (DE) phase, and finally the *zero-integrated* (ZI) phase (see Figure 3-10). The ICL7136 does the measurement process in about three times during the 1 second, or 4,000 clocks/time = 4,000 $\times$ ($^{4}/_{48,000}$ Hz) = 0.33 seconds/time.

During the auto-zero phase three things happen. First, the input high and input low are disconnected from the pins and internally shorted to analog COMMON. Next, the reference capacitor is charged to the reference voltage. Third, a feedback loop is closed around the system to charge the auto-zero capacitor C_{AZ} to compensate for offset voltages in the buffer amplifier, integrator, and comparator.

During the signal integrate phase, the auto-zero loop is opened, the internal short is removed, and the internal input high and low are connected together to the external pins. The converter then integrates the differential voltage between input high and input low for a fixed time. This difference in voltage can be within a wide common mode range. At the end of this phase, the polarity of the integrated signal is determined.

Next, during the deintegrate phase, the input low is internally connected to analog COMMON, and input high is connected across the previously charged reference capacitor. Circuitry within the analog-to-digital chip ensures that the capacitor will be connected with the correct polarity to cause the integrator to return to zero.

The final phase is the zero-integrator phase. First, input low is shorted to the analog COMMON. Second, the reference capacitor is charged to the reference voltage. Finally, a feedback loop is closed around the system to input high to cause the integrator output to return to zero.

Note that the ICL7136 has the high and low input terminals on pins 31 and 30 respectively. Also note, the chip has a separate common ground connection

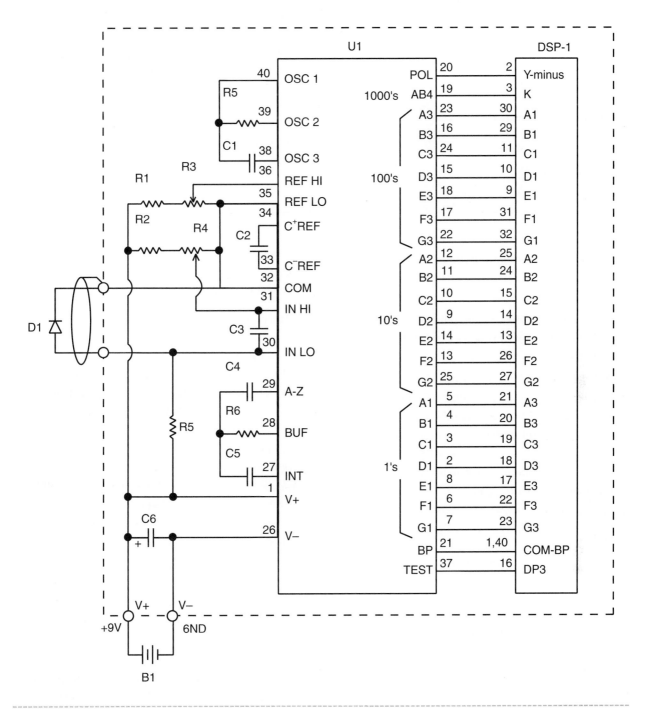

Figure 3-9 *LCD temperature monitor block diagram*

on pin 32. The COM ground is different from the negative power supply connection on pin 26. The analog-to-digital converter sampling is referenced via the clock formed by the oscillator pins 38, 39, and 40.

The potentiometer R3 is used as a zero adjustment on the input of the analog-to-digital chip, while potentiometer R4 is used to adjust the scale factor of the analog-to-digital's input.

The backplane LCD driver occupies a large portion of the chip's output pins as shown. The A1 through G1 or the 1's, digit output is shown from pins 2 through 8, while the 10's digits A2 through G2 are

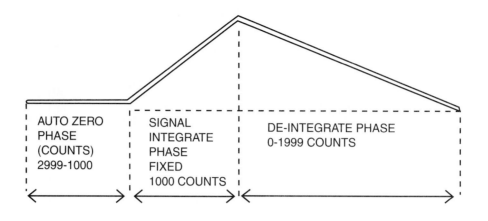

Figure 3-10 *Integrator amplifier output waveform*

found on pins 9 through 14 and pin 25. Finally the 100's digits from QA3 through G3 are found on pins 15 through 24.

A 3½ digital AND brand FE0203 LCD display used in this project is a TN, or twisted nematic type, LCD device. The 40-pin, 3½-digit LCD display is powered by a 5-volt source. In our project we use only the three digits and the decimal point to display temperature. The pinouts diagram for the LCD display is shown in Table 3-1.

Table 3-1

AND Brand Model FE0203 LCD Display Pinouts

LCD Pin Assignments

Pin #	Seg.	Pin #	Seg.	Pin #	Seg.	Pin #	Seg.
1	BP	11	C1	21	A3	31	F1
2	Y	12	DP2	22	F3	32	G1
3	K	13	E2	23	G3	33	NC
4	NC	14	D2	24	B2	34	NC
5	NC	15	C1	25	A2	35	NC
6	NC	16	DP3	26	F2	36	NC
7	NC	17	E3	27	G2	37	NC
8	DP1	18	D3	28	L	38	LO
9	E119	19	C2	29	B1	39	X
10	D1	20	B3	30	A1	40	BP

The digital thermometer consumes little power and can be powered from a 9-volt transistor radio battery for about 3 months or more. Construction of the LCD thermometer is pretty straightforward and can be completed in about 2 hours with little difficulty on a small printed circuit board, as shown in the figure. All the components are mounted on the circuit board including the analog-to-digital chip, all its supporting components, as well as the LCD. The diode sensor is not mounted on the PC board. It is highly advisable to use an IC socket for the 40 pin analog-to-digital chip, so in the event of later failure, you can easily replace the chip if need be.

The input measurement capacitors are all polyester film types, because they can affect the accuracy of the meter. The clock generator capacitor that oscillates near 50 KHz is a ceramic capacitor with good high-frequency characteristics. A multilayer ceramic capacitor is used for the bypass capacitor at C6. Most of the capacitors in this circuit are not polarity sensitive except for C6, connected near the battery. All the resistors in the thermometer circuit are 1% so accuracy of the meter is kept high.

When installing the analog-to-digital chip, make sure you orient the chip correctly to avoid damaging the chip when power is applied. ICs generally have a marking to assist in orientation and installation. The chip may have a small cutout at the top of the package. Pin 1 is to the left of the cutout. Some ICs have a small indented circle just to the right of pin 1.

When installing the LCD, be careful not to apply to much pressure to the front or between the front and the back of the display to avoid damaging it. If you can locate a socket for the LCD, it would advisable to use one. Be very careful when wiring the LCD, because many pins need to be wired, and the possibility for making a mistake is great.

After soldering all the components in the proper place, carefully look over the underside of the circuit board for shorts or solder bridges left from cut components leads. Also look for cold solder joints, which can come back to haunt you later with intermittent operation.

Next, you will need to make a sensor cable from the diode sensor to the analog-to-digital circuitry. First, insulate the diode leads by coating them with spray-on insulation or rubber coating. When you are satisfied that the board looks clean and ready, you can solder in a 9-volt battery clip.

Next, prepare a small diameter coax cable, for example an RG-174, between the sensor and the circuit. The coax cable should not be too long or they will prevent error signals from changing the readings. A 2- or 3-foot cable should work fine.

Now that the LCD thermometer is now complete, you move on to the calibration phase. Note, the diode sensor leads must be insulated from the water when calibrating the thermometer so they won't damage the circuit. You must take precautions for insulating the leads properly. Spray insulation on the leads or coat the leads with a rubber coating.

Fill a small or medium-size bowl with ice and water. Allow a few minutes for the ice to cool the water. An alternative is to calibrate by using a can of instant freeze, available from a local electronics supply house. Apply power to the circuit and place the diode sensor atop an ice cube or spray the instant freeze across the sensor for a minute. Adjust R4 for a 0°C reading. Conversely, you can instead calibrate the LCD thermometer using boiling water by adjusting R3 to 100°C.

You may want to install the LCD thermometer inside a plastic box to protect the meter from damage. Find a suitable plastic box in which to house the

LCD thermometer circuit including the battery. If you choose to mount the LCD thermometer inside a plastic enclosure, you will want to make provisions for an on-off switch and screw terminals or an RCA-type jack for connecting the temperature sensor.

You'll be surprised just how handy your new LCD thermometer will become around your house or shop once you've built one!

LCD Thermometer Parts List

Required Parts

R1 1 megohm, 1/4-watt, 1% resistor

R2 470K ohm, 1/4-watt, 1 % resistor

R3 100K ohm potentiometer (trimpot)

R4 200K ohm potentiometer (trimpot)

R5 100K ohm, 1/4-watt, 1% resistor

R6 390K ohm, 1/4-watt, 1% resistor

C1 47 pF, 35-volt polyester capacitor

C2 0.1 uF, 35-volt polyester capacitor

C3, C5 0.047 uF, 35-volt polyester capacitor

C4 0.47 uF, 35-volt polyester capacitor

C6 0.1uF, 35-volt ceramic capacitor

U1 ICL7136 or TC7136 Harris or equivalent

DSP-1 FE0203 31/2-digit LCD display (AND Co., available from Purdy ElectronicsB1 9-volt transistor radio battery

Miscellaneous PC board, sockets, wire, connectors, enclosure, etc.

Night Scope Project

The intriguing night scope project shown in Figure 3-11 is a device capable of seeing in total darkness. Unlike conventional devices requiring the minute light from the stars or other ambient background light, this system contains its own infrared source, allowing covert viewing of the desired subject. Assembly is shown in two parts, the high-voltage power supply and the final enclosure with optics and an illuminator.

The night scope or infrared viewer project can be used to view a subject for recognition or evidence gathering without any indication to the target that he or she is under surveillance. It is an invaluable device when used for detection, the alignment of infrared alarms, invisible laser gun sights, and in communications systems. This technology can also be used to detect diseased vegetation in certain types of crops from the air, to serve as an aid to nighttime varmint hunting, and to view high-temperature thermographic scenes where heat is used to produce the image.

The night scope viewer project is built using readily available parts for the enclosure and basic optics. The batteries are enclosed in the housing and do not require side packs and cables. The range and field of viewing are determined by the intensity of the integrated infrared source and the viewing angle of the optics. Readily available and low-cost optics are usable, but they may have spherical aberration and other adverse effects. However, using these materials keeps the basic cost down for those not requiring actual viewing of detailed scenes. Improved optics will eliminate these negative effects and can be obtained at most video supply houses.

Assembly focuses around common polyvinyl chloride (PVC) tubing as the main housing and a specially designed, patented, miniature power source for energizing the image tube. The tube is a readily available image converter being used by most manufacturers of similar devices. This tube establishes the limits of viewing resolution and is suitable for most applications but may be limited if one desires video perfection.

The viewing range is determined mainly by the intensity of the infrared source and can be controlled by varying this parameter. The basic unit is shown utilizing a 2-D cell flashlight with an integrated filter placed over the lens to prevent the subject from seeing the source. This provides a working range of up to 50 feet and can be increased to several hundred feet using a more powerful light source such as a 5- or 6-cell flashlight. Infrared LEDs or lasers can also be used as illumination sources.

The unit can also be operated using external sources such as superintense Q-beam handheld lamps with an added filter extending the range to 400 to 500 feet and providing a wide field of illumination. Note that viewing of active infrared sources such as lasers does not require the internal infrared source.

A subminiature high-voltage power supply produces approximately 15KV at several hundred microamps from a 7- to 9-volt rechargeable nickel cadmium or alkaline battery (see Figure 3-12). This voltage is applied to the infrared viewing tube or IR16 with the positive (+) going to the viewing end and the negative (-) to the objective end. A focus voltage is taken from a tap in the multiplier circuit and is approximately $\frac{1}{6}$ of the total potential.

Figure 3-11   *Night scope*

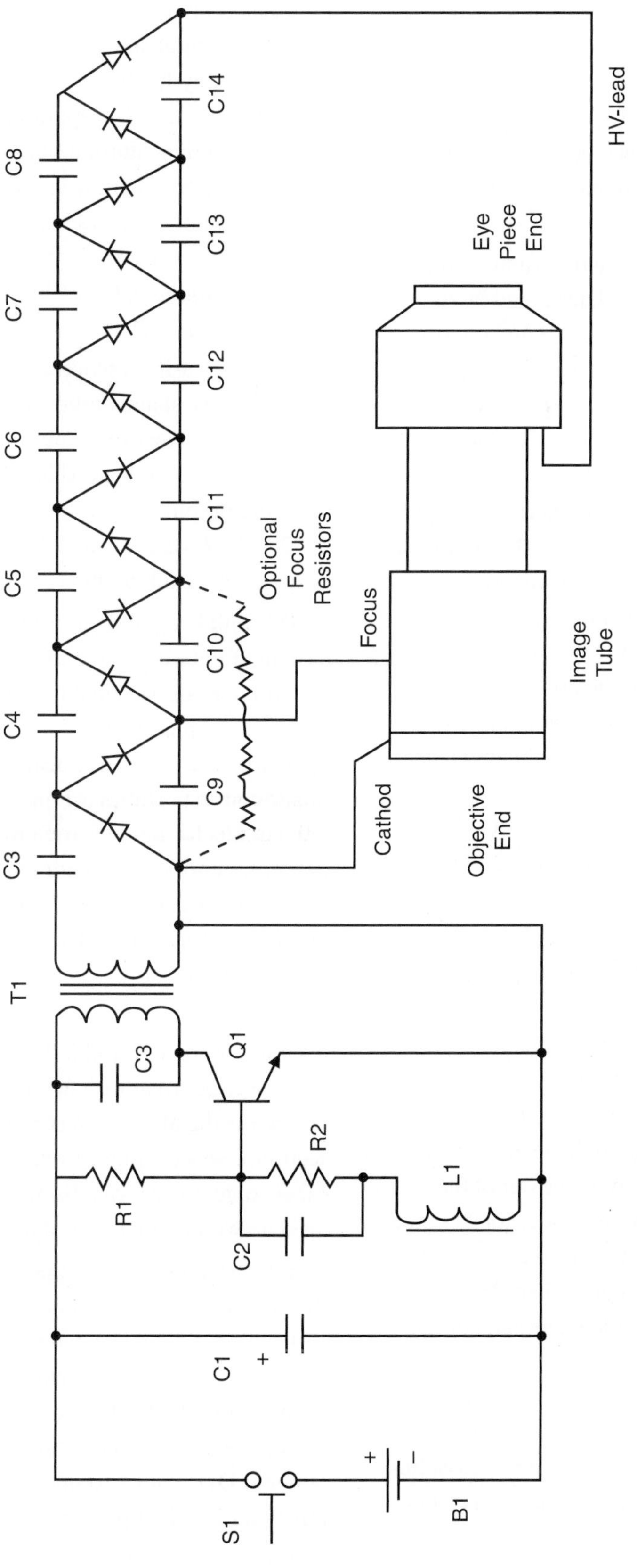

Figure 3-12   *High-voltage power supply*

An objective lens (LENS1) with an adjustable focal length gathers the reflected image, illuminated by the infrared lens, and focuses this image at the objective end of the tube. The image is displayed on the viewing screen of the tube in a greenish tinge. The viewing resolution is usually adequate to provide object identification at a distance of 50 feet or more, depending upon the infrared source as mentioned earlier.

Transistor Q1 is connected as a free running resonant oscillator with a frequency determined by the combination resonance of capacitor C3 and the primary winding of the step-up transformer T1. This oscillating voltage is stepped up to several thousand in the secondary winding of T1. Capacitors C4 through C15, along with diodes D1 through D12, form a full-wave voltage multiplier where the output is multiplied by six and is converted to DC. Output is taken between C5 and C15, as shown, and may be either positive or negative depending on the direction of the diodes. The base of Q1 is connected to a feedback winding of T1, where the oscillator voltage is at the proper value to sustain oscillation. Resistor R2 biases the base into conduction for the initial activation. Resistor R1 limits the base current, whereas capacitor C2 speeds up the deactivation of Q1 by supplying high-frequency energy. The input power is supplied through switch S1 via a snap-in battery.

Assemble the high-voltage power supply board. Start to insert components into the board holes. Be sure to start and proceed from right to left, attempting to obtain the layout as shown. Certain leads of the actual components will be used for connecting points and circuit runs. Do not cut or trim at this time. It is best to temporarily fold the leads over to secure the individual parts from falling out of the board holes for now. Note that the solder joints in the multiplier section, consisting of C4 through C15 and D1 through D12, should be globular shaped and smooth to prevent high-voltage leakage and corona. The solder globs size should be that of a BB. Run your fingers over the solder globs and verify the absence of sharp points and protrusions. Also note that T1 is lying on its side and uses short pieces of bus wire soldered to its pins as extensions for connections to the circuit board.

To test the high-voltage power supply circuit, follow these steps, Separate high-voltage output leads approximately 1 inch from each another. Connect 9 volts to the input, and note the current draw of approximately 150 to 200 milliamperes when S1 is pressed. Decrease the separation of the high-voltage leads until a thin, bluish discharge occurs, using a separation of approximately $^1/_2$ to $^3/_4$ of an inch. Note the current input increasing. The increased value depends upon the length of the spark and should not exceed 300 milliamperes. Check the collector tab of Q1 and add a small heatsink.

If you are using a scope, it may be interesting to note the wave shape at the collector tab. Notice this occurs without any sparking. Note also the takeoff point for the focus lead. This point is approximately at $^1/_6$ the output voltage. The unit may be powered up to 12-*volts direct current* (VDC) but will positively require a heatsink on the tab Q1.

This unit is capable of producing 10 to 20 KV from a small, standard 9-volt battery. It is built on a printed circuit board or a small piece of perforated circuit board and can easily be housed or enclosed, as the application requires. Applications include powering image converter tubes for night vision devices, ignition circuits for flame-throwing or flame-producing units, capacitor charging for energy storage, electric shocking fence, insect eradication, Kirlian photography, ion propulsion electric field generators, ozone producing, and more.

To begin the assembly of the night scope viewer, refer to Figure 3-13 and follow these steps. Assume the power board as outlined is properly operating. Check for the absence of corona in the high-voltage section. Use a coating corona dope to reduce electrical leakage if necessary. Remove all sharp points and insulate with corona dope and so on.

Take a window screen and place it flush against the objective end of the image tube, TUB1. Attach it with a piece of clear scotch tape. Secure the tube on the bench with modeling clay, and temporarily connect it to the leads from the power board. Observe the proper clearance of the leads and components. Darken the room and place a source of infrared filter light pointing toward the tube. (Use a

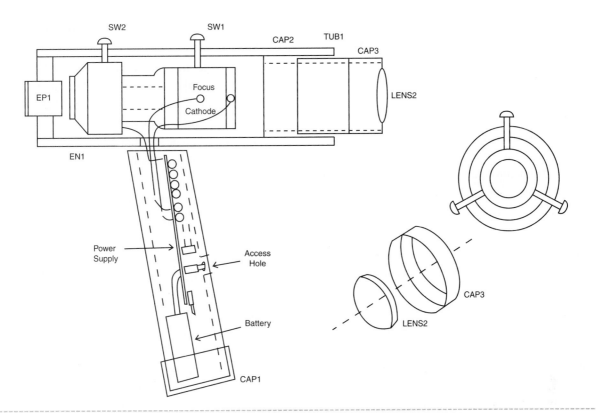

Figure 3-13 *Night vision viewer*

flashlight preferably with an IR filter.) Note the greenish glowing of the tube and the image on the screen appearing either sharp or blurred. If the image is good and sharp, you are in luck. You may further improve the focusing by adding the 22 megohm resistors. This is usually not necessary.

Fabricate EN1 from a 7-inch length of $2^3/8$-inch *inside diameter* (ID) schedule 40 PVC tubing. Note the hole adjacent to the HA1 handle for feeding high-voltage wires to the tube from the power board and $1/4$-20 threaded holes are dimensioned in Figure 3-14 for securing and centering the image tube. These holes are located on a 120-degree radius. Fabricate the HA1 handle from an 8-inch length of $1^1/2$-inch ID schedule 40 PVC tubing. The tube must be shaped and fitted where it fits to the EN1 main enclosure.

Fabricate the BRK 1 and 2 brackets from a half-inch-wide strip of 22-gauge aluminum as shown. Note the holes for #6 × $1/4$ sheet metal screws for securing the assembly together.

Fabricate TUB1 from a $3^1/2$-inch length of 2-inch ID schedule 40 PVC tubing for the objective lens. Note this is only 2 inches long when using the optional optics and C- or T-mount adapter fitting.

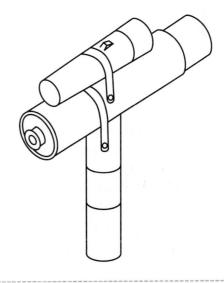

Figure 3-14 *Night vision viewer final assembly*

In order for TUB1 to telescope into the main enclosure EN1, suitable cylindrical shims, CAP2 and CAP3, must be fabricated. These are the $2^{3}/_{8}$ -inch plastic caps. CAP2 has its end removed by cutting out the center, using the wall of tubing as a guide for the knife. CAP3 has a smaller section cut out for LENS1. This method is cheap and works reasonably well. You could substitute the pieces with properly fitted parts fabricated from aluminum or plastic if you desire. This approach is more professional looking but can be much more costly.

The lens shown is a simple, uncorrected convex lens that is adequate for most infrared source viewing. It is not a quality viewing lens such as the optional 50 mm wide-angle lens or 75 mm telephoto lens with C-mount threads. When using this lens, you should either create or purchase an adapter ring that will adapt to the lens threads and fit snugly into the enclosure.

The IR16 image tube has preconnected leads. The negative short lead attached to the objective end must have a 10-inch lead spliced to it. Insert the tube partway into the enclosure and snake the leads through the access hole. Position the tube and gently screw in the retaining screws by hand to secure and center it. Connect the leads from the tube to the power board as shown.

Insert the power board into the HA1 handle. You will have to determine the access hole and drill for the switch S1 once the board is secured in its final position. Wires should be long enough for the complete removal of the assembly when the handle is secured in place via the BRK1 bracket. This allows any preliminary adjustment or service. Leads may be shortened once proper operation is verified. Connect the battery to the power board and you will not have to readjust the focus taps or divider values. Once the operation is verified, check for any excessive corona and eliminate it. Position the board to switch S1 adjacent to the access hole in the handle. It may be necessary to further secure the board in place with foam rubber pieces, a *room temperature vulcanizing* (RTV) adhesive, and so on. Slide a flexible rubber membrane over the access hole and insert the battery and cap CAP1.

To complete the final assembly refer to Figure 3-14 and mount the infrared filtered flashlight atop the night scope viewer housing. You will have to seal any light leaks using coax seal. Adjust the objective and then the eyepiece for the clearest image. The night scope viewer unit is shown with a built-in infrared source consisting of a common, everyday two-cell flashlight fitted with a special infrared filter.

This approach allows total flexibility in viewing sources not requiring infrared illumination, as the light need not be energized or may even be removed. You could replace the two D-cell flashlight batteries with an eight pack of AA Nicad cells providing a 9-volt source to power a bank of infrared LEDs as an alternative light source. A suitable lamp may be substituted, providing several times more illumination. The lamp and battery life will be greatly reduced, as this approach is intended only for intermittent use. Halogen lamps are far more intense and make excellent infrared sources.

Longer-range viewing may be accomplished by using other, more intense sources such as higher powered lights, auto headlamps, and so on. These must be fitted with the proper filters to be usable. A range of several hundred meters may be possible with these higher powered sources. A source capable of allowing viewing from up to 500 feet is referenced in the project parts list. Obtaining maximum performance and range from your system may require the optional lens specified. The viewing of externally illuminated infrared sources will not require the integral infrared source.

You should also note that this device is excellent for viewing the output of most solid-state gallium arsenide laser systems, LEDs, or any other sources of infrared energy in the 9,000 A spectrum. No internal infrared source is necessary when viewing these actual sources. Expect to spend $50 to $100 for this useful infrared imaging system with all the specialized parts available from www.amazing.com.

Night Scope Project Parts List

Required Parts

RI 1.5K ohm, 1/4-watt resistor

R2 15K ohm, 1/4-watt resistor

C1 10 uF, 25-volt electrolytic capacitor

C2 0.047 uF, 50-volt plastic capacitor

C3 0.47 uF, 100-volt plastic capacitor

C4 to C15 270 pF, 3 KV plastic disk capacitor

D1 to D12 6 KV, 100-nanosecond, high-voltage avalanche diodes

Q1 MJE3055 NPN TO 220 case transistor

TI special transformer info #ZBK077

S1 pushbutton switch

PB1 51/2 × 11/2 inch perforated board with 0.1 × 0.1 grid

CLI snap battery clip

WR22 24-inch length of #22 vinyl hookup wire

WRHV20 12-inch, 20 KV silicon wire

IR16 image converter tube

EN1 8 × 2-3/8 inch schedule 40 gray PVC tube

TUB1 3 1/2-inch length × 2-inch schedule 40 gray PVC

BRK1, BRK2 9 × 1/2 inch thin aluminum strip

CAP1 2-inch plastic cap for handle

CAP2, CAP3 2 3/8-inch plastic cap

LENS1 45 / 63 double convex glass lens

SW1, SW2 (6)1/4-20 × 1-inch long nylon screws

SW6 (6) #6 × 1/4-inch sheet metal screws

Optional Parts

PCPBK printed circuit board

CMT1 prefab C-mount adaptor for EN1 enclosure

EP1 small eyepiece

FIL6 6-inch glass infrared filter 99.9% dark

HRL10 200,000 candle power infrared illuminator

Infrared Motion Detector

Infrared radiation exists in the electromagnetic spectrum at a wavelength that is longer than that of visible light. Infrared radiation cannot be seen, but it can be detected. Objects that generate heat also generate infrared radiation, and those objects include animals and the human body, whose radiation is strongest at a wavelength of 9.4 micrometers.

The infrared body-heat motion-detector project, shown in Figure 3-15 is designed to detect the motion of a human or animal both in daylight and at night and to provide a normally open relay output that can be used to activate many types of loads. The motion detector also has terminals for connecting an optional photocell to prevent activation of the load during daylight.

Courtesy of GloLab Corp

Figure 3-15 *Infrared body-heat motion detector*

The heart of the infrared body-heat motion detector is the *pyroelectric sensor*, which is made of a crystalline material that generates a surface electric charge when exposed to heat in the form of infrared radiation (see Figure 3-16). When the amount of radiation striking the crystal changes, the amount of charge also changes and can be measured with a sensitive FET device built into the sensor. The sensor elements are sensitive to radiation over a wide range, so a filter window is added to the TO5 package to limit incoming radiation to the 8- to 14μmeter range, which makes the device most sensitive to human body radiation.

Figure 3-17 illustrates the IR sensor. The FET source terminal pin 2 connects through a pull-down resistor of about 100 K to ground and feeds into a two-stage amplifier having signal conditioning circuits. Each of the two cascaded stages has a gain of 100 for a total gain of about 10,000. The amplifier is typically bandwidth limited to below 10 Hz to reject high-frequency noise and is followed by a window comparator that responds to both the positive and negative transitions of the sensor output signal. A well-filtered power source of from 3 to 15 volts should be connected to the FET drain terminal pin 1.

The PIR325 sensor has two sensing elements connected in a voltage-bucking configuration. This arrangement cancels signals caused by vibration, temperature changes, and sunlight. A body passing in front of the sensor will activate first one and then the other element, whereas other sources will affect both elements simultaneously and be canceled. The radiation source must pass across the sensor in a horizontal direction when sensor pins 1 and 2 are on a horizontal plane so that the elements are sequentially exposed to the IR source.

A Fresnel lens is a Plano convex lens that has been collapsed on itself. As such it forms a flat lens that retains its optical characteristics but is much smaller in thickness and therefore has less absorption losses (see Figure 3-18). The FL65 Fresnel lens is made of an infrared transmitting material discussed in a previous paragraph that has an IR transmission range of 8- to 14μmeters that is most sensitive to human body radiation. It is designed to have its grooves facing the IR sensing element. This causes the smooth surface to be presented to the subject side of the lens, which is

Figure 3-16 *Pyroelectric sensor*

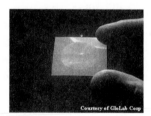

Figure 3-18 *Fresnel lens (Courtesy GloLab Corp)*

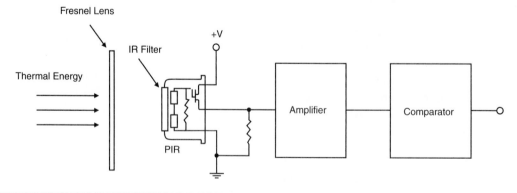

Figure 3-17 *PIR sensor*

usually the outside of an enclosure that houses the sensor (see Figure 3-19).

The FL65 has a focal length of 0.65 inches from the lens to the sensing element. It has been determined by experiment to have a field of view of approximately 10 degrees when used with a PIR325 pyroelectric sensor. When used with a PIR325 sensor and FL65 Fresnel lens, this circuit can detect motion at a distance of up to 90 feet. Figure 3-20 illustrates the direction and range of the IR detection pattern that can be expected with the FL65 Fresnel lens in conjunction with the PIR325 sensor. Note that the distance from the front of the sensing elements to the front of the filter window is 0.045 inch (1.143 millimeters). Mounting can best and most easily be done with strips of Scotch tape. Silicone rubber adhesive can also be used to form a more waterproof seal.

Figure 3-21 is a circuit diagram of the infrared body-heat motion detector. Power is supplied to the circuits and to the relay through a micropower 5-volt regulator U3. The 5 volts from the regulator is further filtered through R2 and C2 and then fed to pin 1 of the PIR325 pyroelectric sensor. The signal output at pin 2 of the sensor is bypassed to ground by a 100 pf

capacitor C1 to shunt any radio-frequency energy that might be picked up from radio transmitters or cell phones. A 100K load resistor R1 is also connected from pin 2 to ground.

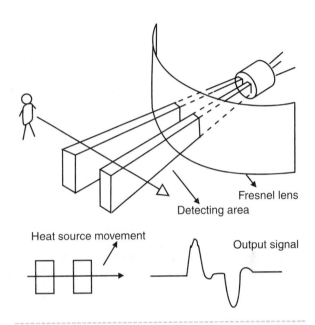

Heat source movement

Output signal

Figure 3-20   *Detection range*

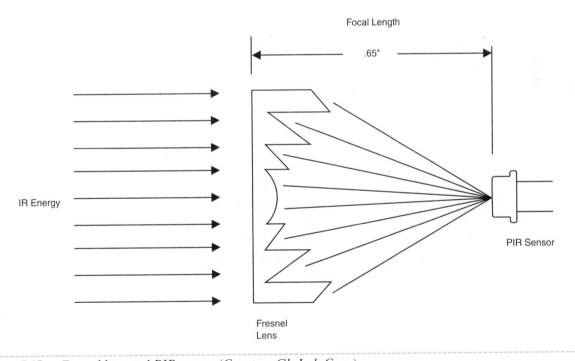

Focal Length

.65"

IR Energy

PIR Sensor

Fresnel Lens

Figure 3-19   *Fresnel lens and PIR sensor (Courtesy GloLab Corp)*

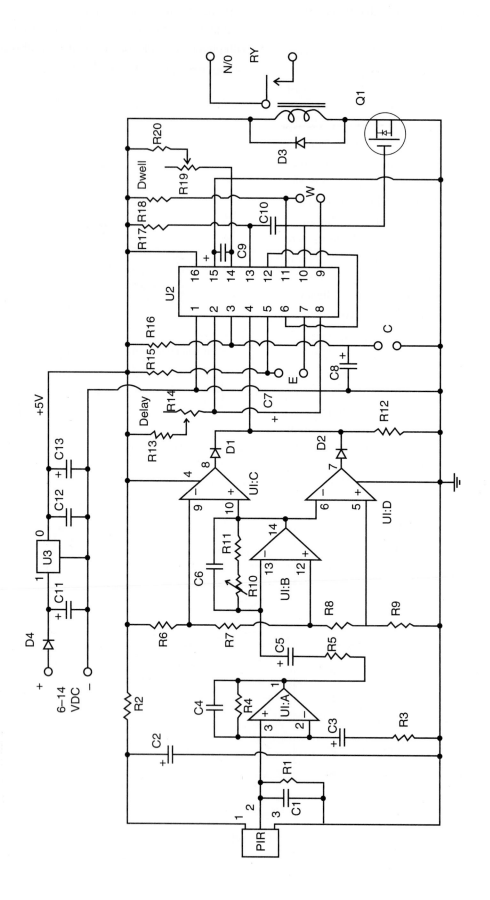

Figure 3-21 *Infrared motion detector (Courtesy GloLab Corp.)*

When motion is detected the sensor will output a very small-voltage transition at pin 2. This voltage must be amplified many times in order to do useful work. Two sections of a LM324 or equivalent quad operational amplifier are used to provide the necessary amplification. Sensor pin 2 feeds into the first stage amplifier U1:A at noninverting input pin 3. This is a very high-impedance input and does not load the sensor. A high-pass filter and feedback network at R4 and C4 connects between U1:A's output pin 1 to inverting input pin 2. A high-pass filter and bias network, C3 and R3, connects from pin 2 to ground. These networks set the amplifier gain and operating point. They also form a bandpass filter that amplifies only signals above DC and below about 10 Hz. The pyroelectric sensor is a thermal device and its response time falls within this band of frequencies. Filtering out signals that are outside its response time eliminates noise from frequencies that are not used anyway and makes the amplifier more stable.

The output of the first amplifier stage U1:A is taken from pin 1. It then feeds through R5 and C5 into the inverting input of the second amplifier stage at pin 13 of U1:B. C5 blocks the flow of DC and, together with R5, forms a high-pass filter to reduce gain at very low frequencies. A feedback network (R10, R11, and C6) connects from U1:B output pin 14 to inverting input pin 13.

R10 is a potentiometer that controls the amount of feedback and therefore the gain of this stage. R11 limits the amount of feedback. The noninverting input to U1:B at pin 12 is biased to $\frac{1}{2}$ of the supply voltage or 2.5 volts by the resistor divider network at R6, R7, R8, and R9. This bias sets the operating point of the amplifier so that its output pin 14 is at 2.5 volts when no motion is being detected.

The output of U1:B pin 14 feeds into a window comparator made of U1:C and U1:D. When an operational amplifier is used as a comparator, it is run at full open-loop gain so that its output switches to a full up or down level when one input is just a few millivolts higher or lower than the other. The purpose of this comparator is to provide a small voltage window or dead zone centered around 2.5 volts that will not respond to small voltage transitions caused by noise or minor fluctuations from the sensor. The inverting

input of U1:C at pin 9 is biased by the voltage at the junction of R6 and R7, so it is about 175 millivolts above the 2.5-volt output level at pin 14 of U1:B, and inverting input pin 10 of U1:C is connected to U1:B at pin 14. U1:C will not turn on until pin 10 goes more positive than pin 9. The noninverting input at pin 5 of U1:D is biased by the voltage at the junction of R8 and R9 to about 175 millivolts below the 2.5-volt output level at pin 14 of U1:B, and inverting input pin 6 of U1:D is connected to U1:B pin 14. U1:D will not turn on until pin 6 goes more negative than pin 5.

When motion is detected and the voltage transition at the output of the second-stage amplifier U1:B pin 14 is positive, it must go more that 175 millivolts above 2.5 volts, or to 2.675 volts. It must do this in order to turn U1:C on so that its output will transition to a high level. If the voltage transition at U1:B pin 14 is negative, it must go more than 175 millivolts below 2.5 volts, or to 2.325 volts, in order to turn U1:D on so that its output will transition to a high level. The window comparator therefore provides a 350-millivolt dead zone centered around 2.5 volts within which it will not respond to voltage-level changes from the amplifier. Any valid motion that is sensed will be amplified enough to generate a transition that will exceed this dead zone and will result in a comparator output. Another characteristic of the comparators is that they each produce a positive output transition when they turn on whether the U1:B pin 14 amplifier output goes high or low. The comparator outputs at U1:C pin 8 and U1:D pin 7 feed into a logic OR circuit consisting of diodes D3 and D4 and pulldown R12. The cathodes of D3 and D4 that are connected together will go high whenever either U1:C or U1:D turns on. This amplifier and comparator circuit will therefore respond to both positive and negative transitions from the sensor.

A CD4538 dual single-shot performs two built-in functions and three optional functions. The built-in functions provide both an adjustable delay between load activation when motion events are sensed and an adjustable dwell time to keep the load active even after motion stops. Optional functions are (1) retriggering of the delay so that repeat motion events will extend the delay between load activation, (2) retriggering of the dwell time so that repeating motion will extend the time that the load remains active, and (3) a day/night function to inhibit activation of a load during daylight.

Each of the two single-shot circuits in U2 has both inverting and noninverting inputs and outputs. When motion is detected, a positive transition at the cathodes of D1 and D2 feeds into noninverting input pin 4 of U2 and then triggers the first single shot on for a period determined by time constants R13, R14, and C7. The noninverting output at pin 6 goes high and remains high during the timeout period. Pin 6 connects to the noninverting input pin 12 of the second single shot, and its positive transition triggers that single shot on for a period determined by time constants R19, R20, and C9. This period is known as the *dwell period*. The noninverting output at pin 10 feeds into NFET Q1 and turns it on, which energizes relay RY1 and closes its contacts.

Both single shots are triggered on by a positive transition and not by a DC level. When the first single shot is triggered on, its output transitions to positive and triggers the second single shot. However, it cannot trigger the second single shot again until its output goes low and then high again. The on time of the first single shot therefore produces a delay between repeat triggering of the second single shot. This is useful for applications where rapid repeat activation of a load is not desired such as when motion is continuously being detected.

When the second single shot is triggered on, it keeps the relay energized for the dwell timeout period. This is useful for applications where it is desired to have a load remain active for some time after motion is detected. Both single shots are retriggerable. If the first single shot is triggered on by a positive transition from D1 and D2, it will stay on for its time constant period and then turn off unless it receives another trigger before it times out. If it does receive another positive transition before it times out, its time constant components will be reset and the delay timeout will be extended by an additional delay period. Retrigger is the default mode. If it is not

desired, terminals E can be jumpered together on the PC board to cause the circuit to ignore trigger transitions that occur during its timeout period.

The second single shot will also retrigger if it receives another positive transition at its input before it times out, however, it can retrigger only if its timeout period is adjusted to be longer than that of the first single shot and if the first single-shot delay is not extended beyond that of the second by retriggering. Retrigger is also the default mode of the second single shot. If it is not desired, terminals W can be jumpered together on the PC board so the circuit will ignore trigger transitions that occur during its timeout period.

A day/night function that inhibits activation of a load during daylight can be added by connecting a resistive cadmium sulfide photo cell to terminals C on the PC board. This type of cell has a high dark resistance and a low light resistance. In the default mode, reset pin 3 of the first single shot is pulled up to Vdd by R16. When a photo cell is connected from pin 3 to ground and its resistance is lowered by light, it pulls pin 3 down and resets the circuits so that they cannot be triggered on. The photocell is not polarized and can be connected in either direction. A switch may be connected in series with the photocell for daylight testing.

One end of the RY1 coil is connected to +5 volts and the other end to the drain of FET Q1. When Q1 turns on, it conducts current through the coil of RY1 to ground and energizes RY1 causing its contacts to close. When Q1 turns off, the collapsing magnetic field of the RY1 coil attempts to produce a large positive voltage spike (a backward *electromotive force*, or back EMF) that would damage Q1 if not suppressed. Diode D3 clamps that spike and prevents back EMF damage. The RY1 contacts are normally open, and they close when motion is detected. The contacts are rated for 3 amperes at 120 volts AC or 32 volts DC.

Current through voltage regulator U3 changes rapidly from 40 milliamperes (drawn mostly by RY1 when Q1 is on) to only a few microamperes (drawn by the amplifier and single-shot circuits when Q1 turns off). The current change produces a positive voltage spike on the +5 power supply that can feed back into the amplifier and cause repeat activation of RY1, especially when the amplifier gain is set high. Filtering of the power system to completely remove such a spike is difficult without using a large capacitor so an alternate method of preventing feedback is used. The reset pin 13 of the second single shot is normally pulled up to Vdd through R17.

Repeat activation of RY1 is prevented by C10, which pulls pin 13 to a down level when the gate of Q1 goes low and Q1 turns off. Pin 13 stays low for a period of about 1 second, determined by the time constant of C10 and R17. If feedback does occur, it will be ignored by the second single shot while it is in a reset state, and all the circuits will have time to stabilize before reset pin 13 goes high again.

The infrared body-heat motion detector consists of an etched, drilled, and screened printed circuit board, and all the board-mounted parts that must be assembled by mounting the parts on the board and soldering them in place. This is a high-density PC board that requires careful assembly and soldering. Figure 3-22 depicts the front side of the circuit board, and Figure 3-23 illustrates the reverse side of the circuit board where the PIR IR heat sensor is installed.

The IR motion detector is fabricated on a 1.7 × 2.4 inch PC board with circuit components on one side and a pyroelectric infrared sensor on the other side. The module should be mounted with the 2.4-inch dimension in the vertical position, for maximum sensitivity to horizontal motion. Four holes on the corners accept #4 mounting screws. A Fresnel lens or other focusing device can be placed in front of the sensor to increase sensing distance by focusing infrared radiation onto the sensor elements.

The component side of the IR motion detector has a sensitivity adjustment R10 that controls amplifier gain and, therefore, range (or detection distance). Turning the adjustment clockwise increases sensitivity and gives a greater range. The pyroelectric infrared sensor in the IR motion detector will detect a human or animal more easily at lower ambient temperatures when a greater difference exists between the human or animal body temperature and surrounding objects.

An optional external resistive photoconductive cell (not supplied) may be connected to the terminals marked C to deactivate the circuit during daylight. By doing this, a load will be activated only at night. This is useful if the IR motion detector is used to turn a light on.

A single-shot circuit controls the amount of time that elapses between motion events that activate a load. This is useful for applications where continued motion might occur and rapid repeat activation of a load is not desired, for example when someone is standing outside a door in view of the sensor. Delay can be varied from 1 second to about 90 seconds by adjusting R14. The single shot is retriggerable so that

Figure 3-22 *Front side of the high-density circuit board (Courtesy GloLab Corp)*

Figure 3-23 *Back side of the high-density circuit board (Courtesy GloLab Corp)*

continued motion will extend the delay. Nonretriggering mode is selected by wiring a jumper across the terminals marked E.

Another single-shot circuit controls the amount of time that a load remains turned on after motion is detected. The on time can be varied from 1 second to about 90 seconds by adjusting R19. This is useful for applications such as keeping a light on until a house is entered. The single shot is retriggerable so that continued motion will extend the on time. Nonretriggering mode is selected by wiring a jumper across the terminals marked W.

A 6- to 14.5-volt battery will power the circuits. Power is connected to the IR motion detector solder pads marked "PWR, + and −." The module circuits draw less than 150 microamperes when no motion is detected and less than 50 milliamperes when motion is detected and the relay is energized. In average use with a short on time, the IR motion detector will draw only 150 microamperes most of the time. A 9-volt alkaline battery should power the module for several months and possibly even much longer.

The power system is reverse-polarity protected so that a reversed power source will not damage the circuits. A DC wall transformer may also power the IR motion detector, however most wall transformers output much higher than their rated voltage when lightly loaded. Therefore the transformer output should be measured to be sure that the 14.5-volt maximum IR motion detector power supply input is not exceeded.

Capacitors in the amplifier and timing circuits require time to charge up to their normal operating voltages before the circuits will operate correctly. The circuits will not respond to motion until about 1 minute after power is applied.

Connect a power source to the PC board power pads marked + PWR −. Connect a load to the PC board relay terminals marked RY. The RY terminals are not a source of power for a load; they simply connect together through the relay contacts when motion is sensed. A load must be powered by an external source that will be switched off and on by the relay.

A thin plastic Fresnel lens that will extend detection range can be purchased separately, mounted in an enclosure, and held in place with tape or silicone rubber. No known adhesive is effective to bond to the lens material without the danger of damage to its surface. Although silicone rubber will not bond to the lens, it can be applied so it overlaps the edge of the lens and forms a captive mount. If a Glolab FL65 long-range Fresnel lens with a focal length of 0.65 inch is used and mounted against the inner surface of an enclosure, four $7/8$-inch (22.225-millimeter) long threaded nylon spacers (Digi-Key product number p/n 1902GK) will mount the IR motion detector PC board so that the sensor is the correct distance from the lens.

To begin building the IR motion detector, first bend the leads of all diodes and resistors close to the diode or resistor body, insert them into the PC boards, and bend the leads against the back side of the board. Cut the excess leads off short enough that they do not touch other connections but long enough to retain the diode or resistor until it is soldered. Be sure to insert diodes with cathode bands in the correct direction, as shown on the boards. Solder all diode and resistor leads. Next, insert each potentiometer (R10, R14, and R19) with its wiper terminal (the one that is different from the other two) in the hole marked by an arrow on the PC board. Press the potentiometer in place and solder. Now, insert small capacitors C1, C4, C6, C11, and C12, bend their leads, cut excess wire off, and solder them into place. Then, insert electrolytic capacitors C2, C3, C5, C7, C8, C9, C10, and C13 with their long positive lead in the hole marked + on the PC board. Bend the leads, cut excess wire off, and solder. Next, insert transistor Q1 and voltage regulator U3 in the direction indicated on the PC boards so they stand at least $1/8$ inch above the board. Solder all pins and cut off the excess leads. Insert U1 and U2 sockets, with their notch as indicated on the PC boards, and hold them against the board while soldering a few pins. Solder all remaining pins. Now, insert relay RY1 and solder. Place the O-ring over the leads of the PIR325 sensor, and insert the sensor leads into the bottom side of the board. Bend the leads over, cut off excess wire, and solder. Do not overheat. Apply only as much heat as necessary for a good solder joint. Next, carefully straighten the leads of U1 and U2 so they extend straight down and will fit into the sockets. This can be done with a pin straightener or by resting the IC on its side on a flat surface. Gently press on the IC while rocking it until the leads face straight down from the top of the IC. Repeat for the other side. Handle the ICs carefully to avoid static discharge damage. Finally, insert U1 and U2 into their sockets with pin 1 near the socket notch.

You are now ready to attach a battery or other power source to the plus and minus power pads and a load to the RY pads. Allow about 1 minute for the circuits to stabilize after power is applied. You can also connect an optional photocell to the C pads for night-only operation. Clockwise rotation of potentiometer R10 increases the amplifier gain; clockwise rotation of R14 increases the delay between motion detection; and clockwise rotation of R19 increases the dwell time when the relay is energized. With the gain control R10 set fully clockwise for maximum gain and with no lens in front of the sensor, it will detect a moving hand at a distance of about 1 foot and a human body at about 3 feet.

Infrared Motion Detector Parts List

Required Parts

R1, R11   100K, 1/8-watt, 5% carbon film

R2, R3, R5, R13, R20 10K, 1/8-watt, 5% carbon film

R4, R12, R15, R16, R17, R18   1 MEG, 1/8-watt, 5% carbon film

R6, R9   2 MEG, 1/8-watt, 5% carbon film

R7, R8   150K, 1/8-watt, 5% carbon film

R10, R14, R19   1 MEG potentiometer

D1, D2, D3   1N914 diode Fairchild 1N914

D4   BAT46 Schottky diode

C1   100 pf, 50-volt ceramic disc

C2, C3, C5   10 uF, 16-volt electrolytic

C4, C6, C11, C12   0.1 uF, 50-volt metalized film

C7, C8, C9, C13   100 uF, 16-volt electrolytic

C10   1 MFD, 50-volt electrolytic

Q1   2N7000 field effect transistor

U1   LP324 or equivalent micropower quad op-amp

U2   CD4538 CMOS dual single shot

U3   Seiko S-812C50AY-B-micropower voltage regulator

PIR   PIR325 pyroelectric infrared sensor (Glolab PIR325)

O-ring   Spacer Polydraulic BUNA-N size 009

RY1   SPST no relay, 5-volt 40 MA coil (P&B T77S1D3-05)

Miscellaneous   IC sockets, transistor socket, wire, connectors, PC board

Optional

FL65   long-distance single-element Fresnel lens (Glolab)

CD-1   CDS photo conductive cell for day/night (PDV-P8001)

Infrared motion-detector kit GLMD

IR motion-detector PC Board GLMDPCB

Liquid Sensing

In this chapter we will explore liquid sensing, a very interesting and important aspect of sensing. Our first project in this chapter is a simple yet useful rain detector. In this chapter you will also learn how to build a liquid sensor, a fluid-level indicator. Weather fans will learn how to construct a humidity monitor to measure humidity around your home or shop. Junior scientists will learn about pH and how to build and use a pH meter to determine whether a liquid is a base or an acid. Nature- and ecology-minded readers will learn about how to build and utilize a stream gauge water-level monitor for studying river and stream flow and runoff.

Rain Detector

The rain detector will enable you to detect the first few drops of rain, which will allow you a few precious moments to roll up your car windows, bring in your laundry, or bring in your possessions. When used as part of a weather-data collection system, the exact time of a shower can be recorded.

The diagram in shown in Figure 4-1 illustrates the sensor portion of the rain detector, consisting of two strips of aluminum foil glued to a piece of plastic. A single square of foil is glued to the plastic with two lead wires underneath, as shown in the figure. The lead wires are stripped back, so that the foil makes good electrical contact with the conductors. But the bare wire should not protrude, allowing the foil to protect the wire from corrosion. A narrow zigzag is cut in the foil to electrically separate the two lead wires. The raindrops bridge the gap causing conduction, which is then sensed by the electronic circuit shown in Figure 4-2. Also note that you could use a scrap piece of circuit board as the sensor, provided that you can etch out or cut out the zigzag pattern with close separation between the plates.

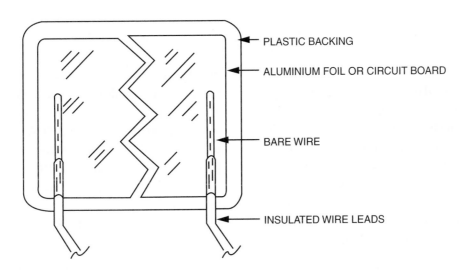

PLASTIC BACKING

ALUMINIUM FOIL OR CIRCUIT BOARD

BARE WIRE

INSULATED WIRE LEADS

Figure 4-1 *Rain detector*

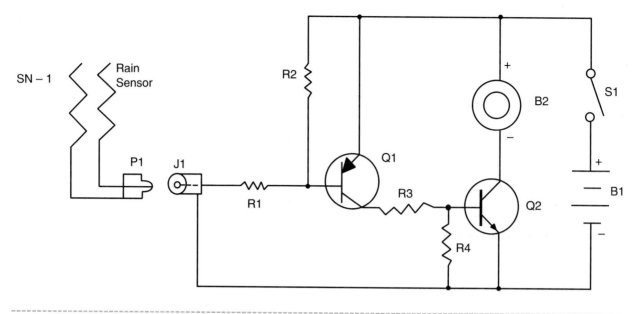

Figure 4-2   *Rain detector circuit*

Figure 4-2 depicts the electronic circuit portion of the rain detector. One end of the sensor is connected to ground as shown, while the other sensor lead is connected through a 1K ohm resistor, which is fed to the first transistor at Q1. Transistor Q1 is a 2N4403 PNP transistor. The output of Q1 is coupled to a second transistor stage at Q2, a 2N4401 NPN transistor, via a 220-ohm resistor. The collector of transistor Q2 is connected to the black or (minus) lead of an electronic buzzer or sonalert. The red or (plus) lead of the electronic buzzer is connected to the positive lead of the battery supply. The prototype uses three AA cells, but a 9-volt battery could be used. A SPST toggle switch was used at S1 to apply power to the circuit.

Construction of the rain detector is quite simple and can be assembled on a perf-board or prototype board. Radio Shack, for example, stocks a few different types of simple low-cost protoboards that can be used for this project. These boards have copper foil around circuit pads with close separation between holes, which is ideal for this project. Nothing critical is in this circuit, so tie point construction could also be used. The components are all readily available and should be available through your local Radio Shack or electronic outlet. When assembling the rain detector, be sure to observe the correct polarity when installing the transistors and the electronic buzzer. Be

sure to identify the pinouts of the transistor before installing them to avoid damaging these components.

Once the rain detector has been assembled, you can simply test the circuit by shorting the two input leads (i.e., connect the free input lead of the 1K ohm resistor to the ground, or minus, of the battery and the buzzer should sound). If all goes well and the circuit works properly, you can then decide how to enclose the circuit board. In the prototype, a plastic box is used to house the rain detector. A small plastic box measuring 4 × 6 × 2 inches is used for this project. A three-cell AA battery holder is mounted on the inside, along the side of the plastic case. The circuit board was mounted on stand-off insulators, which were used to lift the board above the bottom of the plastic case. An RCA-type phone jack is mounted on one side of the side of the plastic case. The RCA jack is used to connect the circuit to the sensor. The center lead of the RCA jack is connected to the free end of the 1K ohm resistor, and the ground, or outside lead, of the RCA jack is connected to the ground of the rain detector circuit at the minus lead of the battery. The toggle switch is mounted on the side of the case near the RCA input jack.

Next, you will need to connect the sensor to a two conductor lead-in wire that has an RCA plug at the opposite end. Finally, you will need to decide the length of the sensor wire. The sensor itself could be

Chapter Four — Liquid Sensing

mounted at the top of a wooden rod (or even a tomato plant stake) and should be placed out in the open on the lawn. Alternately the sensor could be mounted with velcro on the edge of the roof. The sensor should have an unobstructed view of the sky, so rain can fall directly upon the sensor for best results.

Rain Detector Parts List

Required Parts

R1, R4 1K ohm, 1/4-watt resistor

R2 100K ohm, 1/4-watt resistor

R3 220-ohm, 1/4-watt resistor

Q1 2N4403 PNP transistor

Q2 2N4401 NPN transistor

BZ electronic buzzer or Sonalert sounder

S1 SPST toggle power switch

B1 3 AA batteries (4.5 volts DC)

P1 RCA plug

J1 RCA jack

SN-1 rain sensor (see text)

Miscellaneous protoboard circuit board, battery holder, wire, transistor sockets

Fluid Sensor

The fluid sensor project is a variation on a theme of fluid sensing. This project is designed to sense a liquid's level and then close a set of relay contacts. The relay contacts can be used to activate a pump or to activate an alarm system to alert you of a flood condition.

The heart of the fluid sensor is the actual sensor and the CMOS Quad 2-input gates, as shown in Figure 4-3. The circuit operates by detecting resistive conduction through the fluid it is sensing. The conducting fluid decreases the resistance between the sensor probes at C1 and C3, causing the oscillator formed by U1:A and components R1, C2, and C1 to change frequency. As the resistance between the sensing leads decreases the oscillator begins sending

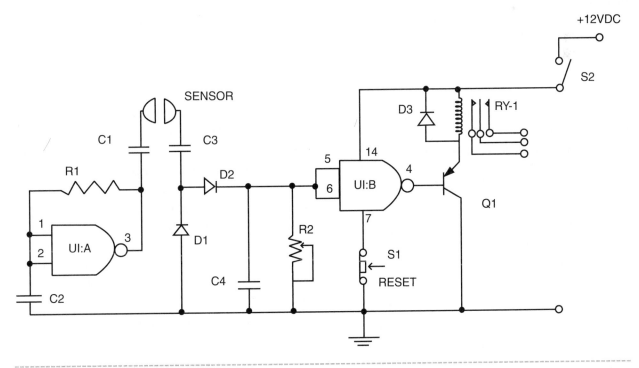

Figure 4-3 *Fluid sensor circuit*

an AC signal to the diodes at D1 and D2. Diodes D1 and D2 are used to rectify the signal, which is used to drive the second quad input gate at U1:b. The DC signal is next smoothed by capacitor C4 and then sent to potentiometer R2. The potentiometer is used to set the sensitivity of the circuit. The Quad 2-input gate at U1:b next drives transistor Q1, which in turn pulls in the relay at RY-1. Diode D3, across the relay, is used to eliminate transients or spikes, which can be caused by the relay coil. A reset switch is shown at S1. Note that the unused input pins 8, 9, 12, and 13 on the quad two-input gates should be tied to ground to eliminate false triggering. Unused output pins should be left unconnected.

The fluid sensor circuit shown uses a *single-pole, double-throw* (SPDT) relay; this allows both a normally closed output and a normally open set of contacts, which can be used to activate a loud bell, an alarm system, or a telephone dialer circuit. The fluid sensor is designed to operate from a 12-volt DC source. The circuit can be operated via a 12-volt battery trickle charged by a 12-volt power supply, or you could elect to use a 12-volt, 1-amp wall wart power supply.

The fluid sensor is basically two contacts that are brought back to the circuit board. The sensor should be constructed from a corrosion-resistant material such as stainless steel. Chrome-plated bicycle spokes removed from an old bike could be used for the sensor. The two sensor probes should be in close proximity to but separated from each other. If two stainless rods are used, for example, a plastic spacer block, plexiglass block, or wooden block could be used to hold the sensor probes together. Depending upon your particular application, you may wish to design a probe assembly such as the one shown in Figure 4-4 that suspends the sensor probes over a flat surface.

Construction of the fluid sensor is quite straightforward. The circuit can be fabricated on a dedicated circuit board or a proto-board or perf-board for quick construction. It is advisable to use a transistor socket as well as an IC socket. In the event that circuit should fail, it is much easier to fix or repair a circuit if sockets are used. Remember that diodes, transistors, and ICs all have particular polarity or orientation, which must be observed. Diodes must be

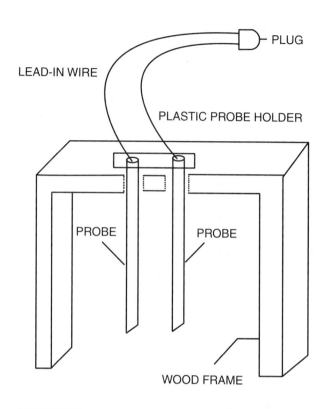

Figure 4-4 *Probe assembly*

installed correctly if the circuit is to function. The arrow points to the cathode of the diode; be sure to carefully observe the polarity prior to installation. Transistors generally have three leads: a base lead, a collector lead, and an emitter lead. The base lead is usually shown on the opposite side from the collector and emitter leads. The emitter lead is usually associated with a pointing arrow, which points either toward the transistor or away from the transistor. A PNP transistor, as used in this circuit, has the arrow pointing toward the center of the transistor. When installing the IC, be sure to carefully observe the orientation when installing. ICs generally have either a cutout or notch at the top of the device, thus pin 1 is usually to the left of the cutout. Other ICs often have a small indented circle near pin 1. After construction of the circuit, check over the circuit board for any stray or cut component leads that may be stuck to the circuit board. Make sure no shorting solder bridges are still left on the board.

Once complete, you can apply a 12-volt power source and test the circuit for proper operation. Place the circuit board on an insulated surface, connect a power source, and then place a jumper wire between capacitors C1 and C3 to simulate fluid at the probes. At this time, you should hear or see the relay change state. If the relay pulls in, the circuit is working. If the relay does not pull in, you will need to check the circuit once again to make sure the wiring and components have been installed correctly.

The fluid sensor circuit is installed in a small metal chassis box. The prototype is built into a 5 × 6 × 3 inch chassis box. The circuit board is placed on stand-offs and mounted to the bottom of the chassis box. A coaxial power jack used for power is mounted on the rear of the chassis. An input jack for the sensor probes is mounted along with the reset and power switch on the front of the chassis box. Depending upon what type of potentiometer you install, you may want to provide an adjustment hole, especially if you used a circuit board trimmer potentiometer. If you used a larger chassis-mounted potentiometer, you will need to drill a hole for the potentiometer control shaft.

Your fluid sensor is ready to protect your home or workshop! Build it and have fun!

Fluid Sensor Parts List

Required Parts

R1   470K ohm, 1/4-watt resistor

R2   15-megohm potentiometer

C1, C2, C3, C4   2.2 nF, 35-volt capacitor

D1, D2   1N4148 silicon diode

U1   MC14093B quad two-input gate integrated circuit

D3   1N4004 silicon rectifier diode

Q1   2N3906 PNP transistor

RY-1   mini 12-volt relay

S1   normally closed pushbutton switch

S2   SPST toggle power switch

Miscellaneous   PC board, IC socket, transistor socket, wire, hardware, connectors

Fluid-Level Indicator

Our next project is the fluid-level indicator circuit shown in Figures 4-5 and 4-6. This circuit not only indicates the amount of water present in the overhead tank but also sets off an alarm when the tank is full. The fluid indicator circuit was designed to show how much liquid is remaining in a tank or container.

The heart of the circuit is a simple sensor, which was built using a scrap piece of circuit board material with five etched lines of different lengths, as shown in Figure 4-5. The prototype fluid sensor is easy to construct and is built using a 1¼-inch wide by 5-inch long piece of scrap circuit board material. Five copper lines are left on the board for the four different fluid levels. Note that there are four sensor lines and a common lead, which is placed at the very bottom of the detector strip. Square or round circle-type copper

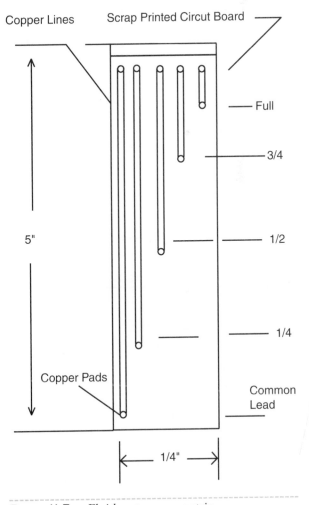

Figure 4-5   *Fluid water sensor strip*

pads are placed at the ends of each of the copper sensor lines; these act as the detection points for the fluid-level sensing. Depending upon your particular sensing situation, you may wish to have shorter or longer sensing strips. By using PC boards in this fashion, it is a simple matter to construct multiple sensing strips for various sensing applications. Also note, that the circuit can be expanded to sense multiple levels as the circuit can be scaled for more sensing levels if desired by adding more CMOS switches and LEDs.

The main circuit for the liquid-level indicator is depicted in Figure 4-6. At the heart of the electronic fluid-level indicator are the CMOS switches at S1 through S4. The circuit uses the widely available CD4066, bilateral switch CMOS IC to display the water level via LEDs. Each of the sensing lines from the PC board sensor are fed back to the CMOS switches at pins 5, 6, 12, and 13, which act as the sens-

ing line inputs. The fluid-level indicator uses a single CMOS CD4066 package for sensing. The ends of each of the CMOS's internal switch contacts are connected together at pins 2, 4, 9, and 11 and are tied to the circuit's common ground connection. The other CMOS switch contacts at pins 1, 3, 8, and 10 are each connected to the indicator LEDs through series resistors, as shown. Power is supplied to the CMOS chip on pin 14, and the ground connection is on pin 7. The fifth or common sense lead at the bottom of the sensing strip is connected to the common anode line of the power supply as a reference lead.

In operation, when the water level is empty, the wires in the tank are open circuited, and the 180K resistors pull the switch low, thereby opening the switch and leaving the LEDs at the off position. As the water starts filling up the tank, the first wire in the tank connected to S1 and the plus (+) supply are

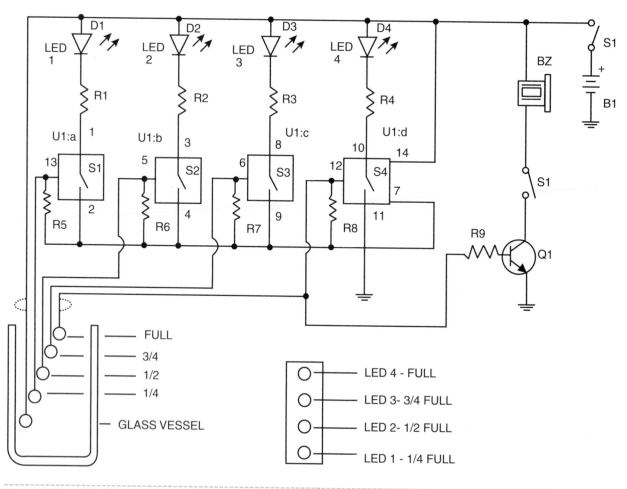

Figure 4-6   *Fluid water level indicator circuit*

shorted by the water. This closes the switch S1 and turns LED 1 to on. As the water continues to fill the tank, LEDs 2, 3, and 4 light up gradually. The number of levels of indication can be increased to eight if two CD4066 ICs are used in a similar fashion.

When the water tank is full, the base of the transistor BC148 is pulled high by the water, saturating the transistor, and thus turning on the electronic buzzer or Sonalert. Note that the SPST switch has to be opened to turn the buzzer off once the tank is filled to the top.

This circuit is simple and straightforward and can be adapted to many applications quite easily. The prototype fluid-level indicator is constructed on a small prototyping board found at any Radio Shack store. These boards are drilled on .010 centers with copper leads at each hole, with connecting copper lines, some of which form common bus leads. These boards are inexpensive and can be readily used for this project. The prototype circuit is quite small because only a single IC, a few resistors, and a single transistor are used. When constructing the circuit board, be sure to use an IC socket for the CMOS chip.

This will greatly help in the event of circuit failure down the road. Be sure to observe proper polarity when installing the IC as well. ICs usually have some form of marking to indicate the pin numbers. Some ICs have a cutout or notch at the top of the IC package. Other ICs often have a small round indented circle near pin 1 of the IC. When installing the LEDs, be sure to insert them properly, observing the correct polarity. If LEDs are mounted backward, they will usually survive, but installing an IC backwards is often a sure-fire way to destroy the IC.

Finally, when installing the transistor be sure to identify the pin before installing it. Transistors have three leads: a base, a collector, and an emitter. Transistors are also very sensitive and can be destroyed by improper installation. Depending upon your particular application, you may wish to install the LEDs on the circuit board along with the other components or separately on the chassis box, with wire leads running

from the LEDs to the circuit board. Installing all the LEDs on the PC board makes a tidy and clean layout with a minimum of connecting wire running all around the chassis box. If you elect to install all the LEDs on the PC board, you will have to precisely determine where the mounting holes will appear on the chassis box enclosure.

The prototype fluid indicator is mounted in a small aluminum chassis box measuring $5 \times 6 \times 3$ inches. The SPST toggle switch and the electronic buzzer are mounted on the top cover of the chassis box. The prototype circuit has all its LEDs mounted on the circuit board, so holes have to be measured precisely in order for the LEDs to line up with the holes drilled into the chassis box. Then the circuit board is mounted on stand-offs, so the LEDs can protrude through the case where the holes are drilled. A nine-pin RS-232 female connector is mounted at the rear of the chassis box, which is used to accept a nine-pin male RS-232 connector, linking the sensor with the electronics package. The fluid-level indicator operates from a 6-volt power source, therefore two AA plastic battery holders are mounted on the bottom of the chassis box. Finally you can make a legend or label the LEDs ¼, ½, ¾, and full.

Once the installation of all the components and circuit is completed, the circuit can be tested to be sure it operates correctly. Wire up your sensor board with a length of five-conductor cable and place a nine-pin RS-232 male connector at the opposite end of the cable. Plug in the sensor connector to the female chassis connector and insert the batteries into the holders, and you are now ready to test the liquid-level indicator circuit. Apply power to the circuit, and slowly insert the sensor strip in a large nonconducting beaker, cylinder, or bowl and the LEDs should begin lighting up. Each LED should light up sequentially and then the alarm buzzer should sound letting you know the container is full.

Your fluid-level indicator is now ready to serve your needs. Your imagination will be the only limit to the applications you may find for the circuit. Have fun!

Fluid-Level Indicator Parts List

Required Parts

R1, R2, R3, R4   330 ohm, 1/4-watt resistor

R5, R6, R7, R8   180k ohm, 1/4-watt resistor

R9   2.2k ohm, 1/4-watt resistor

D1, D2, D3, D4   LEDs

Q1   BC148 or equivalent NPN transistor

U1   CD4066 CMOS electronic switch

BZ   Piezo electronic buzzer

S1   SPST toggle power switch

B1   4 AA cells

Miscellaneous PC board, wire, RS-232 male and female connectors, IC socket, transistor socket, battery holder, etc.

Humidity Monitor

Are you interested in monitoring weather conditions or creating your own weather station? A number of projects in this book can be combined to do just that. This capacitive relative humidity monitor is a good place to start as it is easy to construct, low cost, and has a good long-term stability and reliability.

The heart of the capacitive humidity monitor is the General Eastern G-Cap 2 relative humidity sensor, which is shown as SEN-1 in the circuit diagram in Figure 4-7. The humidity monitor can monitor relative humidity between 0 and 100 percent. The sensing element consists of a proprietary electrode metallization deposited over the humidity sensitive polymer. The structure of the sensor allows for rapid diffusion of water vapor, which speeds up drying and makes recalibration easy. The properties of the sensor film allow the sensor to survive immersion in water. The capacitive sensor changes capacitance from 148 pF at 0 percent relative humidity to 178 pF at 100 percent humidity.

The humidity monitor circuit uses the G-Cap humidity sensor to detect humidity changes by varying the frequency output of a TLC555CP. The CMOS TLC555CP is set up as a stable multivibrator or oscillator employing R3 and SEN-1, which determine the frequency span of the oscillator. As the humidity changes, the oscillator frequency changes from about 13 to 15 KHz. Potentiometer R5 allows the output signal to be adjusted and thus allows calibration of the humidity monitor. The varying frequency output from U3 is converted to a DC signal and coupled to an LM358 op-amp, which provides a varying DC output from 0 to 5 volts DC based on the humidity changes from 0 to 100 percent humidity. The minus (−) input of the op-amp is setup as a reference source, whereas the plus (+) input of the op-amp is the varying signal from the humidity sensor. The output of the LM358 at pin 1 produces a voltage between 0 and 5 volts DC. An Acculex DP652 LCD panel meter is used to display the 0- to 5-volt DC output from the humidity monitor circuit. The output of U4:a is fed to a voltage divider that drops the full scale voltage down to 2 volts for the panel meter input. The minus (−) input of the LCD meter at pin 8 is tied to the common, or ground, at pin 1. The input voltage from the humidity monitor circuit is sent to pin 7 of the LCD meter. The LCD panel meter is powered on pin 1 from the 5-volt regulator at U2. The decimal points at pins 4, 5, and 6 are selected by connecting them to pin 3, which is the *decimal point* (DP) common pin, but the decimal points are not used in this project and are left unconnected.

The humidity monitor circuit is powered from a 12-volt battery at B1. The battery is switched into the circuit via the power switch at S1. The first regulator at U1 converts the 12-volt battery voltage down to 10 volts DC for the op-amp at U4:a on pin 8. The second regulator is fed directly from the 10-volt output from the first regulator. The regulator at U2 is used to supply 5 volts of DC to power both the reference voltage source at R10 and also to power the LCD panel meter circuit.

The humidity monitor should be constructed on a printed circuit board for best results, but you could also elect to build the circuit on an experimenter's prototype board using short length point-to-point

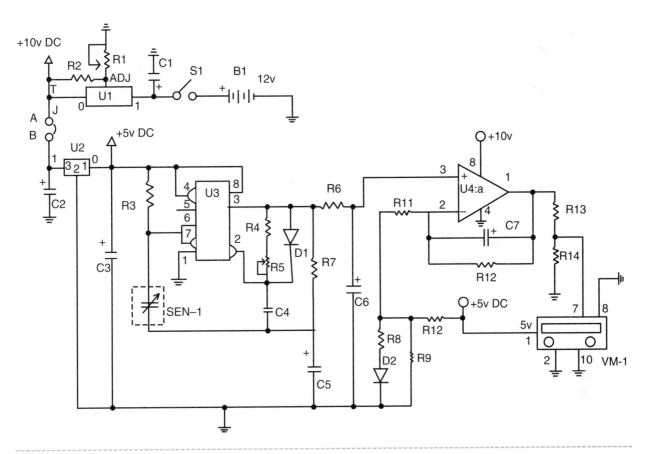

Figure 4-7 *Humidity monitor circuit*

wiring. The 2 ½ × 4 inch glass epoxy circuit board was used for the prototype humidity monitor. When designing a circuit board for the humidity monitor, you will want to consider mounting the sensor SEN-1 at one end of the circuit board to allow the sensor to "look out" the side of the enclosure. You might also elect to mount the sensor remotely, but this is not recommended because the capacitance of the sensor and long lead-in wire could adversely effect the circuit operation.

IC sockets are recommended in the event of circuit problems in the future. IC sockets cost a few cents more at circuit construction time but give you piece of mind for future service work if ever needed. Be careful to observe the polarity of the capacitors when installing them into the circuit in order to avoid damage to the circuit upon power up. The humidity monitor circuit has two diodes, so you will need to pay attention to the polarity of these devices as well when installing them. Before installing the two regulators, identify the input pins from the output pins to

avoid damage to the circuit when applying power to the circuit. You will need to adjust the output of the regulator to 10 volts DV using the potentiometer at R1, prior to installing the ICs to ensure that U4 has a 10-volt source. Once you have ensured that the voltage regulators are producing the correct voltage, you can insert the jumper wire between points A and B.

The humidity monitor circuit prototype is installed in a small metal chassis box. Place the circuit board inside of the chassis box, and align the circuit board to one edge of the chassis box. Drill a half-inch hole in the chassis box to allow the sensor to see outside the chassis box. You can cover the half-hole later with plastic screen material if desired. Drill four holes in the four corners of the circuit board. The circuit board is then mounted on ¼-inch plastic standoffs, to the bottom of the chassis box using ¾-inch 4-40 machine screws. Using the template supplied with the LCD panel meter, a cutout hole is marked on the top of the chassis box for the meter assembly. You will then need to cut out an opening for the LCD meter. You can do this by drilling small holes all around the meter cutout, then punch out the center piece that is left, and finally use a file to make all the edges smooth and straight. Another option is to purchase a nibbler tool to make the rectangular cutout for the meter. The power switch S1 is also mounted on the top of the chassis box. Two four-cell AA battery holders are mounted on the bottom of the chassis box for the batteries. Note that eight AA cells will produce 12 volts of direct current to power the circuit.

Once the humidity monitor is assembled, you are ready to calibrate the humidity meter. Insert batteries into the holders, and switch the power to the circuit. Use a multimeter to ensure that you have 10 volts output at the first regulator and 5 volts at the second regulator's output. At this point you can calibrate the humidity monitor; this can be done in one of two ways. The humidity monitor can be placed next to a pot of boiling water, or the sensor can be taken outside while it is raining. If placed near a pot of boiling water, you'll want to keep the sensor just at the edge of the cloud of steam and only for a short time, long enough to adjust the potentiometer at R5 for full scale of 5 volts while in the cloud of steam. You can also take the sensor outside during a rain storm, but you'll want to keep the sensor protected from the raindrops. If you or a friend have a calibrated humidity instrument, you could also use it to calibrate the humidity monitor.

Your relative humidity monitor is now ready to serve you and can be used to round out your weather monitoring instruments if you are a weather enthusiast.

Humidity Monitor Parts List

Required Parts

R1 5K ohm potentiometer (trimpot)

R2 240 ohm, 1/4-watt, 5% resistor

R3 100K ohm, 1/4-watt, 5% resistor

R4 51.1K ohm, 1/4-watt, 1% resistor

R5 20K ohm potentiometer (trimpot)

R6, R7, R10 30.1K ohm, 1/4-watt, 1% resistor

R8 150K ohm, 1/4-watt, 5% resistor

R9 42.2K ohm, 1/4-watt, 1% resistor

R11 20K ohm, 1/4-watt, 5% resistor

R12 845K ohm, 1/4-watt, 1% resistor

R13 221-ohm, 1/4-watt, 1% resistor

R14 149-ohm, 1/4-watt, 1% resistor

C1 1 uF, 35-volt electrolytic capacitor

C2, C7 0.1 uF, 35-volt tantalum capacitor

C3, C5, C6 4.7 uF, 35-volt electrolytic capacitor

C4 270 pF, 35-volt ceramic NPO capacitor

D1, D2 1N4148 silicon diode

SEN-1 G-Cap 2 capacitive humidity sensor (General Eastern Instruments)

U1 LM317 three-terminal adjustable regulator (National)

U2 LM2936Z-5 three-terminal, 5-volt positive voltage regulator (National)

U3 TLC555CP timer/oscillator IC (Texas Instruments)

U4 LM358 dual op-amp (National)

S1 SPST toggle power switch

B1 8 AA cells or a 12-volt battery

VM-1 DP-652 LCD panel meter; +/-2 volts (Acculex)

Miscellaneous printed circuit board, IC sockets, wire, 4-40 screws and nuts, standoffs, terminals, jumper, etc.

pH Meter

If you are a scientist, chemist, or a pool owner, you are certain to want to know the pH of a solution or of water. So the question is, what is pH? pH is a number that exactly describes the degree of acidity or basicity of a solution. A good analogy can be made with the measurement of temperature. With temperature, we have the terms cold and hot, and we immediately realize that these are very general terms that cannot be used with any degree of accuracy. Accordingly, the temperature scale was developed, and now a specific temperature reading, of say 50°C, means the same to everyone and is scientifically accurate.

In much the same way, the pH scale was developed. Centuries ago, humans discovered that certain materials possessed properties that we called *acid*, whereas others possessed different properties that we called *bases*. Between these two was a neutral area in which the material showed neither acid nor base properties, and we termed this *neutral*.

Now just as in hot and cold, the words *acid* and *base* do not give us a scientific value that we can use. We needed a scale that we all agree on when we discuss the degree of acidity or basicity. But before we study the scale, let's take a moment to find out what makes one material an acid and another a base: An acid must have ionized (or free) hydrogen ions, $H+$; a base must have ionized (or free) hydroxyl ions $OH-$. pH is directly related to the ratio of $H+$ to OH. If the $H+$ is greater than $OH-$, the material is an acid. If the $OH-$ is greater than the $H+$, the material is a base. If equal amounts are present, then the material is a neutral salt (see the pH scale in Figure 4-8).

Lets go back now to the development of the pH scale. A molar solution of hydrochloric acid is about a 3.6 percent solution of the acid. Let's assign this solution a pH of 0. A molar solution of sodium hydroxide (commonly known as lye) is about a 4.0 percent solution of the base. Let's assign this solution a pH of 14.

If we now dilute the acid by adding 1 milliliter of acid to 9 milliliters of pure water, we will have a $1/10$ molar solution. Let's assign this solution a pH of 1. In the same way, let's dilute the molar solution of sodium hydroxide by adding 1 milliliter to 9 milliliters of pure water. We'll assign this resultant solution a pH of 13.

Notice that with the acid, our $1/10$ dilution increased the pH from 0 to 1, while the same dilution of the base reduced the pH from 14 to 13. Now here's the significant point: In both cases, we were going toward 7. As we'll see shortly, a pH of 7 is the exact middle of the scale at which we have neither acid nor base, but neutral.

Continue to dilute both the acid and base by one-tenth, and each time increase the number in the case of the acid and decrease the number in the case of the base. The results look like this.

First, a pH probe (which we'll discuss in more detail shortly) produces a voltage that can be directly related to the pH of the solution in which we place the probe. Secondly, an electronic circuit within the pH meter cabinet receives the voltage from the probe and then presents it to the meter scale. This voltage developed at the probe will cause the meter pointer to move. The value of the number where the pointer stops is the pH of the solution.

The probe can be thought of as a battery whose voltage changes as the pH of the solution in which it is inserted changes. It consists of two parts (in fact many pH measurements are made with two separate probes): (1) the hydrogen sensitive glass bulb, and (2) the reference electrode. The special glass of the bulb has the ability to pass $H+$. This ability then allows the $H+$ molecules inside the bulb to compare themselves with those outside the bulb and to develop a voltage that is related to this difference. This bulb then is a half-cell and will need a companion reference to function.

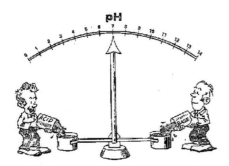

Figure 4-8   *pH scale*

In Figure 4-9 notice that just above the bulb is a reference electrode frit. This is actually a small opening in the glass through which the inside filler solution can very slowly leak out. Now the relationship between the reference electrode and the solution also produces a voltage, and this is the other half cell. Together the pH sensitive bulb and the reference electrode constitute the complete probe.

The value of voltage produced by the probe is fortunately a linear function of the pH. For example, at pH 7.00, the probe produces 0 volts while at pH 6.00, it produces +0.06 volts or +60 millivolts. Notice the plus polarity mark; if the voltage were of minus polarity, the meter pointer would go to the right to a pH reading of 8.00. Generally, a probe will produce about 60 millivolts for each change of 1 pH unit. Thus, a probe voltage of +300 millivolts would cause the meter to read pH 2.00 ($+^{300}/_{60}$ = 5 units, $7 - 2 = 2$).

Because the pH meter and probe are both electronic devices, you might like to have some standard pH solution that you can measure to be sure that everything is calibrated correctly. These solutions are available and they are called standard buffers. You'll soon find that a buffer is a vital part of all pH measurements. A buffer is a solution of a particular pH that has the ability to resist change in pH. (As a side note, buffers in our blood system are what keeps us alive and healthy.)

Generally pH meters for commercial use are rather expensive, but it is possible to build your own pH meter for a small outlay and learn about how pH is measured. Using an ultralow input current amplifier, a CMOS micropower op-amp, and a digital multimeter, you can construct a useful pH meter. A generalized pH meter block diagram is shown Figure 4-10.

The heart of our "volks" pH meter circuit is the low cost *silver/silver chloride* (Ag/AgCl) pH probe at the input of the ultralow current amplifier shown in the diagram in Figure 4-11. The signal from a pH probe has a typical resistance between 10 megohms and 1,000 megohms. Because of this high value, it is very important that the amplifier input currents be as small as possible. The LMC6001, with less than 25 fA input current, is an ideal choice for this pH meter circuit. The theoretical output of a standard Ag/AgCl pH probe is 59.16 mV/pH at 25°C with 0 volts out at a pH of 7.00. This output is proportional to absolute temperature. To compensate for this, a temperature-compensating resistor R1 is placed in the feedback loop. This cancels the temperature dependence of the probe. This resistor must be mounted where it will be at the same temperature as the liquid being measured.

The ultralow input current amplifier, the LMC6001, amplifies the probe output providing a scaled voltage of +/−100 mV/pH from a pH of 7. Overall gain adjustments to the pH meter can be

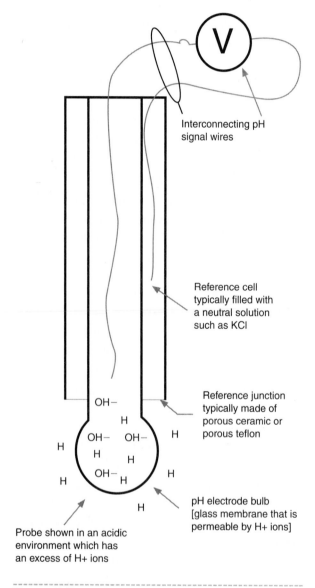

Interconnecting pH signal wires

Reference cell typically filled with a neutral solution such as KCl

Reference junction typically made of porous ceramic or porous teflon

OH−
H
OH−   OH−   H
H
H   H
OH−   H
H
H   H

pH electrode bulb [glass membrane that is permeable by H+ ions]

Probe shown in an acidic environment which has an excess of H+ ions

Figure 4-9   *pH Bulb*

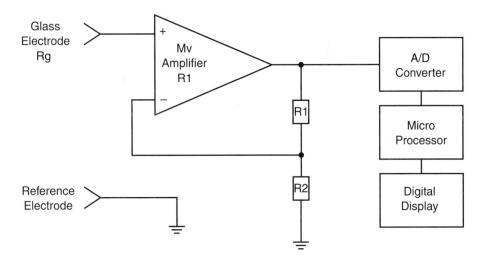

Figure 4-10   *pH meter block diagram*

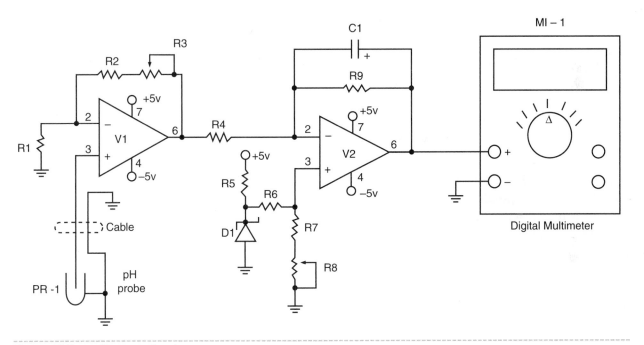

Figure 4-11   *pH meter circuit*

made via potentiometer R3. The second op-amp, a micropower LMC6041, provides phase inversion and offset so that the output is directly proportional to pH, over the full range of the probe. The reference circuit, or offset circuit, consists of a zener diode, two resistors, and the potentiometer at R8. The output of U2 can be directly coupled to a digital multimeter, which can be calibrated to read pH. You could also elect to purchase a low-cost digital panel voltmeter instead of using a multimeter.

Total current consumption of the entire pH meter circuit is only about 1 mA for the whole system. The pH meter circuit requires a 5-volt DC power source, supplied from the dual plus/minus power supply circuit shown in Figure 4-12. The dual plus/minus power supply utilizes two 9-volt transistor radio batteries. Battery B1 supplies a minus (−) voltage to regulator U3, and B2 supplies a positive (+) voltage to the U4. The two regulators provide a stable and accurate 5-volt power source for the two op-amps.

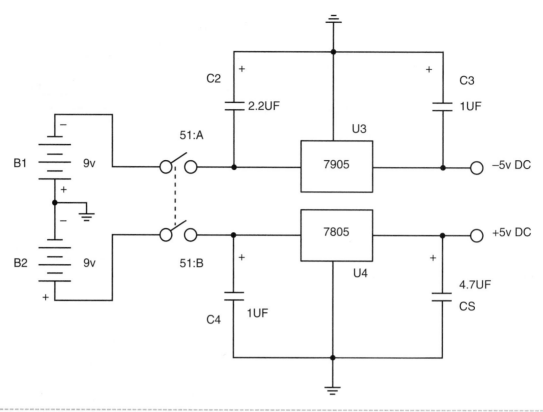

Figure 4-12 *pH meter power supply*

The pH meter circuit can be constructed on a small 2 × 3 inch printed circuit board. The layout should be as short as possible between the two low-current op-amps to remove any signal loss, error, or noise. The use of IC sockets is highly recommended, in case of a circuit failure at a later date. ICs are marked, so you can identify the pinouts of the device. Generally ICs are marked with either a small indented circle on the left side of the IC package or a small cutout on the top of the IC. Pin 1 of the IC is always to the left of either the cutout or the indented circle. The pH meter circuit contains a zener diode, which must be installed correctly in order for the circuit to work properly. The black band on the diode corresponds to the cathode of the diode. The band should be pointing toward resistor R5 and R6. Note that capacitor C1 is a nonpolarized capacitor, and that the resistors are 1% values to ensure accuracy of the meter.

After constructing the circuit board, you'll want to inspect the foil side of the board for any possible shorts, either between copper pads or between cop-

per pads caused by solder blobs or from stray discarded component leads.

Locate a 4 × 6 × 2 inch metal chassis box to house the pH meter circuit. Obtain two 9-volt plastic transistor radio battery holders and mount them to the bottom of the chassis box. The circuit board is then mounted atop plastic standoffs to ensure the circuitry does not short out to the metal chassis box. Power switch S1 can be mounted on the top side of the chassis. A dual banana jack or two single, red and black banana jacks for the multimeter connections can be mounted on top of the chassis. You will need to locate a matching connector for the pH probe; it too can be mounted on the top of the chassis box.

The prototype volks pH meter utilizes a low-cost FastGlass model TSP60001 pH probe purchased from EyeThink Corporation. The pH probe is a glass bulb Ag-AgCI-type, which can measure from 0 to 14 pH.

Once you have the circuit enclosed and the batteries connected, you can begin the setup and calibration. The calibration process is quite simple with little possibility of problems or interactions. First, turn on the power switch, then disconnect the pH probe and, with R3 set to about midrange and the noninverting input of the LMC6001 grounded, adjust R8 until the output is 700 mV. Next, apply −414.1 mV to the non-inverting input of the LMC6001. And finally adjust R3 for an output of 1,400 mV. This completes the calibration. As pH probes may be slightly different, minor gain and offset adjustments should be made by trimming while measuring a precision buffer solution.

pH Meter Parts List

Required Parts

R1 100K, +3,500 ppm/C° (see note)

R2 68.1K ohm, 1/4-watt, 1% resistor

R3 100K ohm potentiometer

R4 9,100K ohm, 1/4-watt, 1% resistor

R5 36.5K ohm, 1/4-watt, 1% resistor

R6 619K ohm, 1/4-watt, 1% resistor

R7 97.6K ohm, 1/4-watt, 1% resistor

R8 10K ohm potentiometer

D1 LM4040D 1Z, 2.5 voltage reference (National)

C1 2.2 uF, 35-volt non-polarized capacitor

C2 2.2 uF, 35-volt capacitor

C3, C4 1 uF, 35-volt electrolytic capacitor

C5 4.7 uF, 35-volt electrolytic capacitor

U1 LMC6001 ultralow input current amplifier (National)

U2 LMC6041 CMOS micropower op-amp (National)

U3 LM7905 negative 5-volt regulator IC

U4 LM7805 positive 5-volt regulator IC

S1 DPST toggle power switch

B1, B2 9-volt transistor radio battery

PR-1 Ag/Ag/CI pH probe (EyeThink)

MI digital multimeter or digital panel meter module

Miscellaneous printed circuit board, IC sockets, battery clips, hardware

Stream Stage Water-Level Measurement

Scientists such as hydrologists often have a need to measure the depth of a stream, creek, or river during or after a flood or during or after spring run-off. They need an accurate way to measure the depth of the water so they can study the flow of rivers and streams due to flooding or run-off or to study the transport of chemicals or particles in water.

Water height is measured by pressure, because pressure at depth is a function of water height. However, the total pressure at a fixed depth is a combination of the water and the overlying atmosphere. A differential pressure gauge detects the height of water above the pressure port and isolates it from the overlying pressure from the atmosphere and its day-to-day variability. You can readily measure the depth of water in a river or stream using the stream gauge system shown in Figures 4-13 and 4-14.

The heart of the stream gauge water-level measurement system is the Honeywell ASCX01DN, the 0 to 1 psi differential pressure sensor shown in Figure 4-15. The smaller psi rating decreases the maximum water depth detected but increases the sensitivity of the measurement. A 1 psi differential pressure gauge has a maximum depth range of 0.72 meters. Assume a voltage range of 0 to 5 volts spread over an 8-bit data point (restricted to a range of integers from 0 to 255) yields a theoretical maximum resolution of 0.28 centimeters for each logger value unit. A 5 psi differential pressure gauge, however, has a maximum depth range of 3.6 meters and minimum theoretical resolution of 2.8 cm/logger value. Honeywell also makes

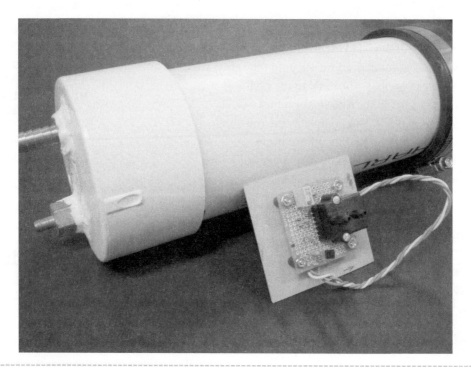

Figure 4-13   *Stream gauge system*

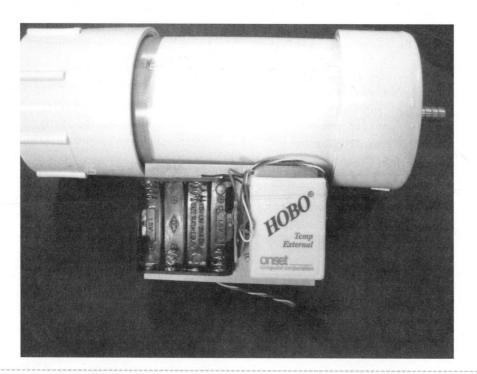

Figure 4-14   *Stream gauge system*

5 psi and larger differential gauges. The Honeywell sensor was selected due to competitive cost, simplicity, and linearity of its response.

The differential gauge actually detects the difference between the pressure applied to the pressure port minus the pressure applied to the background

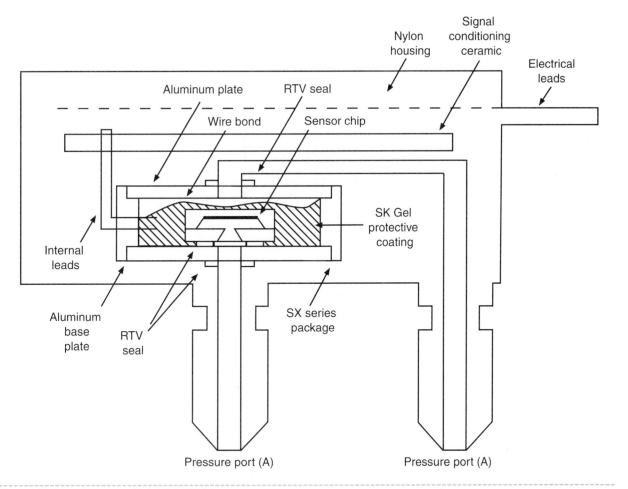

Figure 4-15 *Honeywell ASCX01DN pressure sensor*

port. In this project, the active pressure port B is connected to a length of plastic tube in a fixed location underwater, and the background port A is connected to the atmosphere, via the receiving unit, which is above water.

The stream gauge water-level monitoring system is shown in the overall circuit diagram shown in Figure 4-16. The sensor is powered from a 9-volt source consisting of six AA batteries housed in two battery holders. The regulator provides a constant 5 volts DC to power the pressure sensor. The 5-volt output from the regulator is sent to pin 2 of the pressure sensor. The resistor network, consisting of resistors R1 through R4, provides a zero or balance adjustment for the pressure sensor on pin 1 of the sensor. The trimpot should be set midpoint when constructing the circuit. Pins 5 and 6 on the pressure sensor are not used. The output of the pressure sensor at pin 3 is fed

to a resistor divider network consisting of resistors R5 and R6. The output of the pressure sensor is coupled to an ONSET HOBO 8-bit data logger for recording the pressure sensor values.

HOBO data loggers are available in 8-bit and 12-bit versions. The HOBO H 08-002-02 is a two-channel, small low-cost data logger, which is powered by a small button battery for 1 year of data collection. The two-channel data logger has a built-in temperature sensor and a second free port, which can be used to record the pressure sensor reading. The HOBO series data loggers come in two input configurations, a 4 to 20 ma input and a 0- to 2.5-volt input. Our stream gauge project utilizes the 0- to 2.5-volt input logger. The HOBO data loggers have a small LED at the side of the plastic housing that indicates when the data logger is collecting data. The external input connector is a 2.5-millimeter, two-circuit jack. The data

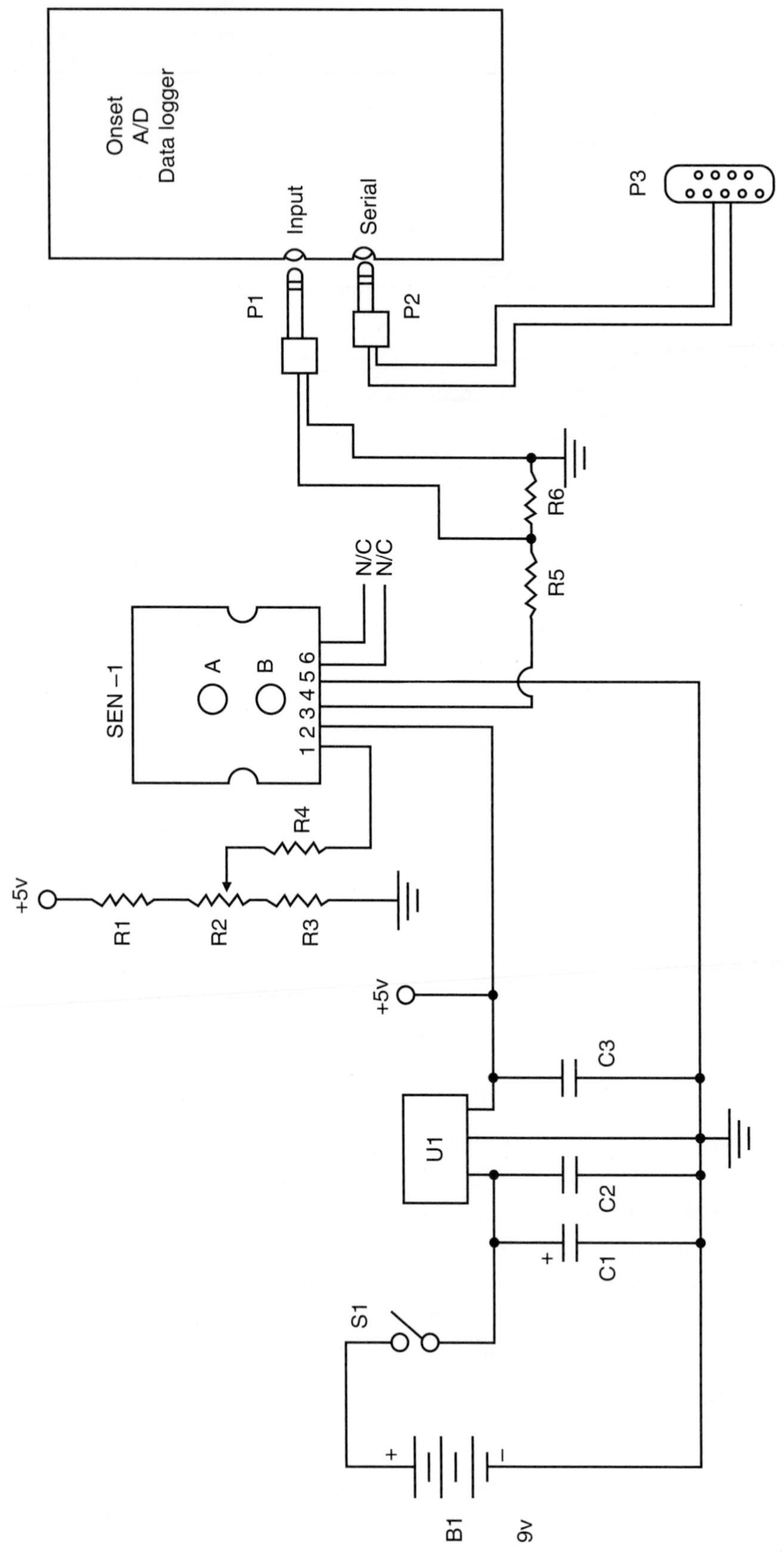

Figure 4-16   Stream stage logger circuit

logger has a second ⅛-inch, two-circuit jack, which is used as the serial output jack. A low-cost ⅛ mini plug to a nine-pin RS-232 serial output cable is available from Onset. The 8-bit HOBO data logger is available for $59.00.

The stream gauge measurement system as mentioned consists of two units: a sending unit and a receiving unit. The sending unit is laid on the bottom of a river or stream and consists of a 5-volt regulator, pressure sensor, sensor offset adjustment, and output voltage divider, as seen in Figure 4-17. The sensing-unit electronics are housed in a 3-inch PVC cylinder, with a three-wire cable connecting the sensing unit to the receiving/recording unit. The receiving/recording unit is also housed in a 3-inch PVC tube, and it houses the ONSET data logger and the 9-volt battery power supply powering the remote sensor in the sensing unit. The three-wire cable between the sensing and receiving units contains the signal wire from the sensor, a common ground wire, and a plus 9-volt power lead. This dual-package system allows the sensor or sending unit to be placed in the stream, while the receiving/recording unit is housed on the banks of the stream or river, allowing easy access to the data logger to retrieve data.

The pictorial diagram shown in Figure 4-18 illustrates how the two enclosures are designed to accommodate the electronics packages to form the stream gauge monitor. The pressure sensor SEN-1, the regulator, the zero adjust circuit, and the output voltage divider circuit are all mounted in the sending unit's PVC-1 enclosure, a 10-inch long, 3-inch diameter piece of PVC pipe. The pressure sensor and electronics are built on a small piece of perf-board and mounted to a $3 \times 2\frac{3}{8}$ inch piece of blank stock circuit board. Circuit board guides are epoxied to the inside of the 3-inch *inside diameter* (ID) PVC pipe. At one end of the PVC-1 pipe, a Fernco 3-inch Quick Cap QC-103 rubber cap is used to waterproof the enclosure but allow access to the circuit board if needed. Two holes are drilled and tapped in a 3-inch PVC flat-end cap to accept two brass fittings. A ¼-inch *outside diameter* (OD) brass fitting is used to couple a length of small-diameter plastic tubing from port B of the pressure sensor to the wall of the PVC end-cap; the other end of the fitting has a brass nipple on which a length of small-diameter tubing is fitted to extend into the stream bed. A compression hose clamp is placed over the tubing. A second brass fitting is mounted on the flat-end cap and is used to

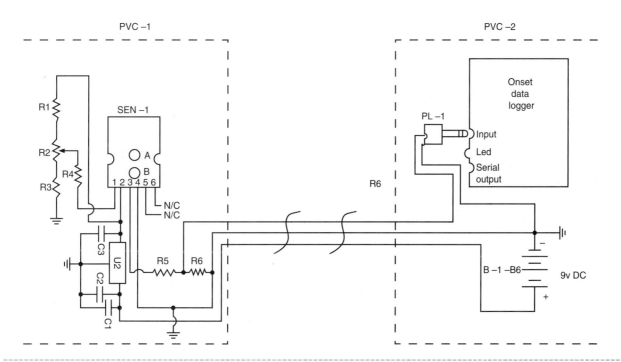

Figure 4-17   *Stream stage pressure sensor*

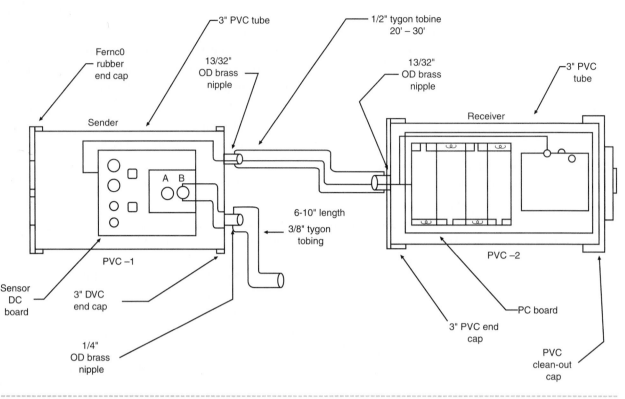

Figure 4-18 *Stream stage sensor recorder system*

run a length of ½-inch OD tubing between the PVC-1 housing and the PVC-2 housing. The ½-inch OD diameter tubing between PVC-1 and PVC-2 allows the power/signal wires to pass between the sensor sending unit and data-logger assembly. A compression hose clamp is secured over the ½-inch tubing to secure the tubing to the brass nipple. The ³⁄₈ -inch Tygon tubing also allows atmospheric air pressure from the outside world to enter the PVC-2 enclosure, which in turn is allowed to pass to the PVC-1 housing and form the differential pressure reference to the sensor. The PVC-1 housing sits on the bottom of the stream, measures the pressure, and sends it to the PVC-2 enclosure sitting above water tied to a tree or affixed to a stake high on the stream bank to allow easy interrogation of the data logger in housed in PVC-2.

The HOBO data logger and the 9-volt battery power supply, consisting of six AA batteries are all mounted on a 6 × 2½ inch piece of blank stock circuit board. Circuit board guides are epoxied to the inside of the PVC-2 enclosure to accept the circuit board. A 2½ inch piece of blank circuit board stock is secured to the main circuit board as a handle to slide the circuit board into the PVC-2 housing.

You will need to locate a second 3-inch ID PVC flat-end cap for the PVC-2 enclosure. Drill and tap a hole in the center of the flat-end cap for a brass fitting with a ¼-inch nipple on the outside of the cap, which will connect the power and signal wires and will allow atmospheric air pressure between the PVC-1 and the PVC-2 enclosure. Place some Teflon tape around the brass fitting and secure in the tapped hole. Clean the outside of both pipe ends and then glue the flat-end cap to one end of the PVC-2 housing. Next you will need to locate a 3-inch clean-out cap and glue it to the opposite end of the PVC-2 housing. The clean-out cap will permit you to slide the circuit board containing the batteries and the data logger into and out of the housing, when interrogating the data logger.

The two power wires from the sending unit at PVC-1 enter the PVC-2 tube and connect to the six-cell AA cell holders. The signal wire from the pressure sensor's voltage divider and ground are routed to a 2.5-millimeter plug, which plugs into the input jack on the HOBO data logger. The ½-inch Tygon tubing between the PVC-1 and PVC-2 assembly allows the normal atmosphere pressure to be used as a reference to port A on the pressure sensor. Remember to attach a compression hose clamp around the ½-inch tubing to secure it to the nipple on the brass fitting.

Stream Gauge Deployment

Basically two options are available for deploying the pressure sensor. Both meet the criteria of exposing the pressure port to a fixed location underwater and exposing the atmospheric port to the atmosphere. Option 1 is to place the entire stream gauge field housing underwater. That is say, place the sensor electronics, the data logger, and batteries all in the same enclosure underwater, and then route a hose from the atmospheric port to the atmosphere, leaving the pressure port exposed to the overlying water column. A hose would come out of the stream and be attached to a nearby tree or stake on the stream bed. Option 2 is to place the sensor housing underwater and route a hose from the pressure port to a fixed location with the remote data-logger housing above water on the stream bank, thus permitting the atmospheric port to be exposed to the atmosphere. Both of these options offer advantages and disadvantages.

The totally underwater deployment is "out of sight and, therefore, out of mind" and less likely to be tampered with or stolen. It allows deployment under surface ice during winter and measures water pressure more accurately because the compression of air inside the hose that connects from the pressure port to some depth in the water column is much less significant in the shorter hose. However, the underwater

deployment is more exposed to the corrosive effects of water, algae, and other underwater threats like flash floods. It is harder to service, because every time you want to interrogate the data logger, you must haul up the entire stream gauge assembly.

We selected option 2 for our project. By doing so, we were able to interrogate the data logger on the dry stream bank without getting wet or entering the stream. Option 2 allows you to walk up to the above-water PVC-2 housing, open the clean-out cap, slide the data logger out just a bit, and plug the ⅛-inch serial port cable into the data logger. You simply attach the ⅛-inch plug adaptor, thereby converting ⅛-inch plug to a nine-pin serial plug, and then connect the cable to your laptop and download the data. Then simply restart the logger, close the clean-out cap, and begin to collect data once again.

Onset supplies a low-cost software package for about $14. This software is easy to use. Simply configure the serial port to talk to the HOBO. Determine the HOBO data logger that you have and select the model and voltage input configuration. Then you will need to determine if you want to start recording immediately or if you wish to have a delayed starting time. You then launch the HOBO by downloading the necessary parameters to the data logger, and you are ready to begin recording your data. When you are ready to interrogate the data loggers, you simply attach a laptop and download the data. Once the data is downloaded, you can use the software to analyze the data or transport the data into a program like Excel to plot your results. Data logging was never so easy.

Stream Gauge Parts List

Required Parts

R1, R3	5K ohm, 1/4-watt resistor
R2	50K potentiometer (trimpot)
R4	200K ohm, 1/4-watt resistor
R5, R6	10K ohm, 1/4-watt resistor
C1	50 uF, 35-volt electrolytic capacitor

C2 0.1 uF, 35-volt electrolytic capacitor

C3 10 uF, 35-volt electrolytic capacitor

U1 LM78L05 5-volt regulator

SEN-1 ASCX01DN 0 to 1 psi Honeywell differential pressure sensor

B1 6 AA cells

P1 2.5 mm mini plug input

HOBO H08-002-02 datalogger Onset computer

Housing (2) 10-inch lengths of 3-inch ID diameter PVC, thick-wall pipe

3-inch diameter flat-end PVC caps (×2)

Fernco 3-inch rubber Quick Cap QC-103 end cap

fittings 2 compression brass fittings, 3/8-inch barb end, and 1 compression brass fitting 1/4 inch

Miscellaneous four-cell AA cell holder, two-cell AA cell holder, PC boards, wire, connectors, 6-inch × 1/8-inch Tygon tubing, etc.

3/8 ID and 1/2 OD Tygon tubing from PVC-1 to PVC-2

Chapter Five

Gas Sensing

Air and gas sensing generally seems so ephemeral, because air and many gases often cannot be seen or smelled. But once again modern electronics comes to our assistance and can help us to sense both air and various gases in our atmosphere. In this chapter we will discuss how to build an air pressure sensing switch, which can be utilized to detect doors moving or approaching vehicles. Our next project is an electronic sniffer, which can detect a number of different gasses and then sound an alarm. With the bargraph pressure sensor, you can monitor water or air pressure changes and display their levels on an LED bargraph display. In this chapter you will also learn about how the new *pellistor* (or pellitized resistor) combustible gas sensors operate and how to construct

your own toxic gas sensor. Weather enthusiasts will be eager to discover an electronic barometer project, which can be used to build a weather station.

Air Pressure Switch

The air pressure switch is a useful project that can be used to detect unwanted intruders entering a private space. Actually this project is a great demonstration project for showing how a change in air pressure can set off an alarm. A sensitive differential pressure sensor is connected to a long length of ¼-inch *outside diameter* (OD), 5/16-inch *inside diameter* (ID) Tygon

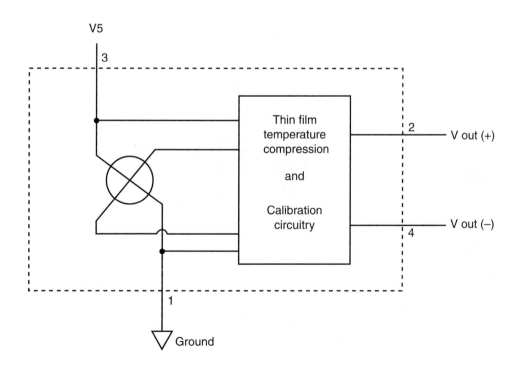

Figure 5-1   *MPX 2100 pressure sensor block diagram*

tubing that can be buried in shallow ground to detect persons or vehicles driving or walking over the tubing. The differential air pressure sensor has two input ports. One port is used as a reference port, and the second port is actually used for sensing air pressure changes fed from the length of Tygon tubing placed in or on the ground. This project could also be used to detect water pressure levels and alert you via a buzzer alarm.

The heart of the air pressure switch is the Motorola MPX2100DP differential pressure sensor, as seen in the block diagram in Figure 5-1. The pressure sensor is a silicon piezoresistive type, which provides a highly accurate and linear output voltage that is proportional to the differential pressure applied. The pressure sensor has two ports, the pressure port at (P1) and the vacuum port at (P2).

The air pressure sensor switch circuit is shown in the diagram in Figure 5-2. The pressure sensor is a four-wire device, where pin 1 is connected to ground, and pin 3 is connected to the 5-volt power source. Pins 2 and 4 of the air pressure sensor are connected to the input of a two-stage op-amp amplifier at U1:a and U1:b. The gain path of U1:a is set by resistors R3 and R4, whereas the gain path of U1:b is formed by R5 and R6. The op-amp section at U1:c is configured as a comparator. The voltage signal from the air pressure sensor is fed to the minus input of the comparator at pin 9. A resistor voltage divider network consisting of R7 and R8 provides a reference voltage to the plus

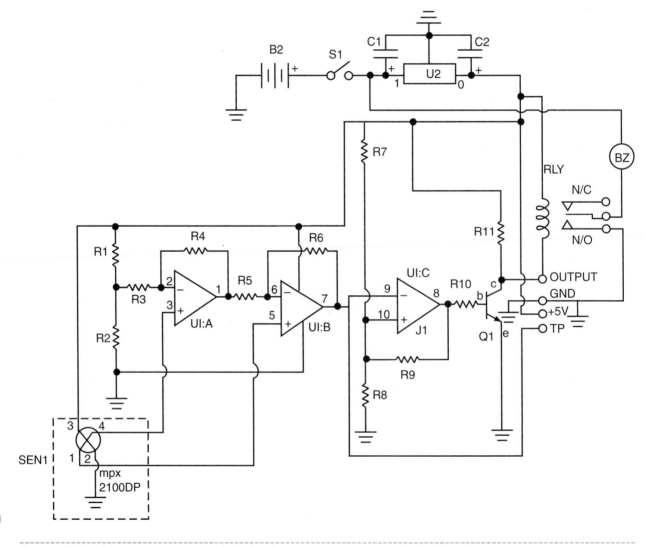

Figure 5-2 *Air pressure sensor switch circuit*

input of U1:c. This network sets the trip point for the comparator. For increased range and control, you could insert a 5K potentiometer between resistors R7 and R8. The center wiper of the potentiometer would then connect to the plus input of the U1:c op-amp.

The output of the comparator at pin 8 is coupled to resistor R10, which in turn is used to drive the transistor at Q1. Transistor Q1 is a relay driver used to control the mini relay at RLY. The relay provides a normally open and normally closed contacts for controlling sirens, buzzers, or warning lamps.

The air pressure switch circuit is powered by a 9-volt battery supply consisting of six C cells. The power switch connects the battery power source to the 5-volt regulator shown at U2. The voltage regulator at U2 is a 7805 three-pin 5-volt regulator and is used to supply a stable power source to the air pressure circuit. Note that the battery power source can be replaced with a 9-volt wall wart power supply for stationary applications.

The air pressure switch circuit can be constructed on small printed circuit board or on a small experimenter prototyping board. All the components including the pressure sensor are mounted on the printed circuit board. Placing the sensor at one end of the circuit board and the relay at the opposite end is generally a good layout for the circuit. An IC socket for the op-amp at U1 and a transistor socket for Q1 are a good idea in the event of a circuit failure down the road. Note that some 1% values resistors are used in this circuit, otherwise none of the other components have to meet particular specifications. A few electrolytic capacitors are used in the power supply portion of the circuit, so be sure to observe polarity when installing them. The LM324 quad op-amp at U1 and the voltage regulator at U2 are easily located. When installing the semiconductors, for example, the transistor and op-amp, be sure to carefully observe their orientations. The transistor has three leads: a base, a collector, and an emitter lead. Make sure you consult the transistor data sheet to be sure you are familiar with the pinouts before inserting it into the circuit. Generally op-amps have either a small square cutout in the top center of the package or a small indented circle at the top left of the package. If your op-amp has a center cutout, then pin 1 on the

op-amp will be to the left of the cutout. If the op-amp has a small indented circle to the top left, then pin 1 of the op-amp will be just to the left of the circle.

After completing the circuit board, carefully recheck your soldering and look for cold solder joints on the foil underside of the board. Next, inspect the underside of the board for stray components leads that may have stuck to the underside of the board. Finally look the board over for shorts or bridges between solder pads.

Now that your air pressure switch circuit has been completed, you can move on to locating a chassis box or enclosure in which to mount it. The prototype air pressure switch is mounted in a metal chassis box measuring $5\frac{1}{2} \times 7 \times 2\frac{1}{2}$ inches. The circuit board was mounted at one end of the bottom of the chassis box. The circuit board was mounted on standoffs, using 1 inch machine screws. If you elect to use batteries, you will need to mount three C cell holders on the bottom or side of the chassis box. If you plan on using a 9-volt wall wart power supply, you will need to install a coaxial power jack on the side of the chassis box. You will have to drill a small hole on the side of the box near the pressure sensor to allow the Tygon tubing to connect to the pressure sensor. Depending upon where your measuring pressure and vacuum are, you will need to observe the pressure port markings. In this application, as in an air pressure switch, you will want to connect the Tygon tubing to port P1, the pressure port.

The power switch and the buzzer, if used, can be mounted on the top of the chassis box to provide ease of use. If desired you could eliminate the buzzer and connect a siren, motor, or other sounding device at the relay contacts. Remember if you select a higher voltage/current load, then the power for the sounding device will have to be provided from a different source.

Now that your air pressure switch has been constructed, you can move on to testing the circuit. Insert the batteries or connect your external power supply and then apply power by turning on the power switch at S1. Take the end of the Tygon tubing and blow into it; the buzzer should sound immediately. The pressure switch is pretty sensitive and will indicate 1 psi of pressure.

Experimenting with the air pressure switch is fun. For demonstration purposes, you can place the Tygon tubing under a carpet or bury it at a place where it can detect vehicles. You can also experiment by using the circuit to detect vacuum conditions or insert the Tygon tubing into water to detect water levels. Have fun!

Air Pressure Switch Parts List

Required Parts

R1 12.1K ohm, 1%, 1/4-watt resistor

R2 15K ohm, 1%, 1/4-watt resistor

R3 20K ohm, 1%, 1/4-watt resistor

R4, R5 100-ohm, 1/4-watt resistor

R6 20K ohm, 1/4-watt resistor

R7, R8 10K ohm, 1/4-watt resistor

R9 121-ohm, 1/4-watt resistor

R10 24.3K ohm, 1%, 1/4-watt resistor

R11 4.75K ohm, 1%, 1/4-watt resistor

C1 1 uF, 35-volt electrolytic capacitor

C2 10 uF, 35-volt electrolytic capacitor

Q1 MMBT3904LT1 transistor

U1 LM324 quad op-amp

U2 LM7805 5-volt regulator

SEN1 MPX2100DP Motorola differential pressure sensor

RLY 5-volt mini SPDT relay (Radio Shack)

BZ 9-volt piezo buzzer

B1 6 C cells

Tubing 1/4-inch OD, 5/16 ID Tygon tubing

Miscellaneous printed circuit board, wire, connectors, screws, standoffs, etc.

Electronic Sniffer

The electronic sniffer is a useful and an interesting project with many potential applications. It can help determine the concentration of various gases in the air to determine if an area is safe for occupation. The electronic sniffer can be used to detect combustible gases, toxic gases, organic solvents *chlorofluorocarbons* (CFCs) by substituting different sensors into the electronic sniffer circuit. The photo shown in Figure 5-3 illustrates the various gas sensors available from Figaro Sensors, Inc.

Figaro TGS sensors are a type of low-cost, thick film metal oxide semiconductor that offer long life and good sensitivity to target gases while using a simple electrical circuit. The Figaro sensors use micrograins of sintered *tin oxide* (SnO_2) that have been heated to a high temperature. This causes the oxygen to be absorbed on the grains and provides a positive potential in the space charge between the molecules. When a gas such as acetone, alcohol, propane, and carbon dioxide passes over the sensor, the sensor's resistance drops. The resistance is linear logarithmically over ranges of a few *parts per million* (PPM) to several 1,000 PPM. The electronic sniffer circuit shown in Figure 5-4 is based on the Figaro TGS826 toxic gas sensor (SEN-1) and is sensitive to carbon dioxide gas.

Figure 5-3 *Figaro sensors*

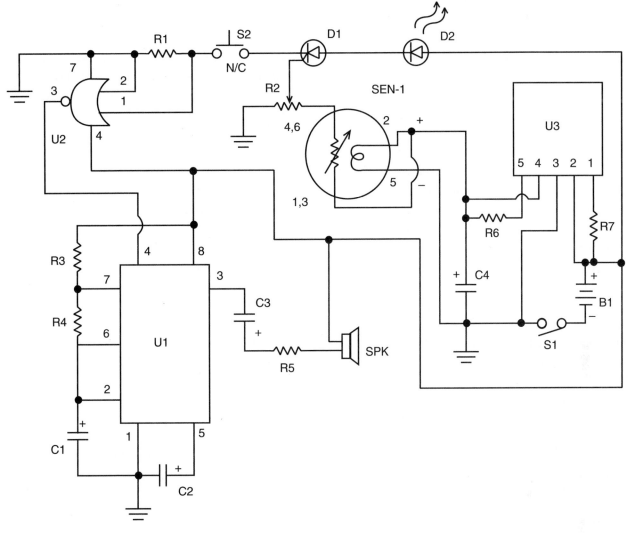

Figure 5-4 *Toxic gas detector circuit*

Most of the Figaro gas sensors have a heater coil that draws a current of about 130 milliamps and must be powered by a 5-volt regulated source. The six-pin toxic gas sensor used in this project is powered by a low-voltage, drop-out National LP3850 regulator, which requires the voltage from four D cells to power the regulator. The regulator provides a stable voltage and current source for the gas sensor for it to operate correctly.

As mentioned previously, the gas-sensitive semiconductor acts like a variable resistor in the presence of toxic gas, thus decreasing its electrical resistance when gaseous toxins are absorbed from the sensor surface. A 25,000-ohm potentiometer R2 connected

to the sensor serves as a load-dividing, voltage-dividing network and as a sensitivity control and it has its center tap connected to the gate of D2. When toxic fumes come in contact with the sensor, decreasing its electrical resistance, current flows through the load potentiometer R2. The voltage developed across the wiper of R2, which is connected to the gate of D1, triggers the *silicon controlled rectifier* (SCR) into conduction. And as conduction occurs, the LED at D2 lights up to indicate that a gas has been detected.

With the SCR now conducting, the input to the CD4071 gate in turn applies a trigger voltage to the LM555 timer/oscillator. Once the gas has been detected and the trigger signal has been activated, the

LM555 oscillator begins driving the speaker at SPK, and the speaker will come to life indicating the presence of a toxic gas. To silence the electronic sniffer, a normally closed momentary pushbutton switch is provided to reset the circuit.

Note that it is normal for the sensor to have a large drop in resistance when it is first turned on. Once the heater warms up, the unit will function normally, and the sensor will begin responding within seconds. The Figaro sensors are quite rugged, however, the sensor element should not be exposed to water. If the sensor does get wet, be sure it does not freeze, because it will allow the sensor to crack and be destroyed.

The electronic sniffer can be constructed on a small printed circuit board or a prototype board available from Radio Shack. The regulator and sensor should be placed near each other to keep the current loss between the two to a minimum. It is advisable to mount the sensor at one edge of the circuit board, so that when the circuit board is placed in a chassis box, a hole can be drilled in the case to allow the gas to enter the sensor. The regulator power supply is connected to the heater element at pins 2 and 5. The sensing portion of the sensor is connected so pins 1 and 3 are connected back to the plus supply of the regulator. Pins 4 and 6 are connected to the potentiometer at R2. When wiring the CD4071, be sure to keep all the unused input pins tied to ground to avoid false trigger. The LM555 oscillator's output frequency is controlled by R4 and C1, and this can be changed if desired by placing a potentiometer at R4.

When constructing the electronic sniffer, you will need to pay particular attention to the orientation of certain components such as the diodes, the SCR, the ICs, as well as the capacitors. Diodes usually have an arrow pointing to a straight line that indicates the cathode. Capacitors, if they are polarity sensitive, are marked with a plus or minus sign to help with orientation. IC sockets are highly recommended and will save you a lot of grief if you have to service the circuit at a later date. When installing ICs, you will notice either a square cutout at the top center of the package or a small indented circle at the top left corner. Generally pin 1 is the left of the cutout or to the left of the indented circle.

Once the circuit board has been built, you will need to recheck your component placement to make sure the components have been installed correctly. Also take notice to make sure that no previously cut components leads could cause a short circuit and possibly damage the circuit once powered up. Lastly, look for cold solder joints, which could prevent the circuit from working correctly.

Now that the circuit board has been built, you are ready to place the circuit into an enclosure. Our prototype electronic sniffer is enclosed in a small metal chassis box measuring $4 \times 5 \times 6$ inches. The power switch at S1, the reset switch at S2, the LED at D2, and the speaker are all mounted on the top front of the chassis panel for easy operation. The prototype circuit utilizes a panel-mounted potentiometer, also mounted on the front top of the chassis box. The circuit board is placed near one end of the chassis box so that the toxic gas sensor can look out the end of the enclosure through a hole drilled at the side of the box. Optionally, you could elect to mount a sensor socket atop the chassis in order to change sensors on a moment's notice. The circuit board is mounted to the chassis on standoffs using ¼-inch long 4-40 machine screws. The two D-cell holders are mounted to the bottom cover of the chassis box so they can be easily replaced.

You will now need to test your electronic sniffer unit for proper operation. First, you insert the D cells into the holders, then apply power via the power switch S1. Adjust the potentiometer to its lowest setting, usually counterclockwise. When the power is first turned on, you may hear the speaker squeaking. Let the electronic sniffer warm up for about 2 minutes (remember the heater coil). After 2 minutes has passed and the sensor has stabilized, next you will need to gradually adjust the potentiometer R2 upward or clockwise to its threshold point. You may have to depress the reset switch and readjust the control for best sensitivity. Your electronic sniffer is now ready to assist you in sniffing out toxic gases. As mentioned earlier, this electronic sniffer can also be used to detect other gases as shown in Table 5-1. Prices of the various gas sensors range from $14.50 for the LP/propane, natural gas, and carbon monoxide sensors, to about $24.00 for the CFC sensors, and finally about $50.00 for the hydrogen, hydrogen sulfide, and ammonia gas sensors. Most of these gas sensors operate in a similar manner, and many of these sensors can be substituted for one another with only small changes required. This of course opens up a large range of gas detection options by merely changing the sensors.

Table 5-1

Figaro Gas Sensors

Gas	Sensor	
Combustible Gases		
LP gas/Propane (500–10,000 ppm)	TGS813	TGS2610*
Natural gas/Methane (500–10,000 ppm)	TGS842	TGS2611*
General combustible gases (500–10,000 ppm)	TGS813	TGS2610*
Toxic Gases		
Carbon monoxide (50–1,000 ppm)	TGS826	
Ammonia (30–300 ppm)	TGS825	
Hydrogen sulfide (5–100 ppm)	TGS825	
Organic Solvents		
Alcohol, tolune, zylene (50–5,000 ppm)	TGS822	TGS2620*
Other volatile organic vapors	TGS822	TGS2622*
Indoor Pollutants		
Carbon dioxide	TGS4160	TGS4161*
Air contaminants (<10 ppm)	TGS800	
CFCs (HCFCs, HFCs)		
R22, R112 (100–3,000 ppm)	TGS830	
R21, R22 (100–3,000 ppm)	TGS831	
R134a, R22 (100–3,000 ppm)	TGS832	

Note that some conditions may exist that truly must be avoided (such as exposing the TGS sensors to silicone vapors like adhesives and hair grooming materials). You should also avoid placing the sensors in highly corrosive environments; high-density exposure to corrosive materials such as H_2S, SOx, C_{12}, HCl for extended periods may cause corrosion or breakage of the lead wires or heater material. You should also ensure that the sensors do not come in contact with water and that the sensors are not exposed to freezing temperatures; this would crack the sensor substrate.

Electronic Sniffer Parts List

Required Parts

R1   470-ohm, 1/4-watt resistor

R2   25K ohm potentiometer

R3, R4   47K ohm, 1/4-watt resistor

R5   10-ohm, 1/4-watt resistor

R6, R7   10K ohm, 1/4-watt resistor

C1   0.006 uF, 35-volt ceramic disc capacitor

C2   0.01 uF, 35-volt ceramic capacitor

C3   2 uF, 35-volt tantalum capacitor

C4   10 uF, 35-volt tantalum capacitor

U1   LM555 timer oscillator IC

U2   CD4071 quad two-input OR gate

U3   LP3875 5-volt regulator

D1   SCR

D2   LED

S1   SPST power switch

S2   normally closed pushbutton switch

B1   4 D cells

SPK   8-ohm mini speaker

SEN-1   TGS826 Figaro toxic gas sensor

Miscellaneous   PC board, IC sockets, wire, battery holders, hardware

Table 5-3

Bargraph Pressure Sensor Applications

Bargraph Pressure Sensor

Pressure sensing devices are used to measure and control many different functions in our modern lives. Many people may not realize in fact that pressure sensing, measurement, and control techniques are found in many devices, appliances, and machines used in our homes, offices, and factories with out us even knowing. Pressure sensing is often little understood, and measurements revolve around three basic types of pressure sensing methods (see Table 5-2).

Table 5-2

Types of Pressure Measurement

Type of Pressure	Description
Absolute pressure	Absolute pressure sensors measure an external pressure relative to a zero-pressure reference point (A vacuum) sealed inside the reference chamber of the die during manufacture. This corresponds to a deflection of the diaphragm equal to approximately 15 psi or 1 atmosphere. Measurement of external pressure is accomplished by applying a relative negative pressure to the pressure side of the sensor.
Differential pressure	Differential pressure sensors measure the difference between pressures applied simultaneously to opposite sides of the diaphragm. A positive pressure applied to the pressure side generates the same (positive) output as an equal negative pressure applied to the vacuum side.
Gauge pressure	Gauge pressure readings are a special case of differential measurement in which the pressure applied to the pressure side is measured against the ambient atmosphere pressure applied to the vacuum side through the vent hole in the chip of the differential pressure sensor elements.

The bargraph pressure sensor project is a fun project that will demonstrate how you can build, experiment, and measure pressure changes in the environments around you. It can be used as a sensitive diagnostic device to sense and measure air pressure, vacuum pressure, or differential pressure (see Table 5-3 of potential pressure sensing applications).

Table 5-3

Bargraph Pressure Sensor Applications

Monitor Application	Example
Monitor pressure drops across industrial liquid filters	Food processing plants, chemical processing plants, sewage and filtration plants
Monitor pressure drops across industrial gas filters	Air conditioning, gas separation, flow monitoring high-vacuum systems
Other monitoring applications	Filtration equipment, computer filters, clean rooms, medical instruments

This project is based on a sensitive low-cost differential piezo pressure sensor that provides an output proportional to the applied pressure. Differential pressure sensors have two pressure ports, allow application of pressure to either side of the diaphragm, and can be used for gauge or differential pressure measurement. The SenSym (a division of Honeywell) IC pressure sensor is a four-pin basic pressure sensor without signal conditioning, as shown in Figure 5-5. The two pressure ports can be seen at port P1 and P2. Port P1 is shown as the high-pressure port. The piezo sensor has four internal 500-ohm resistance elements, which are configured as a Wheatstone bridge. Connected across the output of the sensor are two 550-ohm resistors.

The diagram shown in Figure 5-6 illustrates a block diagram of the bargraph pressure sensor circuitry. Shown to one side of the pressure sensor is the pressure calibration circuitry, and at the other end of the sensor is the temperature compensation circuitry. Pressure is applied to the sensor. The sensor output is amplified by the amplifier section, which is then sent on to the display driver circuitry and finally to the LED display module at the far right of the diagram.

The bargraph pressure sensor circuitry shown in Figure 5-7 begins with the 0 to 7 psi SenSym SPX50D sensor. The pressure sensor is powered from a regulated 5-volt power source, followed by a series of three diodes connected as a span temperature compensation source to pin 3 of the sensor. Pin 1 of the sensor is connected to ground. The positive (+) output of the pressure sensor at pin 2 is connected to both a reference potentiometer (R8) and to the input of the second op-amp at U1:B on pin 5. The minus

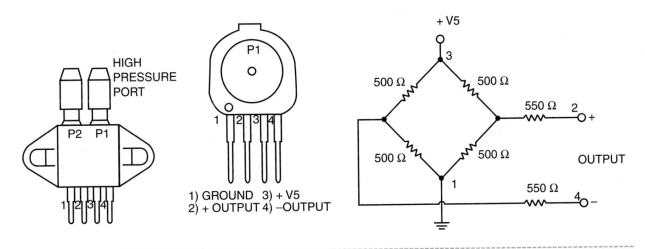

Figure 5-5   *Honeywell SenSym SPX 50D pressure sensor*

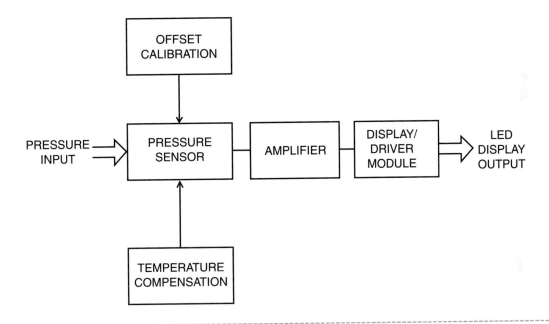

Figure 5-6   *SPX 50D pressure sensor block diagram*

(−) output of the pressure sensor at pin 4 is coupled to the input of the first op-amp on pin 3. The gain of the first op-amp section is set up by the gain resistors R6 and Rp (see Table 5-4). By changing the input resistors, you can select a pressure range of 0 to 1 psi or 0 to 10 psi if desired. The output of the first op-amp is coupled to the second op-amp stage via the resistor at R3. The output at U1:B at pin 7 is sent to the input of the display driver, an LM3914N, on pin 5. The display driver will drive 10 LEDs to a full-scale reading of 0 to 1 psi or 0 to 10 psi. The display driver has a full-scale adjustment control at potentiometer at R13. A zero adjustment of the display can be set by adjusting potentiometer R5. The circuit illustrates two test points: TP1 and TP2. Test point 1 provides a full-scale output voltage, whereas test point 2 provides 5 volts. The LED at D14 provides a power on light for the system. The 10 LEDs are wired to the display drivers on pins 10 through 18 as well as on pin 1.

The bargraph pressure sensor circuit is powered from a 5-volt regulator at U2. The regulator provides 5 volts for the LM 358, the LM3914N display driver, and the sensor. A 9-volt battery supplies power to operate the regulator through switch S1. You could also elect to use a wall wart power supply instead of a battery if desired.

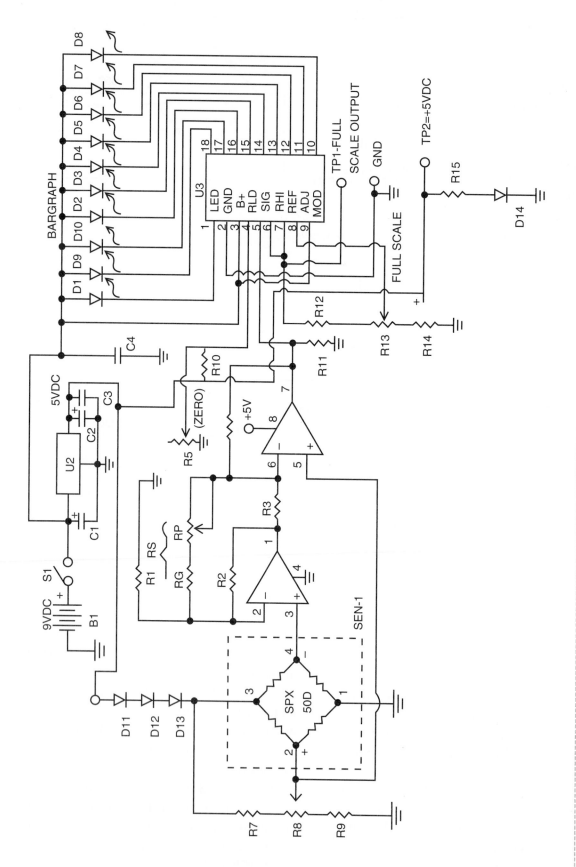

Figure 5-7   *Bargraph pressure sensor circuit*

The bargraph pressure sensor prototype is constructed on a small printed circuit board, as shown in Figure 5-8. When laying out the circuit board design, you should consider mounting the pressure sensor at one edge of the circuit board, as shown, to allow the pressure ports to exit your enclosure. The sensor offset resistors R7, R8, and R9 are all 1% resistors, as are the amplifier gain resistors. All other resistors in the circuit are 5% resistors. The circuit employs only three capacitors. Capacitors C1 and C2 are electrolytic types, and C3 is a ceramic disk type. Potentiometers used in the bargraph pressure sensor are all multiturn circuit board (trimpot) types.

Table 5-4

Gas Bargraph—Full-Scale Gain Resistor Selection

Pressure Range	Device	Rs	
		R6	Rp
0–1 psi	SPX50DN	909 ohms	2K ohms
0–10 psi	SPX50DN	909K ohms	20K ohms
0–15 psi	SPX100	6.9K ohms	20K ohms
0–30 psi	SPX200	6.9K ohms	20K ohms

When installing the pressure sensor, be sure to carefully observe the orientation of the four pinouts on the device. It is important to use IC sockets when building the circuit, in the event of a circuit failure at some later point in time. ICs are sensitive to orientation so particular attention must be paid to proper orientation. Each IC will usually have either a small indented circle on the top of the plastic package or a square cutout in the plastic at the top of the package. If the IC has a small indented circle, pin 1 will be just to the left of the circle. If the IC has a square cutout on top of the package, then pin 1 will be to the left of the cutout.

When installing the LEDs, pay particular attention to their polarity to avoid having to install them twice. The prototype pressure sensor utilizes a 10-segment LED display, which is easier to mount and handle than using separate LEDs. If you use the 10-segment LED display, you can use an IC socket for ease of wiring. The three diodes at D11, D12, and D13 are all small signal diodes and can be identified by the band at one end. This band indicates the diode's cathode. Check the data sheet for the regulator to ensure that you know the difference between the input and output pins of the regulator device.

Once you have completed soldering in all the components, you should carefully inspect the foil side of the circuit board to make sure that no shorts or bridges can be seen between copper pads or lines. Also look the board over to ensure that no cold solder joints exist. Cold solder joints can cause you many headaches later if not caught quickly, so take the extra time to look for them before power is applied to the circuit. Finally check the circuit board once more to make sure that no cut component leads are stuck to the bottom of the circuit board, which may cause short circuits.

After inspecting the circuit board, you are ready to install the circuit board into some sort of chassis box. You can select either a metal or plastic enclosure for this project. You will want to consider mounting the circuit board at one end of the chassis to allow the pressure ports to have easy access to the outside world. Also consider elevating the circuit board on standoffs, so that you can cut a hole for a 10-segment LED display at the top or front of the chassis box, if you decided to use this type of display. This mounting method will allow a clean mounting scheme. The power switch and power on indicator can be mounted on the top or front panel of the enclosure. If you are going to power the circuit from batteries, you can mount a battery holder at the bottom of the enclosure. If you elect to use a wall wart power supply, you will want to mount a coaxial power jack at the rear of the chassis to allow for a power connection.

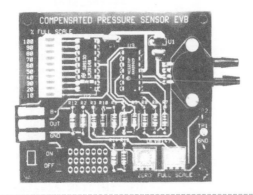

Figure 5-8 *Bargraph pressure sensor prototype printed circuit board*

The pressure sensor has two ports, because it's a differential type sensor. You will need to obtain a length of ¼-inch OD, 5/16-inch ID diameter Tygon tubing. Remember that the pressure port P1 is the high-pressure port. Your application will determine which port you will want to attach the Tygon tubing to. If you want to measure force of incoming air pressure, attach the tubing to port P1; if you want to measure a vacuum, attach the tubing to port P2.

Calibration of the bargraph pressure sensor is pretty straightforward. Only two simple noninteractive adjustments are needed to set the initial offset and full-scale span. To adjust the offset, apply zero differential pressure to the sensor, then adjust R8 for 0 volts at Vd (i.e., at the sensor output between pin 4 and pin 2 of the SPX50DN). To adjust the span, apply 10 psi to the sensor and adjust Rp until the tenth LED lights up. That completes the calibration procedure for the bargraph sensor circuit. The bargraph pressure sensor circuitry is now ready to sniff out air or vacuum pressure changes. Table 5-5 lists a number of different pressure sensor conversions that you may encounter in pressure sensing. Have fun with your new bargraph pressure sensor.

Table 5-5

Pressure Conversion Table

Convert from	Multiply by	Convert to
psi-VG	2.3	feet of water
feet of water	.434	psi-SG
psi	.06895	bars or kPa
bars or kPa	0.1450326	psi
bars of kPa	4.0147	inches of H_2O
bars or kPa	0.2953	inches of Hg
bars or kPa	10.000	mbars
bars or kPa	10.1973	cm of H_2O
bars or kPa	7.5006	mm of Hg
inches of H_2O	0.036127	psi
inches of H_2O	0.073554	inches of Hg
inches of H_2O	0.2491	bars or kPa
inches of H_2O	2.419	mbars
inches of H_2O	2.5400	cm of H_2O
inches of H_2O	1.8683	mm of Hg
inches of Hg	0.4912	psi
inches of Hg	13.596	inches of H_2O
inches of Hg	3.3864	bars or kPa
inches of Hg	33.864	mbars
inches of Hg	34.532	cm of H_2O
inches of Hg	25.400	mm of Hg

Bargraph Pressure Sensor Parts List

Required Parts

R1, R2, R3, R4 100K ohm, 1/4-watt, 1% resistor

R5 200-ohm potentiometer (trimpot)

R6 (see Table 5-4)

Rp (see Table 5-4)

R7 24K ohm, 1/4-watt, 1% resistor

R8 10K ohm potentiometer (trimpot)

R9 15K ohm, 1/4-watt, 1% resistor

R10, R11 1K ohm, 1/4-watt, 5% resistor

R12 1.2K ohm, 1/4-watt, 1% resistor

R13 1K potentiometer (trimpot)

R14 2.7K ohm, 1/4-watt, 1% resistor

R15 470-ohm, 1/4-watt, 5% resistor

C1 1 uF, 35-volt, electrolytic capacitor

C2 10 uF, 35-volt electrolytic capacitor

C3 0.1 uF, 35-volt ceramic disc capacitor

U1 LM358 op-amp (National)

U2 LM7805 regulator (National)

U3 LM3914N display

```
driver IC (National)
D1 through D10   red LEDs
D11, D12, D13   1N914
   silicon diodes
D14   red LED
SES-1   SPX50DN SenSym
   pressure sensor
B1   9-volt battery
S1   SPST toggle switch
Miscellaneous printed
   circuit board, wire,
   sockets, male header
   connectors, battery
   holder, chassis box,
   etc.
```

Pellistor Combustible Gas Sensor

Catalytic combustion has been the most widely used method of detecting flammable gases in industry since the invention of the catalytic *pelletized resistor* (or pellistor) in the mid-1960s.

A pellistor element is simply a platinum wire coil, coated with a catalytic slurry of an inert base material (e.g., alumina) and a metal catalyst that accelerates the oxidation reaction. This type of element is known as the *sensitive element*. A number of catalyst materials are available and the precise type and mix is carefully chosen to optimize sensor performance. Figure 5-9 shows a cross-section of a typical catalytic element.

In addition to the sensitive element, a *nonsensitive element* is manufactured. Two types of nonsensitive element exist. The first is used primarily in high-power devices; in this case the catalytic slurry is replaced with glass, which will not oxidize in combustible gases. The second type is used for low-power devices; this type of nonsensitive element starts life as a sensitive element and as such is manufactured in exactly the same way. However, to prevent oxidation, the catalyst is "poisoned" during production using a suitable substance such as potassium hydroxide.

Pellistors are always manufactured in pairs. The active *catalyzed element* is supplied with an electrically matched element containing no catalyst and is treated to ensure no flammable gas will oxidize on its surface. This *compensator element* is used as a reference resistance to which the sensor's signal is compared, to remove the effects of environmental factors other than the presence of a flammable gas (see Figure 5-10). Pellistors' pairs are conveniently mounted in TO4-size headers or as complete flameproof gas detection heads for use as field devices within fixed gas detection systems. The advantage of using this technique when detecting flammable gases is that it measures flammability directly.

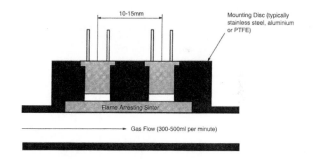

Figure 5-10   *Pellistor dual sensor mount*

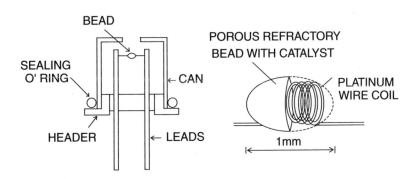

Figure 5-9   *Pellistor gas sensor*

In operation, a pair of beads is always used, and this pair comprises a sensitive and a corresponding nonsensitive element. During manufacture individual pairs are closely matched by voltage and current requirements and therefore do not need any compensation (i.e., a trimming resistor) when used together in equipment. This pair of elements is installed into a Wheatstone bridge, as illustrated in Figure 5-11, which heats them to between 400 and 500°C. When no gas is present the resistances of the two elements are balanced and the bridge will produce a stable baseline signal. Combustible gases are then oxidized on the sensitive element, causing its temperature to rise. Accordingly the resistance of the element also increases. This results in an out-of-balance signal across the bridge and a corresponding change in output voltage, which can be measured easily.

The bridge power supply is typically 2 to 3.5 volts, depending on the sensor. Voltage V1 is set by the potential divider of the sensitive and nonsensitive elements, and, when no gas is present, this is nominally half the bridge supply voltage. Voltage V2 is also half the bridge supply and is predominantly determined by the potential divider of R1 and R2. RV1 is used to fine-tune V1 and V2 so that they are the same value when no gas is present and to enable the bridge output to be set to zero. When gas is present, the resistance of the sensitive element rises, causing V1 to decrease, and the output (being the difference between V1 and V2) to rise also. The function of the nonsensitive element is to eliminate any effects that environmental conditions may have on the resistance of the two elements. Any changes will be the same for both the sensitive and nonsensitive elements, and so the bridge voltage V1 and output voltage remain unchanged.

Pellistor resistors are optimized for use either at a constant voltage or a constant current. Constant current is predominantly used in high-power fixed systems and ensures that the elements are correctly powered, even if the power source is a number of meters away. Where battery life is a concern, constant voltage is preferred, and as such all low-power elements are optimized for constant voltage.

The circuit shown in Figure 5-12 illustrates a combustible gas sensor using the pellistor sensor technology. The heater power supply shown at the left of the circuit is utilized to supply power to the pellistors in the Wheatstone bridge circuit. The heater power supply consists of a 3- to 4.5-volt battery power source that feeds the adjustable voltage regulator, shown at U1, via switch S1:a. The potentiometer at R2 is used to adjust the output of the power supply to the required voltage of 3 volts. The plus (+) output of the heater power supply is coupled to the Wheatstone bridge at V2, whereas the minus (−) of ground of the power supply is fed to the Wheatstone bridge at V1. The Wheatstone bridge consists of four resistances. Resistors R3 and R4 are fixed 1K ohm values, R5 is the reference part of the pellistor sensor, and R6 is the active sensing element of the pellistor at SEN1.

The output of the Wheatstone bridge at E1 and E2 is a voltage, which results when the bridge becomes unbalanced as the sensor element detects a combustible gas. The bridge output at E1 is connected to the common ground of the detector portion of the circuit. The bridge output at E2 is coupled to the minus (−) input of LM393 op-amp comparator. The plus (+) output of the op-amp at pin 5 is connected to feedback circuit consisting of resistors R9 and R10. Pin 5 of the op-map is also connected to a

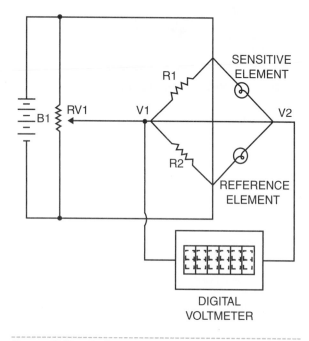

Figure 5-11   *Pellistor gas sensor bridge circuit*

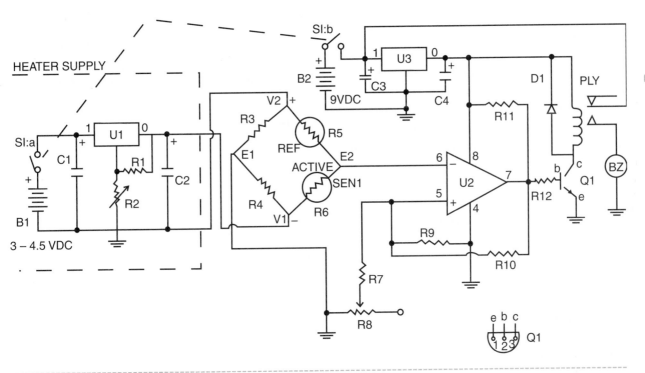

Figure 5-12   *Pellistor toxic gas sensor circuit*

potentiometer at R8, which is used to adjust the reference or set-point of the comparator circuit. The output of the comparator at pin 7 is coupled to resistor R12, which is fed to a PNP transistor at Q1. Transistor Q1 is used to drive a small SPDT relay, which can be used to drive a heavy load. In this example the relay is used to drive a piezo buzzer that sounds when the Wheatstone bridge becomes unbalanced. The normally open contacts of the relay are wired from the 9-volt source through the buzzer to ground. You could use the relay to drive a larger load, such as a pump or siren. The comparator circuit is powered from a 9-volt power source consisting of a 9-volt battery and regulator U3. Switch S1:b applies power to the regulator and, in turn, to the comparator circuit. Note the load, a buzzer in this example, is connected ahead of the regulator and is supplied with 9 volts whereas the comparator is powered from 5 volts.

The combustible gas sensor is constructed on a 3 × 4 inch printed circuit board. No component specifications are critical except for the matched pair pellistor sensor package. All the components are mounted on the circuit board, including the regulators, IC, and transistor and relay. Heatsinks should be placed on both of the regulator packages. The pellistors should be mounted in a heatsink, because they will get a bit warm as power is supplied to heat the element at all times. The pellistor assembly should be mounted at one end of the circuit board, so that the sensor can be mounted near one end of the chassis box. An IC socket is recommended in the event of the circuit failure at some later point in time. The IC must be oriented correctly for the circuit to work properly. Observe the orientation carefully, when installing the IC. ICs usually have some form of marking to denote their orientation. On some ICs, you may find either a small indented circle at the top left of the package or a square cutout at the top center of the package. If the IC package has a small indented circle, pin 1 of the IC is just to the left of the indented circle. If the IC has a square cutout at the top center of the package, pin 1 is to the left of the cutout. Capacitors C1, C2, and C3, C4 are all electrolytic types and must be installed correctly with the positive marking placed correctly with respect to the plus voltages on the power supply. Consult with the data sheets for each of the regulators to be sure you install them correctly and avoid damage to the circuit. The transistor at Q1 has three pins, marked 1, 2, and 3 as shown on the bottom of the diagram (see Figure 5-12). Pin 1 on the transistor corresponds to the emitter, whereas pin 2 is the base, and pin 3 is the transistor's collector pin.

After installing and soldering the components into the circuit, be sure to carefully recheck the foil side of the board for shorts and bridges between circuit pads. Also look for cold solder joints; it's better to identify cold solder joints before applying power to and using the circuit. Finally, examine the foil side of the board to locate stray component leads that may have attached themselves to the board after components were cut. These stray leads may act as shorts and could damage the circuit when power is supplied if not removed. Once you have thoroughly inspected the circuit board, you are now ready to install the circuit board into a chassis box.

Locate a suitable chassis box to enclose the combustible gas detector. The prototype circuit is installed in a $5\frac{3}{4} \times 7 \times 2\frac{1}{2}$ inch metal chassis box. A $\frac{3}{8}$-inch hole should be placed on one end of the chassis box, this should line up with the gas sensor on the circuit board. The circuit board is mounted atop plastic standoffs so that the sensor can see outside the box. If you elect to use batteries at B1, install three C cell holders on the bottom of the chassis. You could also elect to use a 6-volt, 500 ma wall wart power supply if desired. The power source at B2 can be supplied by batteries by installing holders for three AA or three C-cells. The battery holder can be mounted on the bottom of the chassis box.

The DPDT power switch was mounted along with the piezo buzzer on the top cover of the chassis box. If you elected to use a panel-mounted potentiometer for R8, this control could also be mounted on the top front cover of the chassis box.

Once the combustible gas detector has been completed, you are now ready to calibrate the circuit. The combustible gas detector is sensitive to a number of combustible gases (see Table 5-6). You will notice that the sensor is most sensitive to methane, hydrogen, ethylene, and methanol. To calibrate the detector for a particular gas, turn on the combustible gas sensor and expose the sensor to the gas of interest; this will adjust the set-point of the circuit using potentiometer R8. Each gas will require a slightly different set-point for maximum sensitivity. SixthSense manufactures two different catalytic pellistor gas sensors: a CAT16 and a CAT25 (see Table 5-7).

Table 5-6

Pellistor Gas Sensors—Catalytic Sensor Relative Response

Gas/Vapor	Relative Response %
Methane	100
Hydrogen	107
Ethane	82
Propane	63
Butane	51
Pentane	50
Hexane	46
Heptane	44
Octane	38
Ethylene	81
Methanol	84
Ethanol	64
Propane-2-ol	49
Acetone	50
Butane-2-one (MEK)	48
MIBK	–
Cyclohexane	–
Di-Ethyl Ether	40
Ethyl Acetate	46
Toluene	44
Xylene	31
Acetylene	47

Table 5-7

Pellistor Catalytic Sensor Operating Performance

Attribute	CAT 16 Sensor	CAT 25 Sensor
Model number	2111B2016	2111B2125
Operating principle	Constant current	Constant voltage
Gas detected	Most combustible gases	Most combustible gases
Measurement range	0–100% lower explosive limit (LEL)	0–100% LEL
Operating voltage	2.7 volts +/− 0.2 volts	3.3 volts +/−0.02 volts

Operating current	200 mA	70 mA
Max power consumption	580 mW	230 mW
Output sensitivity	>12 mV % methane	>25 mV % methane
Response time	<10 seconds	<10 seconds
Linearity	+/-110% LEL to 100% LEL	+/-110% LEL to 100% LEL

We chose the CAT25 sensor, which requires significantly less current consumption. Two gas exposures must be avoided: exposure to hydrogen sulfide and to hexamethyl disiloxane.

Pellistor Combustible Gas Detector Parts List

Required Parts

R1 240-ohm, 1/4-watt resistor

R2 5K ohm potentiometer (trimpot)

R3, R4 1K ohm, 1/4-watt, 1% resistor

R5 reference element pellistor CAT25 #2111B2125 (Sixth-Sense)

R6 active element (pellistor) CAT25 #2111B2125 (Sixth-Sense)

R7, R9, R10 1 megohm, 1/4-watt resistor

R8 50K potentiometer (trimpot or panel mount)

R11 3.3K ohm, 1/4-watt resistor

R12 1K ohm, 1/4-watt resistor

C1, C3 1 uf, 35-volt electrolytic capacitor

C2, C4 10 uf, 35-volt electrolytic capacitor

Q1 2N2222 transistor (PNP)

U1 LM117T adjustable regulator IC

U2 LM393 comparator op-amp IC

U3 LM7809 9-volt regulator IC

RLY 6-volt mini relay SPDT (Radio Shack)

BZ piezo buzzer

B1 3 C cells; 4 to 5 volts DC

B2 6 AA cells; 9 volts

S1 DPDT toggle switch

Miscellaneous printed circuit board, wire, IC sockets, standoffs, screws, nuts, etc.

Electronic Barometer

Everyone is interested in the weather. And while we constantly see weather reports on TV and often barometric reading is usually specified, to most people, this specific weather-related parameter is usually quite a mystery to most people. So what is barometric pressure?

The weight of the air that makes up our atmosphere exerts a pressure on the surface of the earth. This pressure is known as atmospheric pressure. Generally, the more air above a land mass, the higher the atmospheric pressure; this, in turn, means that atmospheric pressure changes with altitude. For example, atmospheric pressure is greater at sea level than on a mountaintop. To compensate for this difference and facilitate comparison between locations with different altitudes, atmospheric pressure is generally adjusted to the equivalent sea-level pressure. This adjusted pressure is known as barometric pressure. In reality, the weather station measures atmospheric pressure.

Barometric pressure also changes with local weather conditions, making barometric pressure an extremely important and useful weather forecasting tool. High-pressure zones are generally associated with fair weather, while low-pressure zones are generally associated with poor weather. For the purposes of forecasting, however, the absolute barometric pressure value is generally less important than the change in barometric pressure. In general, rising barometric pressure indicates improving weather conditions, and falling pressure indicates deteriorating weather conditions.

Building the electronic barometer shown in Figure 5-13 will help you keep track of the barometric pressure and will easily perform as well or better than most expensive aneroid barometers you are likely to buy. The resolution of the digital display is 0.01 inches of mercury, which is greater than what could be obtained from an analog barometer.

A barometer measures absolute ambient air pressure, usually specified in inches of mercury. At sea level under standard conditions set by the World Meteorological Organization, the accepted absolute air pressure is equal to 29.91216 inches of mercury. This translates to 14.696 pounds per square inch absolute (PSIA). The actual level of air pressure is always changing, and such movements provide an excellent method of predicting the weather. Normal pressure readings fall within the range of 29 to 31 inches of mercury.

One important aspect of the barometric reading is the direction of change: that is, rising or falling. The electronic barometer described here has a hold feature, which allows the present reading to be frozen. Then, at a later time, a new reading may be taken. The change of direction therefore indicates if there the pressure is rising or falling. The new reading may then be frozen until another check of the pressure is made.

About the Circuit

The heart of the electronic barometer is an absolute pressure sensor that has a detection range of 0 to 15 PSIA. This piezoelectric device is composed of four resistors implanted on a silicon substrate, which acts as a diaphragm. One side of the substrate is exposed to a chamber that has been totally evacuated to an almost perfect vacuum. The other side is exposed to ambient air pressure.

The four resistors connected to the sensor are equal in value and are connected in a Wheatstone bridge configuration. The substrate is physically stressed by the action of ambient air pressure on one side pushing against the zero pressure chamber on the other, and as a result two of the resistors have increased resistance, while the other two are lower than normal. This produces an unbalanced bridge output between pins 2 and 4 of the sensor, which represents the actual absolute value of air pressure (and barometric reading).

The pressure sensor is a linear device that, in this circuit, is driven by a 5-volt regulated supply. It develops about 20 millivolts output between pins 2 and 4 at normal barometric pressure conditions. Its sensitivity to pressure changes equals about 0.678 millivolts per inch of mercury.

Analog Amplifier

The pressure sensor bridge circuit is driven by the 5-volt regulated power source and provides a nominal differential output voltage of about 20 millivolts, which will vary by only 0.678 millivolts for each change of one inch of mercury barometric pressure. This minute change in bridge output voltage must be amplified before it can become useful. This is accomplished by U2, a dual operational amplifier.

U2A and U2B are connected as a differential amplifier. The gain of the amplifier is determined by the resistance values of R7 through R13 and is equal to:

$$\text{GAIN} = 2 \times \{1 + 100K/R\}$$

where 100K is the value of R9, R12, and R13, and R is the combined value of R10 and R11 connected in series. Additionally R7 and R8 form a voltage divider that provides 1.5 volts DC offset to the amplifier, and the resistance pair has a Thevinin equivalent of 100K to match R9, R12, and R13.

Using the voltage gain formula shown and the resistor values shown in the schematic diagram, the gain of the amplifier is 1.48. With a sensor output voltage swing of 0.678 millivolts for each one inch of mercury change in air pressure, it can be seen that the output voltage of the amplifier will change 10 millivolts for each inch of mercury. The actual output voltage of the amplifier will be 1.5 volts as provided by offset resistors R7 and R8, plus the bridge differential output voltage multiplied by the gain of the amplifier. Normal output is about 1.8 volts at pin 7 of U2.

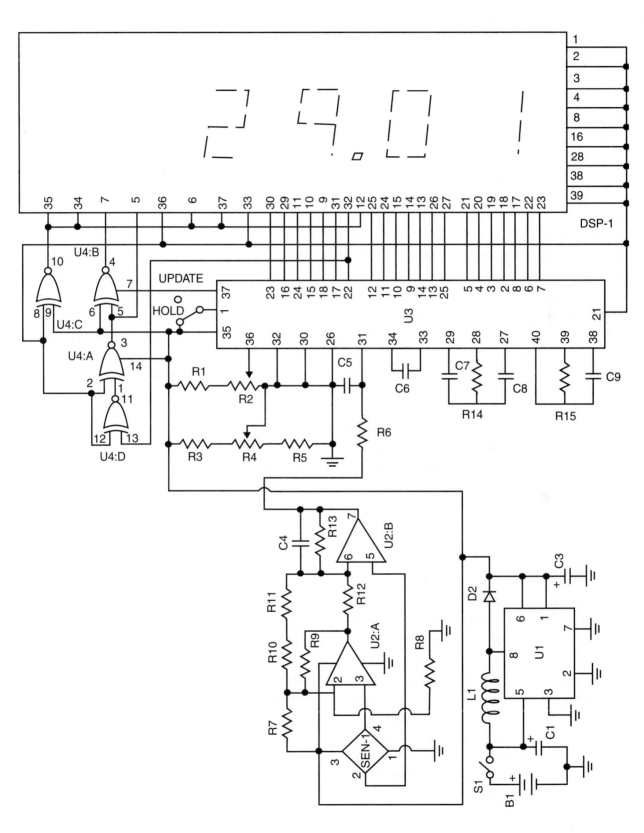

Figure 5-13   *Barometer circuit*

Analog-to-Digital Converter

IC U3 and its associated components form a complete 3½-digit voltage measurement system that drives a *liquid crystal display* (LCD). Only the three least significant digits are used. The half digit 1 is not required because the barometer must display either a 2 or a 3 as its most significant digit. A special driver circuit (U4) is used to accomplish this.

The analog voltage input to U3 is applied to positive input terminal 31 through isolation resistor R6. The negative analog input of U3, terminal 30, is driven by potentiometer R4, which acts as a calibrating adjustment. This takes into account variations in pressure sensors and allows the display to be set to the correct barometric pressure reading as determined from a known, accurate source.

The sensitivity of the A/D converter is determined by the reference voltage appearing between pins 32 and 36. In this circuit it is necessary that the A/D converter has a full scale (1999) sensitivity of 199.9 (200) millivolts. This is accomplished by using potentiometer R2 to set the reference voltage to 100 millivolts (0.100 volts).

With a reference voltage of 100 millivolts, U3 will generate a display of 1,000 when its differential analog input voltage is equal to 100 millivolts. More over a positive or negative change of 10 millivolts as caused by a change of one inch in the barometric pressure will cause the generated display to be either 1,100 or 900, respectively.

Barometric pressure levels range from 28.00 to 31.00 inches in 1-inch increments, and the three least significant digits of the display will be 8.00, 9.00, 0.00. It will be 1.00 when the decimal annunciator (pin l2) is hard-wired for a display resolution of 0.01 inches of mercury.

The range of the barometer display has to cover the range from 28.0 inches to 31.99 inches. Therefore, the most significant digit is required to be either a 2 or a 3. A clever one-chip quad exclusive OR circuit (U4) is used to generate the proper digit. An exclusive OR gate produces a logic 1 output only when its inputs are at opposite logic levels.

Each of the digits of the display is composed of seven segments, identified by the letters a through g. The key to generating a 2 or a 3 is by examination of the g segment of the second most significant digit. If that segment is active because the digit is 8 or 9, the most significant digit must therefore be a 2. If the second most significant digit is a 0 or 1, its g segment is inactive and the most significant digit must therefore be a 3.

The LCD is operated by a square-wave backplane signal generated at U3, pin 21, and is applied to the common terminal, pin 1, of the display. Any segment of the display will be extinguished when it is driven by the backplane waveform and will be activated when it is driven by the same waveform that has been inverted by 180 degrees. IC U4C is used as an inverter that activates the most significant digit segments that are common to a 2 or a 3. These are a, b, d, and g. The inverted backplane signal at pin 10 of U4C is hardwired to pins 34, 35, 36, 37, and 12 of the LCD to permanently illuminate those segments, plus the decimal point.

Segment f of the most significant digit is never activated; therefore pin 36 of the LCD is connected directly to the backplane. U4D simultaneously examines the g segment of the second most significant digit of the display and the backplane waveform. When the g segment of the LCD is active (for digit 8 or digit) the output of U4D pin 11 is high. Otherwise, it is low for digits 0 and l.

IC U4A is used as a conditional inverter so that its output is identical to backplane when the second most significant digit is 0 or 1 and is inverted when the digit is an 8 or 9. U4A feeds the e segment of the most significant digit of the display (pin 5) to illuminate part of the 2 digit, and it also feeds one input of inverter U4B. The c segment of the most significant digit is driven by U4B so that it will always be opposite to the e segment. The logic circuit of U4 always generates the correct most significant digit of the barometric reading, but it cannot produce any digit other than a 2 or a 3.

Barometer Memory

U3 normally updates its reading about three times a second in response to changes in the differential analog input voltage fed to pin 31. However it is designed with a hold feature that allows the display reading to be frozen. This allows the user to store the current reading and update it at any time to provide information regarding the trend of the barometric pressure: rising, steady, or falling. This is accomplished by operating S1, the hold update switch.

Power to drive the circuit is obtained from three AA batteries connected in series to provide about 4.5 volts DC. U1 is a switching regulator chip that is capable of operating with an input voltage range of 0.8 to 5 volts; it delivers a regulated output voltage of 5 volts at pin 6. This voltage stays constant and is used to power the entire circuitry of the barometer.

The circuitry of the barometer is contained on two single-sided printed circuit assemblies called the analog board and display board. The analog board contains the pressure sensor amplifier analog-to-digital converter and power supply. The display board contains the LCD module. The two boards are mounted inside a suitable enclosure that has a rectangular opening cut out for the display module. The batteries are secured inside the enclosure. The only operating control is the hold-update switch.

The specifications for the circuit wiring are not critical and, if desired, this wiring may be hardwired on a perf-board using good construction techniques. When placing polarized components onto the board, be sure they are properly oriented as shown. Just one part placed backward in the circuit may render the barometer inoperative and may cause damage to one or more components. Sockets should be used for U2, U3, U4, and the display module. To make a socket for the display, cut a 40-pin DIP socket in half. The use of sockets permits ease of service, should it ever be necessary. When installing the ICs, you will need to first locate pin 1 of the chip, which is usually identified by a small indented circle near pin 1 of the chip.

When both printed circuit boards are completed, examine each of them very carefully for open short circuits and bad solder connections, which may appear as dull blobs of solder. Any solder joint that is suspect should be redone by removing the old solder with a desoldering tool, cleaning the joint, and carefully applying new solder. It is easier to correct problems at this stage than it would be later if you were to discover that your barometer does not work.

The analog circuitry contains a number of 1% precision metal film resistors to ensure accuracy and stability. Ordinary carbon resistors are not temperature stable and should not be substituted for the metal film types where specified. Be very careful when handling the pressure sensor. Note that pin 1 is identified by a notch on the terminal.

Locate an enclosure that is large enough to hold all the required components including the solar array. Cut a rectangular opening for the display and secure the boards with suitable screws, nuts, and spacers. The hold-update switch may be mounted at any convenient location on the enclosure.

After the barometer is fully assembled, examine the wiring very carefully for shorts or bridges and for proper connections before applying power to the circuit. A multimeter or DVM is required to test and calibrate the unit. Before installing the battery, measure the resistance between the power input terminals of the analog board, +2.4 volts, and the ground to be sure no short circuit exists. Normal indication, with the positive lead of the DVM connected to the positive side of C1, is essentially high resistance (i.e., about 10K). If zero low resistance is measured, examine the circuit assembly very carefully to locate and correct the fault.

Install the batteries, being very careful to observe proper polarity as indicated in the schematic diagram. Measure the terminal voltage of the battery: Normal indication is 4.5 volts or more. Make a voltage measurement across capacitor C3. Normal indication is 5 volts DC. If you do not obtain this reading, disconnect power and check UI, D2, C1, and C3 for proper orientation. Also check L1. Correct the fault before proceeding. Measure the voltage at pin 7 of U2. Normal indication is about l.8 volts. If you do not have 1.8 volts indicated, there may be a fault in the pressure sensor/amplifier circuit. Check the orientation of the sensor and U2. Check the values of resistors R7 through R13. Look for bad solder connections, and then try a new chip.

Measure the voltage between pins 32 and 36 of U3. Adjust R2 for a reading of 100 millivolts (0.1 volts). Once set, do not readjust R2 again. The display should show a four-digit reading with a decimal point. By slowly adjusting R4 over its range, you should be able to see a display of digits that vary from about 29.00 to 31.00 inches of mercury with the decimal point properly illuminated. Set R4 for the proper barometric reading as obtained from a weather report or another barometer that is known to be extremely accurate. If an airport is located nearby, the control tower can provide you with an accurate current barometric pressure reading.

Check the operation of the hold-update switch. In the update position, the display will change with barometric pressure. Allow sufficient time for the reading to change. Note: it is normal for the digital display to sometimes fluctuate by 0.01 or 0.02 inches. With S1 set to the hold position, the reading should remain constant despite changes in the barometric pressure.

Once R4 is properly set and S1 checked, the barometer test and calibration procedure is completed. If the display has deformed digits, it may be caused by a wiring error between the display and analog board, a short, or an opening in the wiring. The location of the deformed digits will tell you where to look for the problem. Refer to the schematic and check each interconnecting wire with an ohmmeter after disconnecting power to the circuit.

If the display is totally blank, check the orientation of U3, and check all components associated with it. If possible, check pin 21 of U3 and pin 1 of the display with an oscilloscope to verify the presence of the square wave backplane signal. Verify the reference voltage of U3 by measuring the voltage between pins 36 (positive) and 32 (negative). Normal indication is 0.1 volts. If incorrect check R1 and R2. If the most significant digit is not a 2 or a 3 as it should be, check the orientation of U4. Check the circuit for shorts or opens, and try a new chip.

Switch S1 is normally left in the update position so that the barometer will show the existing pressure level. However, the hold position may be selected at any time to freeze the reading until a new reading is taken by switching to the update position of the switch. This way the trend of the barometric reading, rising or falling, may be determined.

Electronic Barometer Parts List

Required Parts

R1   22.1K, 1/4-watt, 1% metal film resistor

R2   1K cermet potentiometer (PC mount)

R3   10K, 1/4-watt, 1% metal film resistor

R4   500-ohm cermet potentiometer (PC mount)

R5   3.74K, 1/4-watt, 1% metal film resistor

R6   1 megohm, 1/4-watt, carbon resistor

R7   332K ohms, 14-watt, 1% metal film resistor

R8   143K, 1/4-watt, 1% metal film resistor

R9, R12, R13   100K ohm, 1/4-watt, 1% metal film resistor

R10   15.4-ohm, 1/4-watt, 1% metal film resistor

R11   220-ohm, 1/4-watt carbon resistor

R14   100K ohm, 1/4-watt carbon resistor

C1, C3   68 uF, 25-volt radial electrolytic capacitor

C2, C6   0.1 uF, 50-volt ceramic disk capacitor

C4   1,000 pF, 50-volt ceramic disk capacitor

C5   0.001 uF, 50-volt ceramic disk capacitor

C7   0.47 uF, 50-volt ceramic disk capacitor

C8   0.22 uF, 50-volt ceramic disk capacitor

D1, D2   1N5817 Schottky diode

U1   MAX856CSA regulator IC

U2   LM358N op-amp

U3   CD4030BE two-input exclusive OR gate

U4   ICL7116 CPL analog-to-digital converter/LCD driver

L1   47 micro-henry inductor (Digi-key M7833-ND)

S1, S2   SPST toggle switch

SEN-1   MPX2100A Motorola 15 psi absolute pressure sensor

B1   2 AA cells

DSP-1   3 half-digit LCD display (Digi-key 153-1025)

Miscellaneous IC sockets, PC boards, wire, hardware, battery holder, etc.

Chapter Six

Vibration Sensing

Vibration sensing can be used to solve industrial problems as well as to detect earthquakes deep in the ground. Engine or motor vibration problems can be monitored over time to study machine problems that may be developing. In the first project in this chapter, you will build a vibration hour meter. This simple yet unique circuit will allow you to record the length of time a machine or natural event occurs, using an off-the-shelf hour meter. As long as the vibration persists the hour meter will record the event.

Our next project is a vibration alarm that will notify you of intruding persons or animals. This project uses a commonly available audio speaker as a sensor to detect vibration. The vibration alarm project could be used to scare away pesky animals from your garden or to warn you of approaching intruders. Our next project, the piezo seismic alarm sensor, uses a gas barbeque piezo sparker mechanism as a seismic sensor. The piezo seismic alarm makes an interesting science fair project for the budding scientist. Our advanced project in this chapter, the AS-1 seismograph, is capable of detecting earthquakes all around the world. This seismograph can be used for serious amateur research and observations. Many schools across the country are currently using the AS-1 for seismic research and display. Qualified schools can actually receive an AS-1 for free (see information at end of chapter).

Vibration Hour Meter

The vibration hour meter is a simple yet unique circuit that will allow you to record the time of a machine or system or natural event by using an off-the-shelf hour meter. It's the perfect science fair project for the budding scientist in the family. Certainly this project has many yet undiscovered applications.

The vibration hour meter circuit illustrated in Figure 6-1 begins with the piezo sensor shown at X1. The piezo sensor can be obtained from a gas-grill sparker element. The piezo sensor element is fed directly to pin 3 of a CD4069UB CMOS HEX inverter IC at U1. The sensitivity of the hour meter can be decreased by lowering the value of resistor R1 if desired. The output of U1 is next coupled to diode D1. The network of R2 and C1 form a pulse shaper that inputs the pulses or vibrations into the next inverter segment at U1:b.

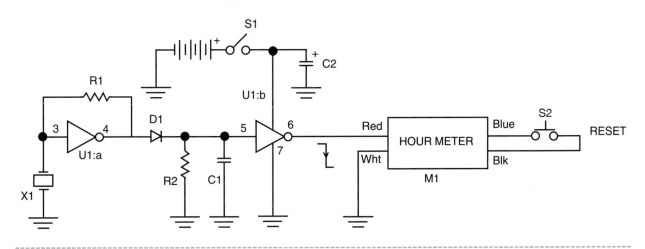

Figure 6-1   *Vibration activated hour meter*

Upon activation of a vibration, a downward pulse is fed directly to the hour meter at M1. As vibration from the sensor continues the hour meter will continue to run, thus recording the length of the event. The hour meter can be reset for the next event by depressing the pushbutton at switch S1. Note that all unused input pins of the CD4069UB should be tied to the 3-volt source.

The hour meter circuit is powered from two AA cells wired in series to form a 3-volt power supply. A 10 uf capacitor is placed across the battery source. Power to the hour meter circuit is applied via the power switch at S1.

The hour meter circuit can be built on a small piece of experimenter board or on a dedicated printed circuit board. A specific layout of the PC board is not critical, however some precautions must be observed. You may wish to have the sensor mounted along with all the other parts on the same PC board or you may wish to have the sensor as a probe device using a coax cable lead-in from the sensor to the actual circuit board, depending upon your particular application.

Certain components require special attention when mounting. For example, a single silicon diode is used. When installing diodes, care must be taken to install them correctly. Most diodes have a black or red band at one end of the diode, which corresponds to the cathode of the diode. On the diagram the cathode is the vertical line facing the input of U1:b. Some capacitors have polarity, and particular attention must be paid to installing these. In this circuit a single electrolytic capacitor at C2 is used. Note the positive lead of the capacitor should be connected to the positive lead of the battery. The hour meter circuit utilizes a single integrated circuit at U1. It is advisable to use an IC socket as assurance against a circuit failure at a later date. ICs must be orientated correctly in order to prevent damage to the IC and usually have some form of marking indicating orientation. Most ICs will have either a small indented circle at the left side of the IC or a cutout on the top of the plastic IC package. Pin 1 of the IC is to the very left of either the small indented circle or the cutout. The hour meter has two mounting tabs, which allow it to be mounted

to the circuit board. And it has four leads; the red and white leads are power and ground respectively. The reset pins are located on the blue and black leads.

The hour meter circuit can be housed in a small metal enclosure if desired. The hour meter and both switches can be mounted at the front top of the enclosure. A two-cell AA battery holder can be mounted to the bottom of the enclosure to hold the batteries. If desired you can mount the piezo sensor as a remote probe. The probe lead should consist of a small diameter coax cable such as RG174/U. The probe lead-in should be kept under 10 to 12 feet if possible to avoid degradation of the signal over a long distance. A two-circuit RCA jack can be mounted on the metal enclosure for the probe cable, and an RCA plug can be placed at the end of the probe cable.

Now the hour meter circuit can be mounted to the bottom of the metal enclosure via $1/4$-inch standoffs and $1/2$-inch 4-40 screws. Once the circuit has been mounted, make sure the power switch is off, install the batteries, and attach the vibration probe to the circuit. Next, you can turn on the power to the circuit and press the reset switch S2. The vibration hour meter is now ready to do its job. Place the probe on a hard surface, rap on the sensor, and the hour meter should begin to operate.

Vibration Hour Meter Parts List

Required Parts

R1	22 megohm, 1/4-watt, 5% resistor
R2	4.7 megohm, 1/4-watt, 5% resistor
C1	0.1 uf, 35-volt ceramic disc capacitor
C2	10 uf, 35-volt electrolytic capacitor
D1	1N4148 silicon diode
U1	CD4069UB CMOS inverter IC
M1	hour meter, Red Lion CUB3, TR-01/A
X1	piezoelectric sensing device
S1	SPST toggle power switch

S2 N/O normally open
    pushbutton switch
    (reset)

B1, B2 AA penlight
    batteries

Miscellaneous PC board,
    IC socket, battery
    holder, hardware, etc.

Seismic Vibration Alarm

A low-cost seismic vibration sensor can serve you in a number of different ways. This unique vibration alarm can be used not only to detect people around your home or campsite, as a burglar alarm or camping alarm does, but it can be used in your garden to detect large animals, such as deer, and scare them away. This simple alarm consumes little power and can be built in a length of PVC pipe that is easy to transport.

The seismic vibration alarm is built around its simple sensor, a 2-inch modified speaker and the op-amp, as shown in Figure 6-2. The project is based on the speaker and the unusual op-amp at U1, a CA3094. This op-amp consists of a programmable transconductance amplifier connected to a darlington transistor. In this circuit the darlington amplifier is combined with a PNP transistor at Q1, to form a monostable timer, which determines how long the buzzer will sound. When the ground shakes, the vibration sensing speaker generates a small voltage that is amplified causing the voltage on pin 1, the darlington amplifier in the op-amp, and the 2N4403 transistor to turn on with regenerative feedback provided with the diode D1. The 2N4401 transistor turns on, thus powering the buzzer at BZ, until the monostable multivibrator timer resets itself.

The seismic vibration alarm circuit can be powered by a 6- to 9-volt power source. A 9-volt battery will work, but will not last too long. For long periods of operation you will want to use 6 C batteries to provide 9 volts or an external power source.

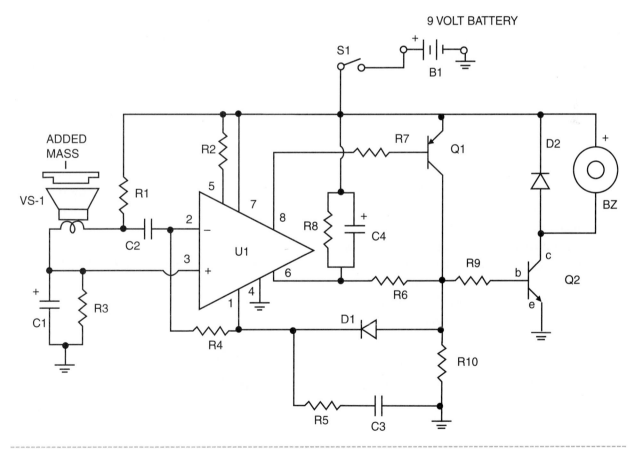

Figure 6-2   *Seismic vibration alarm circuit*

The vibration alarm circuit is used to drive a piezo buzzer, but you could easily substitute a relay to control a larger current, alarm device if desired. You could also use the relay to control a remote transmitter to inform you of an unwanted visitor via remote receiver.

As mentioned earlier the heart of the seismic vibration alarm is the 2-inch speaker sensor. In order to lower the resonance of the speaker to sense low-frequency vibrations of humans and animals, you will need to attach a mass to the speaker cone. To accomplish this, glue an object such as a baby-food jar lid to the speaker cone. The weight added to the speaker will lower the frequencies just enough so that the sensor/speaker can be used to detect larger animals and humans.

Construction of the seismic vibration sensor can be accomplished on either a perf-board or printed circuit board as desired. A small circuit board measuring 2½ × 2 inches was used for the prototype circuit in this project. Components associated with the op-amp should be mounted as close to the op-amp as possible and have their leads kept as short as possible to prevent circuit oscillation. IC and transistor sockets are recommended for the project. In the event of a future circuit failure, the sockets will greatly aid in repairing the circuit. When constructing this project, be sure to carefully observe the proper polarity of the diodes, capacitors, transistor, and the IC. Transistors have three leads and are usually configured as *emitter, base, and collector* (or EBC). The emitter lead on older metal-can transistors was often oriented near the metal tab on the side of the transistor. Newer plastic transistors will usually have no tab but may have a small door and are often oriented by a flat edge on one side of the transistor. Be sure you know the correct pinouts of the device before installing the transistor. ICs are usually identified in two ways. An IC package will have either a plastic cutout at the top edge of the device, with pin 1 to the left of the cutout, or it will have a small indented circle near pin 1 of the IC to help identify the top of the IC and pin 1.

The piezo buzzer and the sensor speaker are both mounted off the circuit board, but all other components are mounted on the circuit board. The vibration alarm circuit board can be built to be placed inside a length of PVC pipe (see Figure 6-3). As a self-contained vibration alarm, you will want to either mount the piezo buzzer on the outside of the PVC enclosure and waterproof it, or mount the buzzer on the inside of PVC tube and have a hole for the sound to come out of the tube. Place a piece of Mylar over the front of the buzzer before mounting it inside to protect the buzzer from the elements. A three-C-cell battery holder can be epoxied to the inside of the PVC tube, but make sure you place them near the clean-out cap so you can easily replace the batteries as needed. If you are using the vibration alarm as a portable camping alarm, you may not need to be concerned with waterproofing the PVC tube. However if you will be using the vibration alarm in your garden regularly, you may want to take extra care to waterproof the PVC enclosure and the buzzer device, especially if it is mounted on the outside of the PVC tube.

A 12-inch length of 3-inch diameter PVC can be used as the main housing. Glue an end cap to one end and a clean-out cap assembly to the other end. The clean-out cap assembly has a screw cap that can be used to pass the circuit board into the PVC tube along with the battery. Once the circuit is housed in the PVC pipe, place Teflon tape around the threads of the screw-on clean-out cap to waterproof the assembly.

A larger detection area may be incorporated by burying a long pole or PVC pipe vertically just below the surface of the ground and locating the vibration sensor above the pole. Vibrations will readily travel down the pole whenever a footstep occurs anywhere along its length. The vibration alarm is now ready to alert you of unwanted two or four legged intruders.

Vibration Alarm Parts List

Required Parts

R1, R3, R4 22-megohm, 1/4-watt resistor

R2 1-megohm, 1/4-watt resistor

R5 100-ohm, 1/4-watt resistor

R6, R7, R8 680K ohm, 1/4-watt resistor

R9, R10 4.7K ohm, 1/4-watt resistor

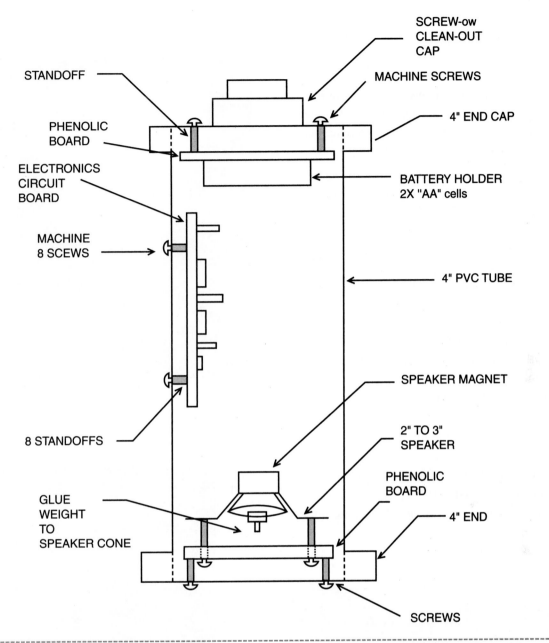

SCREW-ow
CLEAN-OUT
CAP

STANDOFF

MACHINE SCREWS

PHENOLIC
BOARD

4" END CAP

ELECTRONICS
CIRCUIT
BOARD

BATTERY HOLDER
2X "AA" cells

MACHINE
8 SCEWS

4" PVC TUBE

SPEAKER MAGNET

2" TO 3"
SPEAKER

8 STANDOFFS

PHENOLIC
BOARD

GLUE
WEIGHT
TO
SPEAKER CONE

4" END

SCREWS

Figure 6-3   *Seismic vibration alarm mounting*

C1   1 uF, 35-volt elec-
     trolytic capacitor

C2   0.1 uF, 35-volt
     ceramic disc capacitor

C3   220 pF, 35-volt
     capacitor

C4   22 uF, 35-volt elec-
     trolytic capacitor

D1   1N914 silicon diode

D2   1N4001 silicon diode

Q12N4403 transistor

Q2   2N4401 transistor

U1   CA3094 transconduc-
     tance op-amp

BZ   piezo buzzer

VS-1 2-inch diameter,
      8-ohm speaker

S1   SPST toggle switch

B1   6 C cells or a 9-
     volt battery

Miscellaneous PC board,
     IC and transistor
     sockets, wire, PVC
     pipe, PVC end caps,
     PVC clean-out cap
     assembly, PVC glue,
     battery holders,
     speaker mass, etc.

Piezo Seismic Detector

The seismic detector/demonstrator in this section is a great science fair project that won't break your budget. You will want to consider building this low-cost seismic detector made from a commonly available piezo sounder.

The piezo seismic detector project is shown in Figures 6-4 through 6-6. The project revolves around a low-cost Radio Shack piezo sounder and a low-cost op-amp. The piezo sounder has a piezo crystal, used in this project as the seismic sensor. The piezo crystal in the piezo sounder is designed to create sound when voltage is applied to its leads. Piezo crystals are two-way devices (i.e., they can be used to create sound, but they can also generate electricity as well). Piezo crystals, like quartz crystals have interesting properties. You can use an oscillator to drive a crystal to produce sound or voltage, but you can also use the crystal to generate a voltage when the crystal is stressed. In our seismic project, we will use the piezo crystal to generate a tiny voltage when it senses movement.

We will begin by modifying a Radio Shack model 273-060A piezo sounder. This will serve as your vibration sensor once it is modified. Looking at the

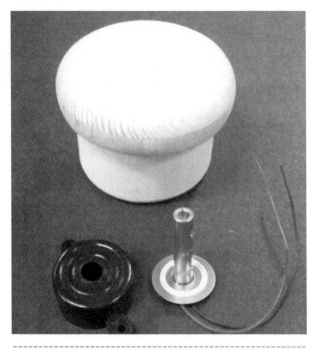

Figure 6-5　*Piezo sensor and mounting components*

Figure 6-4　*Piezo sensor and mounting post*

Figure 6-6　*Mounted piezo sensor housing*

back of the piezo sounder, you will notice two indentations along the recessed bottom cover. Insert a very small screw driver to pry the bottom cover from the piezo sounder case. Once you have opened the case, you will notice a brass-looking plate, which is the piezo element, and you will notice a circuit board attached to the bottom cover. Cut the black and red wires from the circuit board and unsolder the three wires from the piezo element. Now solder the red wire that you removed from the circuit board to the center element on the piezo plate. Next solder the black wire to the outer edge of the brass ring on the piezo element. Notice on the reverse side of the piezo disk a white looking rubber ring. Then look inside the black piezo case; you will find a center cylinder corresponding to the rubber ring on the piezo disk. The cylinder is a reference pressure point, which is used to apply pressure to the piezo element. Do not remove the rubber ring! Locate the back cover and a roll of electrical tape. Tape over the circuit board area with black electrical tape and trim the excess tape.

Next locate a $5/16$-inch diameter, 2-inch long brass dual-shaft coupler. Now, epoxy the brass dual-shaft coupler to the center piezo disk on the rubber ring side of the piezo disk (see Figure 6-4). Make sure the piezo disk surface is clean and dry before attempting to epoxy the shaft coupler. Allow the epoxy to dry overnight.

Once the epoxy has dried, locate a small inexpensive round 2- to 3-inch wooden box from a craft supply store. Take the black plastic piezo sounder case and place it over the bottom of the small round wooden box (see Figure 6-5). Mark where the plastic tabs meet the box, and drill a hole for each mounting tab. Locate a steel rod that is the same diameter as the inner diameter of the shaft coupler (i.e., $1/8$ to $5/32$ inch). You will want to cut the probe shaft to about 8 inches long. One end of the probe should be tapered to a point, and the other end must be sanded to fit inside the free end of the dual-shaft coupler. After the epoxy is dry, loosen the outer set screw on the shaft coupler and insert the steel rod into the open end of the shaft coupler. Locate the plastic piezo disk case and make the center hole larger to accommodate the brass coupler diameter, which you will then place through the piezo case. As you reassemble the

piezo sounder, you will have to first make sure that the rod has been placed through the hole in the case. Then place the piezo disk back into the plastic case, so that the rubber ring mates with the ridge on the case. Seat the piezo buzzer into the case firmly, then place the backer plate firmly against the piezo element and secure the case.

Next you will need to screw the piezo sounder case to the bottom of the small wooden cylinder box (see Figure 6-6). Locate a 20-foot length of shielded microphone or coax cable and solder-splice the piezo wires to the microphone cable. Use heatshrink tubing to insulate each of the wire connections, and place a length of heatshrink tubing over the final assembly to keep the wires dry at all times.

The vibration from the piezo crystal produces a minute voltage, which is fed to a sensitive op-amp at U1, as shown in Figure 6-7. The piezo crystal is connected to a conditioning circuit consisting of a few resistors and capacitors. From the conditioning circuit, the signal is sent to the OPA124P op-amp, which amplifies the seismic signal. The output of the op-amp is coupled to a meter circuit and an LED activity indicator. The meter circuit converts the AC seismic signal to DC to drive the DC milliammeter. The output of the meter display circuit is sent to a 10K ohm potentiometer, which can be fed to a data logger or analog-to-digital converter circuit. The analog to digital converter could be a self-contained data logger such as an Onset mini HOBO data logger or an analog-to-digital converter card in a personal computer.

The piezo seismic detector can be constructed on a Quick-board, a prototype circuit board or a conventional printed circuit board. Good construction techniques should be used, and leads between components should be kept as short as possible. The front end of the circuit consisting of the piezo sensor and the op-amp circuit should be shielded when completed.

When constructing the circuit, be sure to use an IC socket for the op-amp. An IC socket will save you repair time later in the event of circuit failure. The IC must be oriented correctly for the circuit to work properly. At one end of the IC will be either a rectangular cutout at the top of the IC or a small indented circle at the top right of the IC. If there is a rectangu-

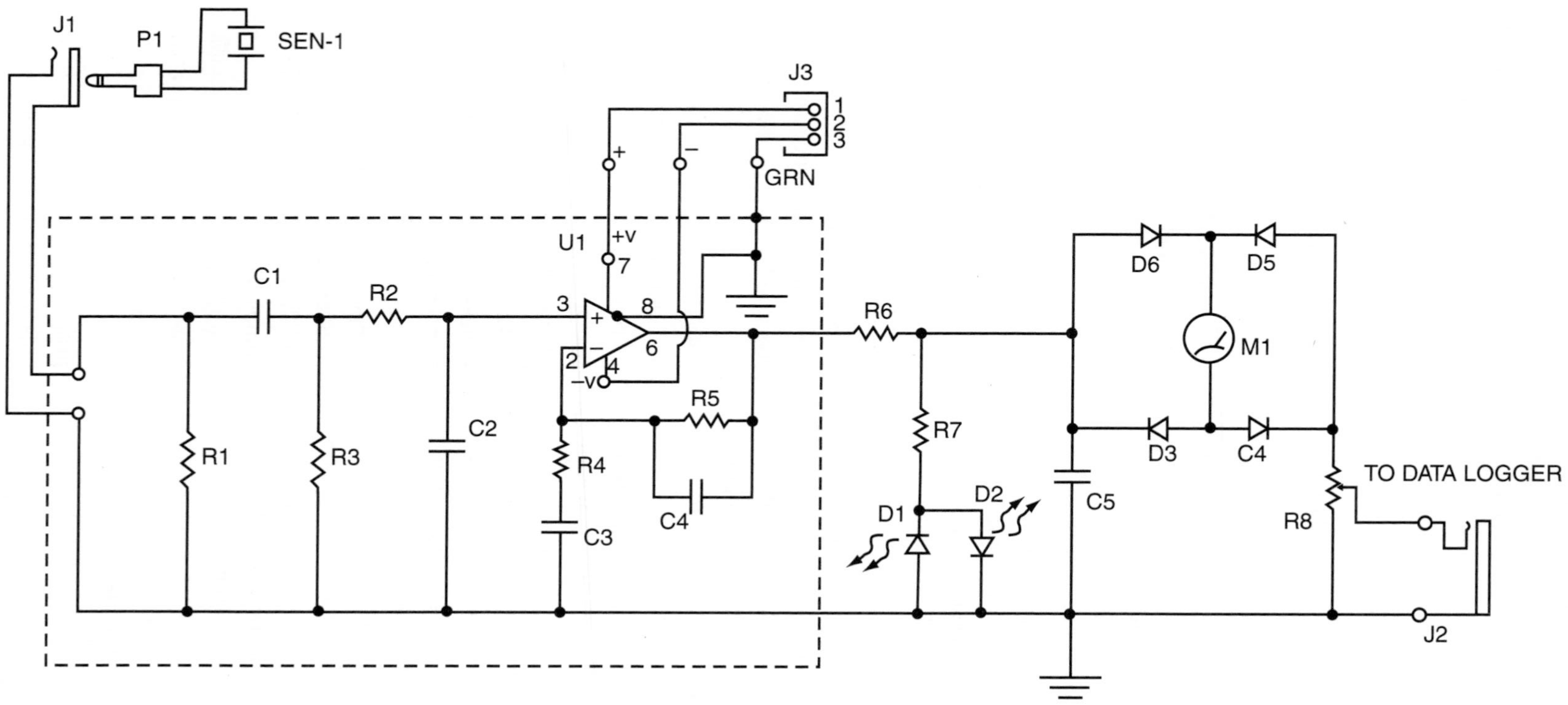

Figure 6-7   High-frequency piezo seismic vibration sensor circuit

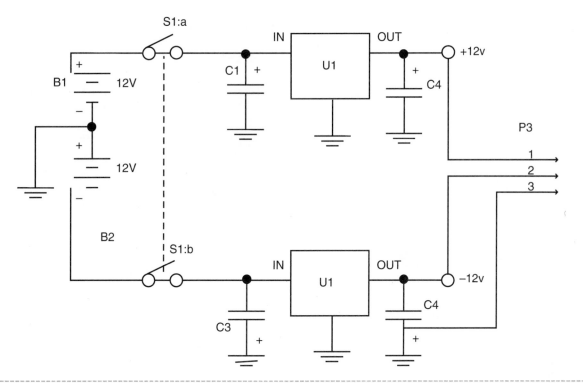

Figure 6-8   *+12/−12 VDC power supply*

lar cutout at the top of the IC, pin 1 will be to the left of the cutout. If the IC has a small indented circle at the top of the IC, then pin 1 will be just to the left of the circle. Be sure also to observe the correct polarity when installing the capacitors, diodes, and the meter. The diodes connected to the meter need to be installed correctly in order for the circuit to operate correctly. The arrow points to the cathode of the diode, which is usually noted as the black band at one end of the diode. The activity indicator LEDs are mounted in both directions; it doesn't matter which one is in what location as long as the bands are opposite each other when you are finished. The milliammeter should be marked with a plus (+) and a minus (−) lead.

When you have completed building the circuit board, be sure to look at the copper side of the board to make sure that no stray or cut component leads are still stuck to the circuit board. Remove anything that is stuck to the board. Look for bridged or shorted circuit leads before applying power to the circuit, to avoid damaging the circuit. Note that the piezo seismic detector circuit requires a plus and a minus 12-volt DC power supply. The dual power supply is illustrated in Figure 6-8. In this dual 12-volt power supply, a 78L12 plus 12-volt regulator and a

79L12 minus 12-volt regulator are connected through power switch S1 to two 12-volt batteries as shown, or they can be connected to a 115 VAC dual 12-volt power supply or dual wall wart power supply. For portable applications, you may wish to power the circuit from two 12-volt lantern batteries. The piezo seismic detector draws very little current in its idle state, so 12-volt lantern batteries should have a long life in a portable setup.

The piezo seismic detector electronics are mounted in a small metal chassis box to eliminate picking up interference from noise or RF sources. The meter and activity indicator LEDs are mounted on the top front of the chassis box along with the power toggle switch S1. The piezo sensor wires from the circuit board are brought out of the chassis box via an RCA jack J1 at the rear of the enclosure. The data logger output wires are connected to an RCA jack, J2 at the rear of the chassis box. A three-conductor microphone jack at J3 is mounted to the rear panel of the chassis box in order to connect the circuit board to two external 12-volt lantern batteries. The diagram in Figure 6-9 depicts the how to mount the piezo seismic crystal to maximize the conduction of vibration from the ground to the piezo crystal.

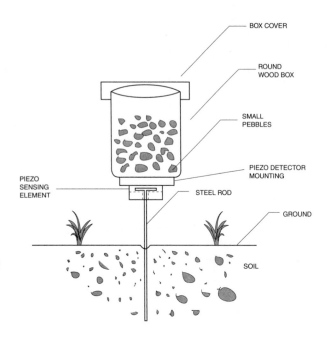

BOX COVER

ROUND
WOOD BOX

SMALL
PEBBLES

PIEZO DETECTOR
MOUNTING

PIEZO
SENSING
ELEMENT

STEEL ROD

GROUND

SOIL

Figure 6-9 *Piezo seismic/vibration sensor mounting*

Your vibration sensor is now complete and ready to install. You are ready to install your new vibration/seismic sensor. Locate an area away from your house and walkways with little or no traffic as a place to install your sensor. Dig a small-diameter hole about 10 inches deep, and insert the metal sensor probe into the hole. Once the probe is secure in the ground repack the dirt around the hole and tamp it firmly around the sensor. The sensor should protrude from the ground only by about an inch or two. Once the sensor assembly has been completed, the last detail is to fill the small box with lead shot or rocks to give the box some mass. Run your microphone cable back to your electronics monitoring circuit and attach the microphone cable wire to the circuit. This can be done with matching spade clips or with some sort of male and female connectors.

Finally, connect the microphone cable from the piezo seismic sensor to the piezo amplifier electronics box. Connect the analog-to-digital converter to the output of the seismic amplifier. Finally connect the power supply, and your piezo seismic sensor monitor is complete and ready to serve you!

Piezo Seismic Detector Parts List

Required Parts

R1, R3, R5   100-megohm, 1/4-watt resistor

R2   3.3-megohm, 1/4-watt resistor

R4   56K ohm, 1/4-watt resistor

R6   2.2K ohm, 1/4-watt resistor

R7   6.8K ohm, 1/4-watt resistor

R8   10K ohm potentiometer

C1   10 nF, 630-volt Mylar capacitor

C2   22 pf, 100-volt Mylar capacitor

C3, C4   100 pF, 50-volt Mylar capacitor

C5   6.8 uF, 50-volt electrolytic capacitor

D1, D2   red LEDs

D3, D4, D5, D6 SB140 Schottky diodes

U1   OPA124R op-amp

M1   0 to 1 Ma DC meter

J1, J2   RCA jacks

J3   three-pin microphone connector

SEN-1   piezo sounder (Radio Shack, 273-060)

Miscellaneous   PC board, IC socket, wire, hardware, screws, nuts, 2-1/2-inch diameter round wooden box, dual-shaft coupler, etc.

Research Seismograph

The AS-1 research seismograph system shown in Figure 6-10 is easy to use and was designed and built to make the science of seismology accessible to individuals and schools at a reasonable cost. With the AS-1 seismometer, you will be able to record, analyze, and archive earthquakes on a personal computer for about $500 if you purchase the complete system and much less if you construct the system yourself from scratch. In the classroom, the instrument will allow students to watch a quake roll in as it is occurring. Students will be able to watch a current event unfold, long before the media reports begin.

The AS-1 system will record local quakes as small as magnitude 3.5 from a distance of 150 kilometers. It will record larger quakes occurring anywhere in the world. The AS-1 seismometer combined with the AmaSeis software allows real-time recording of seismic events and simplifies data analysis. The AmaSeis software automatically stores the data about the ground movements so that the earthquake can be analyzed at a later time.

Although the AS-1 seismograph is a relatively simple and inexpensive seismograph, it is capable of relatively accurate and effective recording of ground vibration from earthquakes and other sources. Modern high-performance seismographs that are used by seismologists for research and earthquake monitoring have very accurate timing and calibration, very low electronic noise, and are capable of recording a

broader range of ground vibration amplitudes and signal frequency ranges. Research seismographs also usually record three components of ground motion (one vertical and two horizontal components) rather than simply the vertical component that the AS-1 seismograph records.

When a very large earthquake occurs anywhere in the world (or a smaller event occurs closer to the seismograph station), the AS-1 seismometer responds to the small ground vibrations that are generated by and propagate from the earthquake location to the seismograph station. The ground vibration is converted by the seismometer to a small electrical current, which is first amplified and filtered, then converted to a digital signal via an analog-to-digital converter, and then sent to an attached personal computer, which is running the AmaSeis software.

Overview

Earthquakes are caused by the sudden movement of blocks of the earth's crust. Such movements involve the fracture of brittle rocks and the movement of rock along the fractures. A fracture in a rock along

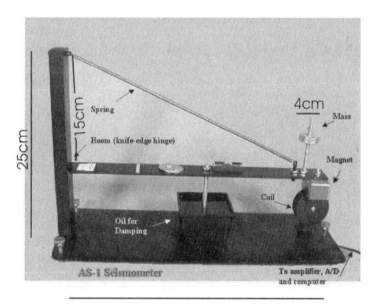

Figure 6-10   *AS-1 research seismograph system*

which this movement occurs is called a *fault*. When an earthquake occurs, it releases energy in the form of waves that radiate from the earthquake source in all directions.

The different types of energy waves shake the ground in different ways and travel through the earth at different velocities. The fastest waves, and therefore the first to arrive at a given location, are *P waves* (or primary waves), also called *compressional waves*. Like sound waves, they move by alternately compressing and expanding material in the direction of motion. The next significant waves to arrive are the *S waves* (or secondary waves). S waves travel through the earth moving material up and down and side to side, perpendicular to the direction of the motion. The last to arrive are *surface waves*, which travel more slowly than P and S waves and move the ground like a rolling ocean wave (*Rayleigh*) or side to side like an S wave (*Love*). The distinction between types of wave motions is important because it determines what will be recorded on a seismometer. The AS-1 is a vertical component seismometer, which means that it records waves that displace the ground in a vertical direction only. Thus, we can expect to record P, S, and Rayleigh waves.

Arrival times of prominent phases, such as P and S waves, can be identified on the extracted seismograms and compared with standard travel-time curves so that the distance from the earthquake epicenter to the recording location can be determined. The operation of the AS-1 along with the AmaSeis software allows easy selection and display of portions of the seismic data. Seismograms corresponding to specific earthquakes or other signals can be enlarged, filtered, viewed on the screen, and saved for further use or additional analysis. Seismogram displays can also be printed. Using the procedures described here, magnitude estimates of earthquakes recorded on the AS-1 seismograph can be calculated from relatively simple equations and from information read from the seismic data.

Figure 6-11 illustrates the AS-1 educational seismometer showing the main components of the instrument. The AS-1 is a vertical component or vertical velocity transducer seismometer. The seismometer has a natural period of 1.5 seconds. Up and down motions of the ground, and therefore of the base and frame of the seismometer, cause the coil to move relative to the magnet that is suspended by the spring and boom assembly. The mass of the seismometer,

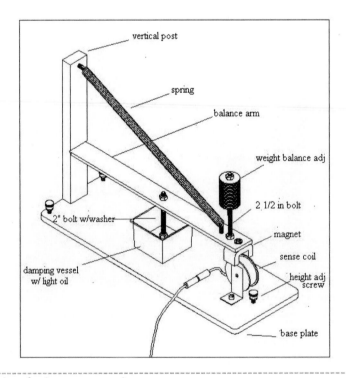

Figure 6-11 *AS-1 educational seismometer main components*

consisting primarily of the magnet, the boom, and the washers tends to remain steady because of inertia when the base moves. The motion of the coil relative to the magnet generates a small current in the coil. The current is amplified and digitized by an amplifier and then connected to the computer for recording and display. The damping reduces the tendency for the mass and spring system to oscillate for a long duration from a single source of ground motion (arrival of seismic waves at the location). Damping is accomplished by placing oil in the container and a washer mounted to a bolt extending downward from the boom into the oil.

Seismometer Assembly

1. If you purchased the AS-1 seismometer as a kit, unpack the box completely and identify the parts from the packing list.

2. Place the seismometer base assembly on a level surface. The base contains three leveling screws. Use the bubble level and adjust the leveling screws as needed. Check the level of the base, both front to back and side to side. The photo shown in Figure 6-12 illustrates the base with the 10½-inch high vertical post at one end of the base plate and the coil attached to the other end of the base plate. A 13½-inch spring is shown attached to the top of the vertical post. If you did not purchase the kit, you will have to construct a base plate measuring 17 inches long by 6 inches wide from metal or wood. The aluminum vertical post measures 10½ inches high by 2 inches wide by 1 inch thick. About 4 inches from the bottom of the vertical post, cut a notch across the 2-inch side

that will face the inside of the seismograph assembly. You will need to fabricate and secure an aluminum vertical post to the base plate with screws, and you will need to attach an eye hook to the top of the vertical post on one of the 2-inch flat sides.

3. If you purchased the AS-1 kit, remove the tape from the knife edge at the end of the boom. One end of the spring should be already connected to the upright support post. Attach the other end of the spring to the boom. Position the boom's knife edge into the slot on the support post. The boom will be positioned at an angle; this is to be expected and will not damage the boom. CAUTION: Before moving the seismometer, disconnect the boom to prevent damage to the knife edge. Figure 6-13 depicts the boom assembly with the magnet attached at one end. From the figure, you can see the boom has a vertical balance bolt with washers as well as a damping bolt with washers in the center of the boom. You will also notice the small post next to the vertical balance bolt; this post has a hole in the center to which attaches the free end of the spring.

If you are building the seismometer from scratch rather than from a kit, you will need to obtain a 14½ long by 2 inch wide aluminum strip to form the boom assembly. Drill three holes in the aluminum boom. In the center of the boom, you will need to drill a $5/16$ hole for a 2-inch long 10-32 machine screw. Next, drill a $5/16$-inch hole for the vertical balance bolt assembly. This hole should be just to the left of the magnet or about $1\,3/8$ inches from the end of the boom. Now you will need to drill a $1/4$-

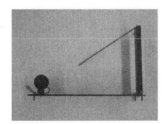

Figure 6-12   *Base plate, vertical post, and spring*

Figure 6-13   *Fulcrum arm and magnet*

inch hole for the small spring end post. This spring end post has a small hole in the center, to attach the free end of the spring. The spring end post is mounted about $1\frac{7}{8}$ inches from the magnet end of the boom. Finally, epoxy a small horseshoe magnet (see parts list) to the end of the boom. The opposite end of the boom has a knife-edge point; grind a sharp edge to this side of the boom. Now install the 2-inch-long bolt though the center hole you drilled in the boom. Place the 2-inch-long bolt through the hole with the head of the bolt on the top of the boom and a nut placed underneath the boom to secure the bolt. At the free end of the bolt, place a nut about ½ inch from free end. Then place a 2-inch diameter washer on the bolt and place a nut on the bolt to secure the washer in place. This bolt with the washer attached is the damping assembly, which will be placed in a jar of light oil at a later time.

4. The upright bolt on the boom arm (shown in Figure 6-14) requires a stack of 1 to 10 washers (depending on the tightness of the spring), approximately 2.5 centimeters (1 inch) from the top of the 2½-long 10-32 bolt. Secure the stack between hex nuts as shown.

5. Tighten the coil to the base plate of the seismometer. The coil should be upright and perpendicular to the base, and it may require tightening and or bending to orient it correctly. Tighten the bolt that holds the magnet to the boom arm. With the added weight of the washer stack, the boom arm should now be suspended over the coil. Center the magnet over the coil (see Figure 6-15) by sliding the

Figure 6-14   *Upright bolt on the boom arm*

knife edge at the support post (shown in Figure 6-16). The boom arm should move freely up and down, and the magnet should not touch the coil. This may require additional bending of the coil.

6. Attach the spring between the small vertical boom post and the 10½-inch vertical support post, as shown.

7. Carefully swing the boom up until it is possible to position the damping fluid container under the damping vane (see Figure 6-17). Do not fill the container before it is positioned correctly. CAUTION: Do not release the boom arm abruptly or allow the magnet to hit the coil. The equipment is sensitive; handle and adjust the parts carefully and gently.

8. The boom arm must be level to accurately record vertical motion. Begin by positioning a

Figure 6-15   *Coil and magnet assembly*

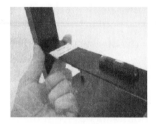

Figure 6-16   *Boom arm*

Figure 6-17   *Completed seismic sensor*

bubble level on the boom. If needed, add one or two washers along the boom until the boom is level. The positioning of the washers and the bubble level will vary. The bubble level and the balance washer(s) stay on the boom arm to keep it level.

9. Rotate the bubble level 90 degrees on the boom. Adjust the base-leveling screws until the bubble is again centered. Rotate the bubble level 90 degrees back to its starting position.

10. Check that the magnet is still centered above the coil. The boom can be horizontally adjusted by pushing the boom at the support post. The boom arm should move freely up and down, and the magnet should not touch the coil.

11. Add either mineral oil or synthetic 10-40 motor oil to the damping fluid container. The fluid level should be 1 centimeter ($^3/_8$ inch) below the top of the container.

Seismic Pre-amplifier and Filter

The sensor coil produces a weak signal, which must be amplified electronically to a level sufficient to drive a chart recorder or computer interface device. A small, 220-volt AC relay coil is disassembled and used in our seismograph as the signal pickup device. You might also elect to purchase the coil (see the parts list). The pickup coil is located within 1 centimeter of the magnet attached to the seismometer's pendulum. Seismic disturbances cause the seismometer's frame and base to move with respect to the pendulum and magnet (and, hence, with respect to the pickup coil that is attached to the frame). The motion is quite small and very slow, and the amplitude of the resultant signal is in the low microvolt region. This weak signal is quite vulnerable to electrical interference and noise; it must be amplified and filtered before it can be used to drive a chart recorder or other instrument. The pre-amplifier should be located as close to the seismometer as possible; shielded cable should be used between the pre-amplifier and the pickup coil.

The signal at the pre-amplifier input connection is applied to the inverting input of a *commutating auto-zero* (CAZ) instrumentation operational amplifier, a Maxim MAX420CPA (shown in Figure 6-18). The CAZ amplifier is unique in its ability to sense its own internal offset voltage and automatically correct for it. Long-term output voltage drift, due to thermal changes and 1/f noise, is completely eliminated, making the device ideally suitable for the amplification of the long-period seismic signals. The amplifier is configured as a single-ended (unbalanced) inverting

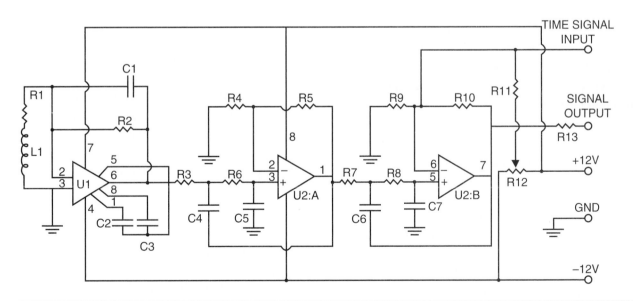

Figure 6-18 *Seismic pre-amplifier and 1 Hz low-pass filter*

amplifier, and the noninverting input is connected to ground potential as shown. A 10-megohm feedback resistor sets the stage-gain to 1,000 (30 dB). In parallel with the feedback resistor, a 0.01-μF capacitor is incorporated to filter out 60 Hz noise, which may have been introduced via magnetic coupling to the pickup coil. The CAZ amplifier IC is relatively expensive at $7.30, but it is indispensable to the excellent overall performance of the seismometer.

The gain is proportional to the value of the resistance and is computed by the formula:

$$g = 10,000,000/(10,000 + \text{DC resistance of the coil})$$

Note that the gain is dependent on the DC resistance of the pickup-coil wire and can also be adjusted by changing the value of the 10K input resistor.

There are two major forms of electrical noise that seem to present the most trouble to amateur seismic recording systems: (1) 60 Hz power-line noise with its harmonics and (2) the very low-frequency noise caused by the characteristics of the pre-amplifier used to amplify the faint signal from the pickup coil. Beyond these two electrical forms of noise is another type of noise that is not electrical in nature, but rather is a true earth motion. It is caused by local, man-made disturbances such as building vibrations resulting from nearby traffic or machinery. This form of noise, however, has a periodicity sufficiently different from those of the earthquake waveforms we are interested in and can thus be filtered without causing too severe a degradation to the waveforms of interest.

The CAZ amplifier is somewhat sensitive to the circuitry that follows it, requiring a load impedance greater than 10 kilohms. The second op-amp, the TL-081(A), is used to isolate the first stage from the load. It acts as the first stage of the fourth-order filter to reduce the response of the seismic output signal for events, such as those that would be caused by people walking past the seismometer, nearby automobile traffic, and remaining electrical-type interference caused by computers, motors, fluorescent lamps, and so on. Thus, any 60 Hz power-line noise will be reduced in amplitude by a factor greater than 1,000 (31 dB to be exact). Because we are interested in seismic signals that have a period greater than 1 second, such a filter has no adverse effect on the quality of our recorded waveforms.

The final op-amp section incorporates a fourth-order filter instead of the second-order filter, as shown on the previous page. Because of its much sharper cutoff characteristics (24 dB per octave), a fourth-order filter can be tuned to a higher cutoff frequency—on the order of 1 Hz. Signals of a longer period would thus pass unattenuated; but unwanted, interfering signals at 3 Hz or faster would be effectively removed.

The inverting input of the third stage is also used to introduce an offset bias if necessary when a particular chart recorder or data acquisition system requires a signal voltage that never drops below 0 volts (i.e., one that requires a zero-signal set-point at some specified DC voltage level). Normally, the offset adjustment is set to produce 0 volts with no seismic signal, but the adjustment is included to provide for a +/− 5-volt offset capability. An optional connection to the inverting input of the third stage is used to accept the time-signal pulses from the optional time-code receiver decoder circuitry.

Time Signal Decoder

A very useful option to the seismometer system is the addition of time-marker signals to the seismic record. These signals provide for the determination of the precise arrival times of seismic events. This is especially important if you are interested in correlating your data with that of other seismic stations for the purpose of triangulating to locate epicenters. Unfortunately, the time base of a chart recorder or even a good computer is completely inadequate to provide true time-of-arrival data. Even the most expensive Pentium IV, 3 GHz computers are still being shipped with a low-cost, inefficient time-of-day clock that simply cannot keep accurate time to better than 1 or 2 minutes per month. Something better is needed; so the following system was developed that provides for unambiguous, always-accurate time marks for our seismic recordings.

The *National Institute of Standards and Technology* (NIST) operates an easy-to-receive shortwave time-signal service; it's broadcast by radio station WWV from Fort Collins, Colorado. The transmitted time signals contain various tones that can be used to automatically insert markers onto the seismogram; as the seismic event is being recorded the time marks are inserted.

Any shortwave radio, such as those available from Radio Shack, that can tune to either 2.5, 5.0, 10.0, 15.0, or 20.0 MHz can be used to receive WWV. Tune the radio to the frequency that gives you the best reception. The audio output of the receiver's recorder or earphone jack is used to feed the following circuitry. The first stage of the time-marker decoder (shown in Figure 6-19) is a simple op-amp that allows us to establish a 200 mV *peak-to-peak* (pp) audio signal for the decoding circuits. This gain of this amplifier stage is adjustable to compensate for the differences in audio levels available from different shortwave receivers.

The output from the op-amp is then fed, simultaneously, to three NE-567 tone-decoder ICs. The first decoder is tuned to 440 Hz (to provide for the 45-second-long marker beginning at 2 minutes past each hour, except the first hour). The second IC is tuned to decode the 1.5 KHz tone during the first second of each hour. And the third IC decodes the 1.0 KHz tone during the first second of each minute other than the first minute. The ICs are amazingly immune to falsing from noise on the received signal. In general, if you can hear the tones from the radio at all, the decoder ICs will properly respond to them. Remember, the transmitted tones are based on *Coor-*

dinated Universal Time (UTC), formerly known as *Greenwich Mean Time* (GMT). The 440 Hz tone, normally sent during the second minute of each hour, is not sent during the first hour of each day, thus indicating (by its absence) the beginning of the next day, UTC.

All three decoded outputs are connected in parallel (a logical OR circuit) and to an amplitude adjustment potentiometer to provide a slight offset pulse whenever any of the special tones are transmitted. The offset pulses are then connected to the output amplifier of the seismic pre-amplifier at the attachment point. Usually, the amplitude of the time pulses is adjusted to appear as about 3 percent of the full-scale seismic signal.

The circuit diagram illustrated in Figure 6-20 shows a triple-voltage power supply, which can be used to power the seismic pre-amplifier and the time-signal decoder circuit. The power supply circuit utilizes a step-down transformer at T1, which is coupled to the full-wave bridge rectifier circuit that converts the 115-volt AC voltage from the transformer to a 24-volt DC voltage. The center-tap of the transformer is connected to ground, and therefore both a plus and minus voltage is derived from the transformer and is then sent to the IC regulators. Regulator U1 provides a plus (+) 9-volt DC voltage to power the time

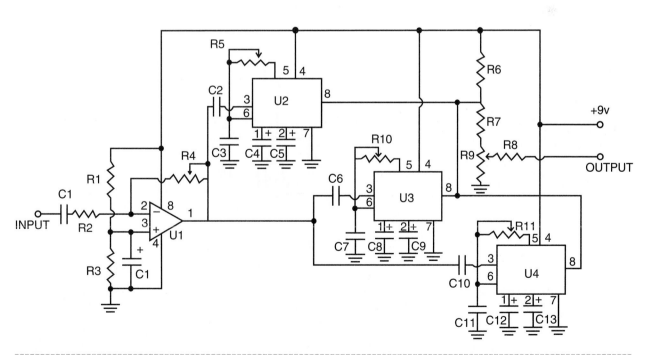

Figure 6-19 *Seismic clock synchronization circuit*

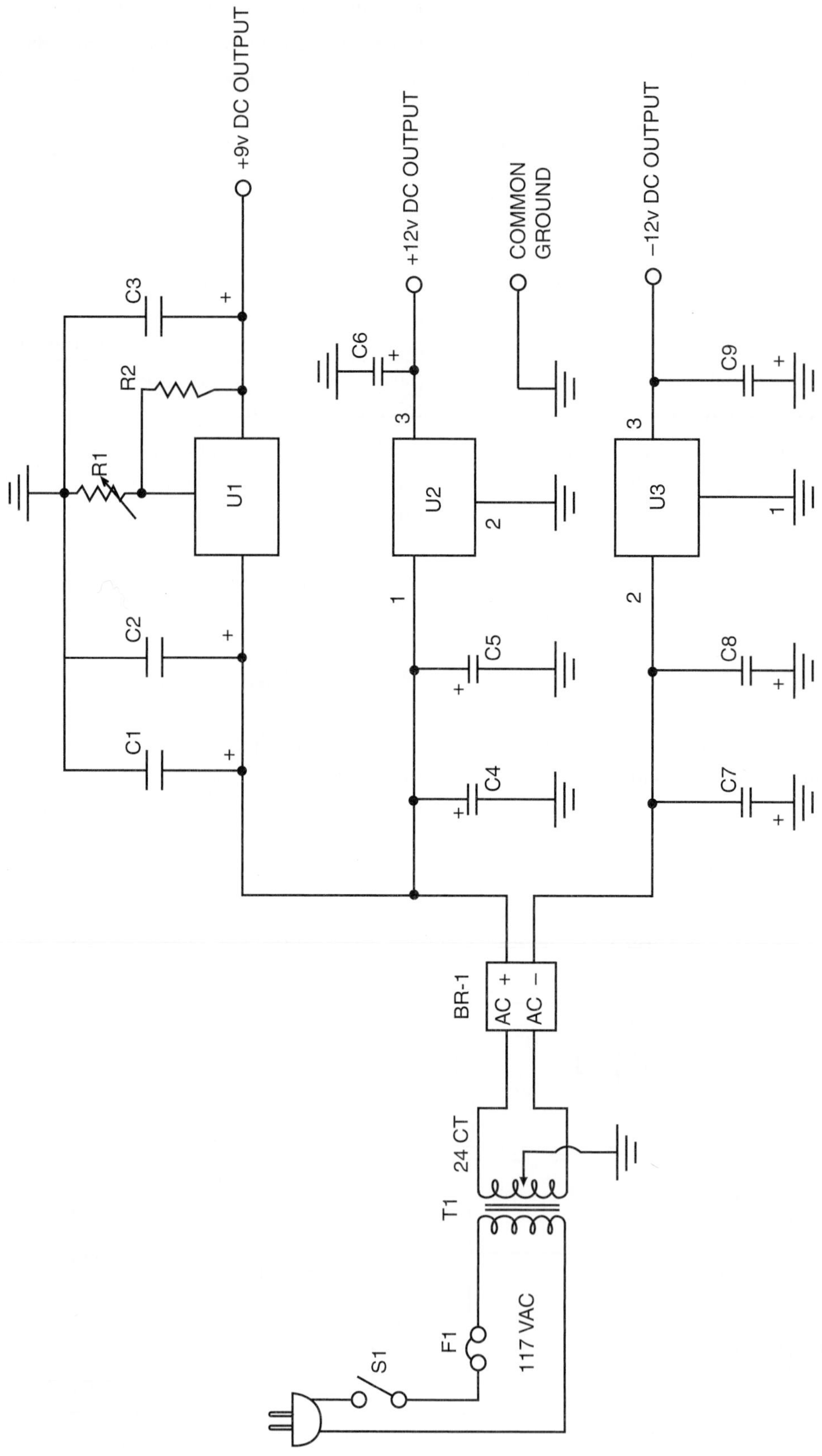

Figure 6-20   +12/−12 and 9-volt DC power supplies

decoder circuit, whereas the regulator at U2 is used to provide a plus (+) 12-volt DC voltage for the op-amps in the seismic pre-amplifier. The regulator at U3 provides a minus (−) 12-volt DC voltage source for the pre-amp's op-amps in Figure 6-18.

Data Logging

The circuitry of the seismometer, once it is amplified by the seismic pre-amplifier, is suitable to directly drive a chart recorder or a conventional drum recorder. A very useful alternative is to send the seismic signals to a computerized data-logging system. In addition to the obvious advantage of saving the costs of chart paper and pens, an enormous amount of data can be saved to disk for later, nearly instantaneous recall. Additionally, the data can be smoothed, filtered, and analyzed in any manner desired by using suitable algorithms in the software.

The following is a small (but growing) list of hardware/software packages that are available at a nominal cost. Numerous computerized logging and data analysis packages are on the market, costing many hundreds of dollars. I've restricted this list to systems costing less than about $100, on the premise that the construction of a homemade seismometer implies that the builder may have limited resources at hand.

- Radio-Sky Publishing: For the technically inclined, a 3-channel analog-to-digital converter with a simple yet flexible software (freeware) package is available from the Radio-Sky Publishing (www.radiosky.com/). This system can be built for less than $30. If you can construct the seismometer preamplifier circuit, you can certainly construct this little beauty! The analog-to-digital converter interfaces directly to your computer's parallel port.

- DATAQ Instruments: One of the best values in commercial, off-the-shelf systems is the four-channel, 12-bit A/D device offered by DATAQ Instruments. It is called the DI-158U Kit, and it contains an elaborate software package with terrific data analysis capabilities.

The price, for both the hardware and software is just $99. It interfaces to your computer's serial port. Currently they are offering a DI-194RS starter kit with a four-channel, 10-bit A/D converter and chart recorder software for $24.95—unbelievable!

Locating the Seismometer

Find a suitable location for the seismometer, preferably where local vibrations from people or machinery are at a minimum. Vibrations can be caused by air-conditioning units, traffic, construction, or weather. The vibrations, called *seismic noise*, can hide seismic signals from smaller events. The best surface for the seismometer is a concrete or stone floor, in a location free from extreme temperature variations. A back room or storage closet out of the way is ideal for optimal recordings. The seismometer can be out on display in a public place so that students can see it in operation, but the trade-off is that fewer earthquakes may be detected because of increased local vibration.

Installing the AS-1 Seismograph

1. Assemble the AS-1 seismometer as described in the instructions. The seismometer will work best in a quiet environment. The most important factor is placing it on a solid floor—a concrete slab basement or first level is best. However, the seismometer will work nearly anywhere at a lower gain setting, including on an upper-level floor of the building and on a table top (for testing and demonstration). Because the instrument is not very sensitive to high-frequency signals, vibrations from walking near the seismometer are considered small unless the person is walking within a couple of meters of the seismometer. For this reason, it is desirable to place the connected computer and display at least 2 meters away from the seismometer.

2. Be sure to level the base with the leveling screws and the boom by moving the level bubble and a washer along the top surface of the boom until the boom is approximately level. Place a drop of the motor oil on the hinge of the boom (be very careful of the sharp edge and do not damage the edge) to help prevent friction on the hinge. Be sure that the magnet and coil assemblies are positioned as shown in the instructions and that the magnet and coil do not touch when the boom (and magnet) are moved up and down. When you are ready to connect the seismometer pre-amplifier to the computer and set up the software, we suggest that you do not use the software provided with the AS-1 seismometer, but rather that you use the AmaSeis software instead (see parts listing). Carefully place the clear plastic cover over the seismometer to decrease air currents on the spring, which can cause significant noise.

3. To test the instrument, be sure that the output is approximately centered. Then walk up close or jump near the seismometer, and you should see a small noise pulse on the screen. Gain levels for reasonably quiet sites (concrete slab or floor installation) should be about 20 to 60. At this level, you may record one the following types of events: (1) sometimes the signal will be visible but no distinct arrivals will be distinguished, and amplitudes will be so small that measuring them for magnitude calculation will be impossible; (2) sometimes very clear arrivals and amplitude will be visible.

4. These rules of thumb will depend on the site and gain level, your geographic location (for example, western or eastern United States), the nature of the earthquake, and the path that the waves have traveled from the earthquake location to your station.

Note: Many choices need to be made when deciding to build this project. First you will need to decide if you want to purchase the seismograph for $275.00 or to build it from scratch. The only compo-

nents that may be difficult to find might be the spring, the sensing coil, and the magnet. Both the magnet and sensing coil can be purchased from Larry Cochrane (see Appendix). Next you will have to decide if you want to purchase the seismic amplifier and analog-to-digital converter from the seismograph supplier for $275.00 of build the seismic pre-amplifier (shown in Figure 6-18). You will also have a decision regarding building and obtaining an A/D converter from Radio-Sky or purchasing the A/D converter from DATAQ. The AmaSeis software is free for the download from Dr. Alan Jones (see Appendix).

The AmaSeis seismic display/recording software is currently widely used by many high schools and universities around the country. The software has many great features and is well supported by Alan Jones (see appendix for contact information).

The Help file for AmaSeis software AmaSeis (found at www.geol.binghamton.edu/faculty/jones/) has a fairly complete help file that includes instructions for setting up the software, options that are available for analyzing and displaying data, and a tutorial. To obtain help information, click on the Help icon (menu) at the top of the screen, and then on Help topics. Select Contents. Double-click on one of the topics and then on one of the question marks that appears before the name of a subtopic. Then use the double-arrow controls (right or left arrows at the top of the Help dialog box) to find information related to that topic. To investigate a specific topic, feature, or setting, select Index from the Help topics dialog box and double-click on the appropriate entry. An online tutorial with detailed descriptions for the use of the AmaSeis software is also available.

The following **settings** for AmaSeis can be found at: Settings Menu > Helicorder.

Lines per hour: usually set to 1. This gives a 24-hour record on screen, which is useful for monitoring because when you observe the screen once per day, you will know immediately if an earthquake has occurred. To go back to look at earlier data, use the scroll bar on the right. Lines per hour can be set to as high as 20 (60 in the latest release of AmaSeis) so that the cursor moves quickly across the screen and the details of the waveform can be seen. This setting is useful for laboratory testing, demonstrations, and make-your-own-earthquake activities (stomp tests, etc.). For lines per hour greater than 1, temporarily set the decimate factor to 1.

Gain: amplification of the signal by the software. Set this to 1 for set-up and initial centering of the trace on the screen, then to a higher number depending on the noise level. In reasonably quiet installations, a gain of 50 or more is possible.

Low-pass filter cutoff: low-pass filter options for the trace on the helicorder screen display. Set this at 3 Hz; if high frequency noise is visible on the screen, try a lower cutoff such as 1 or 0.5 Hz to see if the noise is reduced. Earthquake signals will generally not be greatly affected by this cutoff frequency for this seismometer.

Days to retain record: controls how far back in time one can scroll to view the seismic data on the screen. If you are scrolling back more than a few days, move the scroll bar slider with the mouse, instead of using the up or down arrows or clicking on the bar above or below the slider. Set this to 365 to allow viewing data back as far as one year. The value can be set larger, but if there are old data that you would like to view, they usually would have been saved as individual events (extracted and saved as .sac files, so that data archive doesn't get too large) and then opened under the File menu.

Decimate factor: set to 10, normally. This setting will provide a fairly accurate view of the data on the screen but allow rapid scrolling. A decimate factor of 10 plots only every 10th point; some distortion is caused by this choice, so for large numbers of lines per hour, make-your-own-earthquake experiments, and whenever you want to see the most accurate view of a seismogram on the screen, set the decimate factor to 1. The decimate factor does not affect the plotting of an extracted seismogram, which is always plotted with all of the available points.

The Show data values dialog box and display allows you to center the cursor so that it tracks approximately on the zero line (the blue line) that is defined for each hour. To center the trace, it is convenient to start with a low-gain setting (such as 1). Turn the black knob on the small black interface box

that comes with the AS-1 seismometer that is between the seismometer and the computer in the electric circuit. The knob adjusts the cursor position on the screen. It is very sensitive, so make small adjustments and determine which way to turn the knob in order to cause the cursor to move upward. As you adjust the knob, the cursor should get closer to the zero line and the display numbers should be close to zero. When you have succeeded in getting the numbers close to zero, turn the gain up and repeat, eventually setting the gain to the highest level that is consistent with the background noise (you should see relatively continuous, small, approximately +/− 1- to 2-millimeter movements on the screen) and making the displayed values close to zero. To adjust the knob when the gain is set to a high value (greater than about 10), it is convenient to use a pencil and very, very lightly tap the knob from the side to cause it to move. **This centering process can take a considerable amount of time. Proceed systematically and patiently.**

When you are ready to **archive data**, do the following. Recording the AS-1 seismic data with the Ama-Seis software generates about 1 MB of data per day. If your computer has insufficient disk space to archive data for many months or years, you can copy the files onto a zip disk or to a *compact disk rewritable* (CDR). Alternatively, you can sort through the recorded data and delete data for days in which there was no significant signal.

Be sure to take time for **time synchronization** and **maintaining a seismograph catalog**. It is useful and good scientific practice to develop a catalog for your seismograph station. Entries are made automatically to the catalog every day or two that include absolute time checks or synchronizations. Time checks can be made by comparing the computer's clock to an accurate, absolute time signal from a radio-synchronized clock such as arctime.com or radio station WWV at 5, 10, or 15 MHz frequency on a shortwave radio, or, if the computer is connected to the Internet, the computer clock can be periodically synchronized to absolute time using the software tool AboutTime at www.arachnoid.com/. Catalog entries also include times of arrival, amplitude and magnitude information of earthquakes (recorded by the seismograph), and comments about the weather, background noise conditions, or operational status of the seismograph.

One very convenient radio synchronized clock (Atomic Clock) is the SkyScan wall or desk clock (available from Sam's Club for $20) that has very large lettering. The clock can be set to show GMT time and can be used to visually check the time correction of your computer running AmaSeis and record the drift of the computer clock. Occasionally resetting (synchronizing) the computer clock to GMT time will keep the AmaSeis time display reasonably accurate. A time correction can be applied to saved seismogram files in AmaSeis. The atomic clock makes a nice display for GMT time even if you have an Internet connection on your computer that is used to run AmaSeis and update your computer clock automatically. The SkyScan Atomic Clock is also available ($30) at http://eggshop.net/skysatcloc.html and at http://homestore3.com/skysatcloc.html. Another atomic clock option (with large-digits display) is the WS-8001U wall clock from La Crosse Technology, available for $40 at www.weathermeter.com/.

Relative calibration of the AS-1 seismograph can be conducted using a step function method. A step function (a small, sudden increase or decrease in the position of the mass with respect to the coil) can be applied to the seismometer by lowering (or removing) a small mass onto the boom of the instrument. Because lowering the mass causes an inconsistent input to the seismometer, the method works best if you lift the mass. Because the seismometer is very sensitive, the relative calibration process, although simple, must be performed with a very specific procedure and very carefully. To perform the calibration, use a piece of masking tape or post-it note material to mark a position on the boom of the AS-1 seismometer that is 10 centimeters from the hinge and upright. Place the cover back over the instrument and drill a small hole (approximately 4 millimeters or $1/8$ inch) in the top of the cover above the mark on the boom. Make the mass used for calibration by cutting a 1×2 centimeter rectangle of lightweight poster board. This mass is about 0.063 grams. Fold the poster board rectangle into the shape of a tent, and attach a 3-meter long piece of nylon monofilament thread (very light and available in fabric and sewing stores) to the top of the poster board tent with a small piece of tape. Thread the monofilament through the hole in the cover (from the inside of the cover) so that the small mass is suspended above the mark on the boom by the thread. Because the instrument is sensitive to tilt if one is standing near it, and because the air current noise is large, put the cover on and lower the mass through the small hole in the top of the cover by holding the other end of the thread from a distance of about 3 meters, standing still. It is convenient to attach a piece of tape to the free end of the thread to make it easier to find and hold on to the end of the thread. Lowering the mass gives a calibration pulse, but a better pulse (about 1,400–1,500 counts maximum amplitude) is obtained by letting the instrument stabilize and then pulling on the thread to lift the mass suddenly off the boom. This method appears to be stable and results in calibration pulses that varied only about $+/-10$ percent between tests. The maximum amplitude of this test should provide a relatively linear correction factor to the amplification of different instruments. For more information on calibration of the AS-1 seismometer, see http://quake.eas.gatech.edu/MagWeb/CalReptAS-1.htm. This relative method of calibration has the following uses:

- Performing the calibration tests to see if the seismometer is working properly (the calibration pulse should look approximately like the one in Figure 6-21; the *lifting-the-mass* pulse is used; the calibration pulse is extracted from the standard AmaSeis display to view it in a close-up view). Repeated calibrations suggest that measurable values of the calibration test pulse should be (approximately): amplitude of first trough (negative) = $-1420 +/- 150$ counts amplitude of first peak (positive) = $480 +/- 50$ counts time from beginning of calibration pulse to first zero crossing = $3.0 +/- 0.5$ seconds

- The relative calibration allows you to compare different instruments. AS-1 seismographs having calibration pulses that are approximately the same as the pulse in Figure 6-21

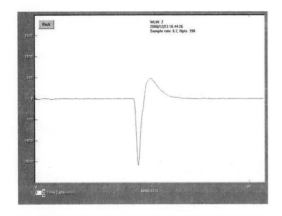

Figure 6-21   *Calibration pulse*

can be assumed to have the same displacement amplification values as given here, and they can be used to estimate magnitudes using the amplification values and procedures described here. If the calibration pulse for a different AS-1 seismometer is similar to the one shown in Figure 6-21 (and parameters given here) but has different maximum amplitude, the amplification factors used can be adjusted in the magnitude calculations by the relative difference between the two AS-1 instruments.

- The calibration provides a check on the polarity of the seismometer. Because lifting the mass from the boom causes the boom to move up relative to the coil that is attached to the base of the seismometer, the first motion of the output trace (the calibration pulse) should be down or negative. A first up motion of the ground from an incoming seismic wave will move the base and the coil up relative to the magnet (corresponding to a relative down motion of the boom) attached to the boom, because of inertia of the seismometer mass suspended by the spring. This up motion of the ground will therefore cause an up or positive signal on the seismograph trace. If the calibration pulse (corresponding to lifting the small mass) has a first positive motion, the polarity of the seismometer is reversed. To correct the polarity, simply switch the input wire connections on the amplifier.

Seismograph Filter Circuit Parts List

Required Parts

R1   10K ohm, 1/4-watt resistor

R2   10 megohm, 1/4-watt resistor

R3, R6, R7, R8   1-megohm, 1/4-watt resistor

R4, R9   130-ohm, 1/4-watt resistor

R5   20K ohm, 1/4-watt resistor

R10   160K ohm, 1/4-watt resistor

R11   220K ohm, 1/4-watt resistor

R12   100K offset potentiometer

C1   0.01 uF, 50-volt ceramic disk capacitor

C2, C3, C4, C5, C6, C7   0.1 uF, 50-volt ceramic capacitor

U1   MAX420CPA CAZ auto-zero instrumentation amplifier

U2   TL082 op-amp

L1   disassembled relay coil (see text)

Miscellaneous   PC board, wire, IC sockets, connectors, enclosure, etc.

Optional Seismic Clock Synchronization Circuit Parts List

Required Parts

R1, R2, R3, R6, R8   10K ohm, 1/4-watt resistor

R4   100K ohm potentiometer (PC board)

R5, R10, R11   20K ohm potentiometer (PC board)

R7   150K ohm, 1/4-watt resistor

R9 10K potentiometer
(PC board)

C1, C5, C9, C13 4.7 uF,
50-volt electrolytic
capacitor

C2, C6, C7, C10, C11
0.1 uF, 50-volt
ceramic disk capacitor

C4 47 uF, 50-volt elec-
trolytic capacitor

C8, C12 22 uF, 50-volt
electrolytic capacitor

Miscellaneous PC board,
IC sockets, wire,
hardware

Power Supply Parts List

Required Parts

R1 5K ohm potentiometer
(PC board)

R2 240-ohm, 1/4-watt
resistor

C1, C4, C7 4,700 uF,
50-volt ceramic disc
capacitor

C2, C5, C8 0.1 uF, 50-
volt electrolytic
capacitor

C3, C6, C9 10 uF, 50-
volt electrolytic
capacitor

BR-1 bridge rectifier,
3 amps, 100 volts

U1 LM7812 plus (+12
volts) DC regulator

U2 LM7912 minus (-12
volts) DC regulator

U3 LM7809 plus (+9
volts) regulator

T1 transformer 24-
volt/3-amp center
tapped

S1 SPST toggle power
switch

Miscellaneous PC board,
wire, connectors, etc.

The AS-1 was designed and built for the IRIS Seismic Consortium to offer low-cost seismic systems for educational purposes. If you represent a qualified school group, you can apply for a grant to obtain one at no cost. Go to the IRIS Web site at www.iris.edu/edu/ASI.htm for more information. Or you can purchase an AS-1 system by contacting Jeff Batten and The Amateur Seismologist.

The Amateur Seismologist
2155 Verdugo Boulevard, PMB 528
Montrose, CA 91020
(818) 249-1759
info@amateurseismologist.com

AS-1 Specifications

Seismometer: Vertical velocity transducer with oil damping

Natural period: 1.5 seconds (filter extends bandwidth to 20 seconds)

Amplifier: 100 db with low-pass filter, with a zero adjustment but no gain adjustment

Digitizer: 12-bit analog-to-digital converter

Bandwidth: 0.1 second to 20 seconds

Dimensions: 17 inches long, 6 inches wide, 12 inches high

Computer interface: serial port to PC

Data format: ASCII numbers from 0 through 2085

Data rate: 6.2 samples per second

Complete system: $550.00 plus California sales tax if applicable, plus shipping and handling

Seismometer only: $275.00

Seismometer electronics only: $275.00

Seismograph plastic cover and plastic damping oil dish: $40.00 from E & A International

AmaSeis software from: www.geol.binghamton.edu/faculty/jones/AmaSeis.html. For details on how AmaSeis reads from a DATAQ AD, go to www.dataq.com/support/techinfo/dataform.htm

Pickup Coil

The pickup coil is extracted from a new 220 VAC relay, with a 5-foot shielded cable and an RCA plug at the end. The coil has 10,000 turns and a DC resistance of around 9,000 ohms. The coil diameter is 23 millimeters ($^{15}/_{16}$ inch) and is 17 millimeters ($^{11}/_{16}$ inches) wide. Cost: $15.00.

Magnet

The magnet is rated at 22 lbs of pull. The dimensions of the magnet are 25 millimeters (1 inch) high, 40 millimeters ($1^{9}/_{16}$ inches) wide. The open end of the magnet is 21 millimeters ($^{13}/_{16}$ inch) wide. This gives approximately 2 millimeters ($^{1}/_{16}$ inch) on each side of 17 millimeters ($^{11}/_{16}$ inch) wide coil when placed inside or near the edge of the magnet. Cost: $20.00. Contact: lcochrane@webtronics.com

The AS-1 Seismograph magnitude determination
www.eas.purdue.edu/~braile/edumod/as1mag/as1mag3.htm

AmaSeis software setup
www.iris.washington.edu/about/ENO/AS1AmaSeis_v3.pdf

Chapter Seven

Detecting Magnetic Fields

In this chapter we will discuss different types of magnetic sensors, from small induction pickup coils that can be used to listen to telephone conversations and to locate hidden electrical conduit, to larger coil detectors that can be used to detect magnetic fields produced by moving cars and trains. We will also construct an electronic compass and an ELF radiation monitor that can be used to survey your home appliances for harmful radiation. Radio enthusiasts will be interested in the ionospheric disturbance receiver, used for radio propagation studies. Science-minded experimenters will discover the research potential for the earth-field magnetometer for detecting solar magnetic storms originating from the sun.

Historical Review

Until 1821, iron magnets were the only form of magnetism known. A Danish scientist, Hans Christian Oersted, while demonstrating to friends the flow of an electric current in a wire, noticed that the current caused a nearby compass needle to move. The new phenomenon was studied in France by André-Marie Ampére, who concluded that the nature of magnetism was quite different from what everyone had believed. It was basically a force between electric currents: two parallel currents in the same direction attract and will repel when they run in opposite directions. Iron magnets are a very special case, which Ampere was also able to explain.

Michael Faraday is credited with fundamental discoveries on electricity and magnetism. He also proposed a widely used method for visualizing magnetic fields. Field lines of a bar magnet are commonly illustrated by holding over a magnet a piece of paper with iron filings sprinkled on it. Similarly, field lines of the earth start near the South Pole of the earth, curve around in space, and converge again near the North Pole. Faraday called them *lines of force*, but the term *field lines* is now in common use.

In nature, magnetic fields are produced in the rarefied gas of space, in the glowing heat of sunspots, and in the molten core of the earth. In the earth's *magnetosphere*, currents also flow through space and modify this pattern: on the side facing the sun, field lines are compressed earthward, whereas on the night side, they are pulled out into a very long tail, like that of a comet. Near Earth, however, the lines remain very close to the *dipole pattern* of a bar magnet, so named because of its two poles.

To Faraday, field lines were mainly a method of displaying the structure of the magnetic force. In space research, however, they have a much broader significance, because electrons and ions tend to stay attached to them, like beads on a wire, even becoming trapped when conditions are right. Because of this attachment, they define an *easy direction* in the rarefied gas of space. Like the grain of a piece of wood, the easy direction is a direction in which ions and electrons as well as electric currents (and certain radio-type waves) can move easily. In contrast, motion from one line to another is more difficult.

Faraday's ideas evolved into the *magnetic field* concept—that space in which magnetic forces may be observed is somehow changed by the force. Faraday also showed that a magnetic field that varied in time —like the one produced by an alternating current (AC)—could drive electric currents if, for example, copper wires were placed in it in the appropriate way. That idea has come to be known as *magnetic induction*, the phenomenon on which electric transformers are based.

So, magnetic fields could produce electric currents, and we already know that electric currents produce magnetic fields. Would it perhaps be possible for space to support a wave motion alternating between the two?

magnetic field → electric current → magnetic field → electric current →

There was one stumbling block. Such a wave could not exist in empty space, because empty space contained no copper wires and could not carry the currents needed to complete the above cycle.

A brilliant young Scotsman, James Clerk Maxwell, solved this riddle in 1861 by proposing that the equations of electricity needed one more term to represent an electric current that could travel through empty space, but only for very fast oscillations. With that term added (the *displacement current*), the equations of electricity and magnetism allowed a wave to exist, propagating at the speed of light. Figure 7-1 illustrates such a wave—(H) is the magnetic part, (E) the electric part—the term Maxwell added. The wave is drawn propagating just along one line. Actually it fills space, but it would be hard to draw that. An electromagnetic wave is comprised of two fields: a magnetic field (H) and an electric field (E). These combined two fields radiate into space.

Maxwell proposed that it indeed was light. There had been earlier hints—as noted, the velocity of light had appeared unexpectedly in the equations of electricity and magnetism—and further studies confirmed it. For instance, if a beam of light hits the side of a glass prism, only part of the light enters or passes through; another part is reflected. Maxwell's theory correctly predicted properties of the reflected beam.

Then in Germany, Heinrich Hertz showed that an electric current bouncing back and forth in a wire (today it would be called an *antenna*) could be the source of such waves. Electric sparks create such back-and-forth currents when they jump across a gap, hence the crackling heard on AM radio caused by lightning. And Hertz in 1886 used such sparks to send a radio signal across his lab. Later the Italian Marconi, with more sensitive detectors, extended the range of radio reception and in 1903 detected signals from Europe and from as far as Cape Cod, Massachusetts.

Electromagnetic waves eventually led to the inventions of radio and television and to a huge electronic industry. But they are also generated in space by unstable electron beams in the magnetosphere, as well as at the sun and far-away in our universe.

Transformer Action

When an alternating voltage is applied to an inductance, such as that of a coil, an *electromotive force* (emf) is induced in the inductance by the varying magnetic field that accompanies the flow of alternating current. Now if a second coil is brought into the same field as that of the first coil, a similar emf will be induced in the second coil. This induced emf may be used to force a current through a wire or other electrical device connected to the terminals of the second coil.

In this manner, the two coils are said to be coupled together, and the pair of coils make up a *transformer*. The coil connected to the source of energy, such as 120-volt AC house current, is called the *primary* coil, and the other coil, is called the *secondary* coil.

The usefulness of a transformer lies in the fact that electrical energy can be transformed from one circuit (the primary) to the other (the secondary) without direct connection. Such is the case with induction pickup (discussed shortly), where it acts as the secondary of a transformer. Whatever we are listening to with the pickup probe is the primary. And as we will see, this can be used in a telephone, an AC motor, or a TV set.

The transformer, of course, can be used only on alternating current circuits, because no voltage will be induced in the secondary, or *winding* coil, if the magnetic field is not changing. When direct current is applied to the primary of a transformer, such as when connecting a battery across the winding, a voltage will be induced in the secondary winding only at the instant of closing or opening the primary because it is only at these times that the field is changing. At other times, the current is constant and no voltage is induced into the secondary because there is no changing field. Of course, when the battery is not connected, there also is no current or change in current, and there is no voltage induced in the secondary winding.

The Radiation Field and the Induction Field

In brief, a magnetic field forms part of an electromagnetic field, which radiates from an antenna, and this radiation is caused by the changing electric and magnetic fields about the antenna conductor or wire. We also know that the changing fields produced a moving field that travel away from the antenna at the speed of light.

The components of this moving field are the *induction* field and the *radiation* field. The induction field acts as if it were permanently associated with the antenna, and its energy is alternately stored in the antenna and removed from it. The induction field components are usually no longer detectable beyond a distance of about two wavelengths from the antenna. For *very high frequency* (VHF) TV channels, this is on the order of 10 to 30 feet. The portion of the electromagnetic field that carries the signals from the antenna, or point of transmission, to the point of reception is called the radiation field. Let us look more closely at the development of the fields surrounding a wire. We will examine the difference between the composition of the magnetic and electric fields.

The Magnetic Field

The magnetic field is composed of two components which are in phase. That is, they both reach a maximum at the same time and are both minimum at the same time. One of these components, however is inversely proportional to the distance from the antenna, $1/r$, while the other component is inversely proportional to the square of the distance from the antenna, $1/r^2$.

The Electric Field

The electric field contains three components, one of which is inversely proportional to the distance from the antenna, $1/r$; a second, which is inversely proportional to the square of the distance, $1/r^2$; and a third, which is inversely proportional to the cube of the distance, $1/r^3$. In the case of the electric field, however, all the components are not in phase. The component that is inversely proportional to the cube of the distance ($1/r^3$) has a 90-degree phase relationship with the other two. Thus, two separate electromagnetic fields are produced. In the first, the radiation field, which propagates the farthest, the electric and magnetic fields are at right angles in space and are in phase as shown in Figure 7-1. In the second, the induction field, the electric and magnetic fields are at right angles in space but 90 degrees out of phase, as shown in Figure 7-2.

Because the electric component in the induction field is inversely proportional to the cube of the distance, it dies out very quickly and therefore may be neglected whenever the distance, r, is more than a few wavelengths. Such is the case when one is listening to a broadcast station in the AM broadcast band, as it is being received via the radiation field and not the induction field. However, at shorter distances, the i/r^3 term does not die out fast enough, and the induction field term must be taken into account. This is the case when using the pickup probe close to a motor field or a radio speaker. The induction field, in which the electric and magnetic fields are 90 degrees out of phase, dissipates no power. Any power delivered to the field during one part of a cycle is returned during another part.

As stated earlier, the effect of the induction field is negligible at distances greater than a few wavelengths from the antenna. Thus, at high frequencies, where

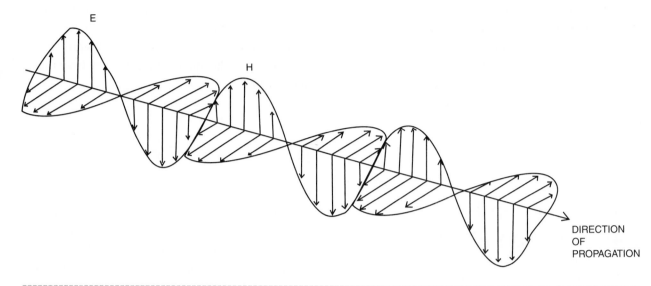

Figure 7-1   *Radiation field propagation*

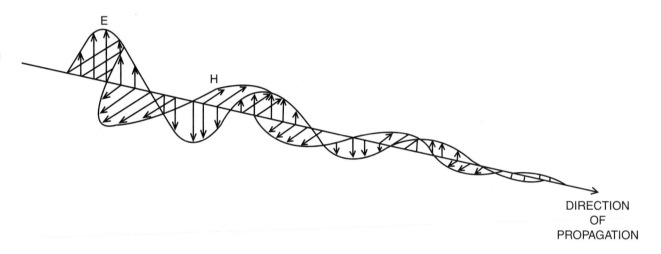

Figure 7-2   *Induction field propagation*

the wavelength is short, the induction field effectively extends only a few feet from the antenna. However at low frequencies, where a wavelength is much longer, the induction field makes itself felt at a considerable distance. With the induction pickup loop as input to the mag-ear project that follows, we will be using the induction field for some of our experiments.

Magnetic Detectors

One of the most simple types of magnetic detection devices is the induction pickup coil. The induction coil can be coupled with a high-gain amplifier and used to listen to many different magnetic phenomena. Induction pickup coils are available from many local radio/electronic supply stores at a cost of only a few dollars,

The induction pickup coil is shown in Figure 7-3. The induction pickup consists of an induction coil and a small suction cup, which is used to clamp the device to the base or earpiece of a telephone. The output of the pickup probe is connected through a 3- to 4-foot shielded cable to a miniature 3½-millimeter male plug.

The pickup coil consists of a large number of turns of very thin wire and has a DC resistance of 400 to 500 ohms. This coil serves as a secondary winding of a transformer, and the source of the signal we are seeking to listen to is the primary winding. The coil may have a soft-metal core, which serves to concentrate the induced lines of flux. Figure 7-4 shows a schematic diagram of the induction pickup coil.

The induction coil or telephone pickup can be coupled together with a high-gain amplifier to form a device called the mag-ear, which can be used to amplify the signal from the induction pickup coil and

Figure 7-3   *Induction pickup coil*

to investigate a number of magnetic phenomena, as we will see. The mag-ear amplifier circuit shown in Figure 7-5 is a high-gain amplifier used to amplify the signals from a magnetic pickup coil. The mag-ear circuit consists of two amplification stages. The first stage consists of a TL082 op-amp acting as a pre-amplifier, followed by an LM 386 high-output audio amplifier IC. The input at J1 is coupled first to a capacitor at C1, followed by a resistor at R1. The gain of the U1, a TL082 op-amp, is controlled via switch S1, a three-position rotary switch that selects the various gain resistors. The network formed by C2, R5, and R6 forms part of a bias network connected to pin 3 of U1. This network is coupled to the 9-volt power source. The output of U1 is coupled to the second-stage audio amplifier, via capacitor C3 and audio volume control R8. The output of the U2, the LM386 audio amplifier is coupled through capacitor C6 to an 8-ohm speaker.

A mini switched ⅛-inch audio jack disconnects the speaker if a headphone is plugged into the jack at J2. The mag-ear amplifier circuit is powered by a 9-volt transistor radio battery through the power switch at S2. The mag-ear amplifier circuit can be constructed on a perf-board or a circuit board, whatever your preference may be. Locate two IC sockets; they will aid in trouble shooting and replacement if the circuit has problems at a later date. There are a few electrolytic capacitors in the mag-ear circuit, which must

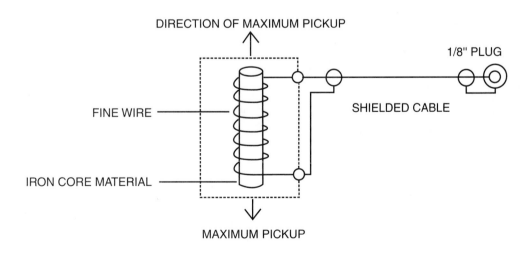

Figure 7-4   *Induction pickup coil diagram*

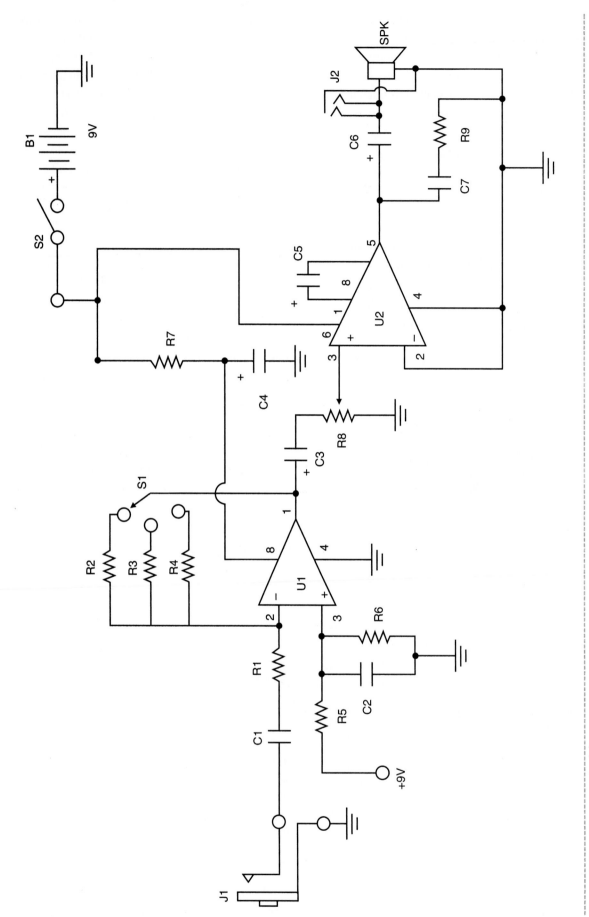

Figure 7-5   *Mag-ear amplifier*

be installed correctly. Therefore it is important to observe the proper polarity. Also be sure that the ICs are correctly installed to avoid damaging the circuit when first powering up. Most ICs have either a cutout notch or indented circle, which generally indicate the top of the IC, with pin 1 just left of the cutout notch or indented circle. After constructing the mag-ear circuit, be sure to check and recheck the circuit against the schematic for possible mistakes in wiring. Also look for any cold solder joints or solder bridges before applying power to the mag-ear circuit.

Once the mag-ear circuit has been built, you will need to locate a metal chassis box in which to install the high-gain amplifier circuit. Drill holes on the front of the chassis for the input and earphone jacks, as well as for the speaker. Both the on-off switch and the gain selector switch S1 can also be mounted on the front of the chassis box. Next install a battery holder for the 9-volt battery, and then mount the circuit or perf-board inside the chassis box on plastic standoffs, using ½-inch long 4-40 machine screws.

Now that the circuit has been built, you can test it to make sure it is working correctly. Plug in a pickup coil assembly to the input jack at J1, and apply power to the circuit via switch S2. You should then bring the sensing coil near a TV set that has been powered on. If the mag-ear circuit is working correctly, you should hear a humming sound as the pickup coil is brought near the TV set. If you hear a humming sound, the circuit is functioning correctly and you can begin using the circuit for your observation and experiments.

Mag-Ear Amplifier Parts List

Required Parts

R1   1K ohm, 1/4-watt resistor

R2   10K ohm, 1/4-watt resistor

R3   100,000-ohm, 1/4-watt resistor

R4   1,000,000-ohm, 1/4-watt resistor

R5, R6   47K ohm, 1/4-watt resistor

R7, R9   10-ohm, 1/4-watt resistor

R8   10K ohm potentiometer

C1, C2, C7   0.1 uF ceramic capacitor

C3, C5   10 uF, 50-volt electrolytic capacitor

C4   100 uF, 50-volt electrolytic capacitor

C6   220 uF, 50-volt electrolytic capacitor

U1   TL082 dual op-amp (one section used)

U2   LM386 audio amplifier

SPK   8-ohm mini speaker

S1   three-position rotary gain switch

S2   SPST on-off switch

B1   9-volt transistor radio battery

J1, J2   ⅛-inch switched mini jack

Miscellaneous   PC board, wire, IC socket, chassis, standoffs, screws, nuts, etc.

Amplifying Telephone Conversations

Now, with your mag-ear amplifier constructed and ready to go, you can experiment with the induction pickup coil. The induction pickup can be used to amplify the sound received over a telephone. It can also be used to provide remote sensing of a telephone ring. In order to amplify voice or telephone conversation, you will need to discover the two locations on a telephone where the induction pickup can be placed to receive signals that will be amplified by the mag-ear. One of these locations is the induction coil, usually in the base of the telephone, and the other location is the front or back side of the earphone on the telephone handset. The induction coil in the base of the telephone may be at different locations depending on the type of telephone you have. The signal picked up from the earpiece will usually be the stronger of the two, and a suction cup can be used to attach the pickup probe to the telephone hand piece. The back of the handset is best as it will be out of the way during normal use of the phone. Adjust the volume control of the mag-ear to a place where there is plenty of listening volume without audio feedback. You can also use the magnetic pickup as a remote telephone ring indicator by placing the induction coil near the telephone's internal ringer coil. On occasion, you might want to be able to hear the telephone ring at a location distant from the phone itself, for example if you are working in the garden. The mag-ear can then be used to amplify the sound of the ringer remotely.

Listening to Magnetized Wire

Some interesting sounds can be heard from magnetized wire. You can use the wire strings of an electric guitar or an ordinary thick wire. The electric guitar, just like an acoustic guitar, uses stretched steel wires that vibrate when plucked. In the electric guitar, a magnet and a coil are associated with each other similar to the induction pickup probe. When the wire vibrates, it causes a changing magnetic field in the induction pickup coil so that a changing voltage is produced, which is then amplified by the guitar amplifier.

The induction probe of the mag-ear can be placed at any position along the length of the guitar wire. However, no amplified sound will be heard in the mag-ear unless a small magnet is placed under the wire. The movement of wire disturbs the field induced by the magnet into the probe. If you put the magnet under the wire anywhere along its length, the wire will become magnetized. You can locate that place by moving the probe along the wire until you find the spot of maximum sound from the plucked wire. In the late 1930s, the wire tape recorder was the only means of quickly recording and playing back sound. The popular magnetic tape recorder was not to come along until the mid-1950s. In the 1960s, the cassette tape recorder was invented, and more recently in the 1980s, the VCR was invented and became extremely popular. Now almost every family household has at least one VCR.

Making a Wire Sound Like Big Ben

A thick wire can be made to sound like Big Ben when it is amplified by the mag-ear. A stiff wire 6 to 8 inches long, clamped at one end, will produce a sound like Big Ben when it is struck with an object such as a wooden pencil, if the inductance pickup is placed near the vibrating end. As with the vibrating guitar wire, a magnet should be placed below the wire to generate a magnetic field to be picked up by the inductance probe. Eventually the wire will become magnetized by the magnet, and the magnet will no longer be needed. The bong sound that is produced by the vibrating wire and amplified by the mag-ear is extremely realistic. Different sounds can be produced by different lengths of clamped wire, as is done in many of the grandfather clock gongs. Those bongs, however, are acoustically produced, whereas the technique we are using is electronic in nature.

Detecting Magnetic Field Leakage

The mag-ear used with the induction pickup coil is a very handy means for the detection of wanted or unwanted magnetic fields. One example of magnetic field leakage is electromagnetic interference.

Locating Electromagnetic Interference (EMI)

You can trace or probe for magnetic fields surrounding electronic equipment. These fields might be generated within the equipment or externally to the equipment and yet affect its operation. Much of the data regarding the shielding effectiveness of various alloys (including seam quality) is measured in the frequency range of 50 Hz to 20 KHz. Because the mag-ear operates over most of this frequency range, it can be used to observe relative shielding effectiveness of various mu-metal configurations (which confine magnetic fields) and to observe magnetic fields radiated from electronic or electrical devices.

Probing Magnetic Fields

If you place the mag-ear near a piece of electronic equipment, its induction pickup will act as an antenna. With an earphone added, acoustic noise can be shut out while you are listening to the magnetic fields. The mag-ear will pick up magnetic fields generated and radiated by electronic wristwatches, electronic clocks, motors, pocket calculators, TV receivers, teletype equipment, computers and their peripheral equipment, power lines and electrical conduit, and switching current as small as the clicks caused by turning a flashlight on and off.

During the integration of various units into a system, the electrical influence of one unit on the operation of the others can be determined with the aid of the mag-ear and the induction probe. Because the mag-ear is battery operated and completely independent of an AC power supply, the introduction of power generated by or conducted through magnetic field paths is avoided. Fields can be introduced into the unit only through the antenna input probe. This makes the mag-ear extremely versatile for probing magnetic fields to determine their strength and direction and their effect on other equipment. It can also indicate how magnetic shielding reduces these quantities.

Listen to TV Receiver Radiation

The mag-ear can easily detect the strong magnetic fields radiated by a TV receiver. Very high currents are used in the deflection circuits of a TV set in order to cause the electron beam of the picture tube to swing properly from left to right, retrace, drop down a line, and swing from left to right, until a complete frame of a picture is presented to the human eye. A complete picture is presented to the eye 30 times a second.

Locating Electrical Conduit

When you are working with electrical wiring and conduit, it is necessary to know the location of existing line, connections, and switches. Even when you are hanging window curtains, Venetian blinds, or suspended potted plants, you should know the location of existing wires or conduit so you don't inadvertently cut, drill, or nail into them. This is most important if the electrical lines are energized at the time.

You can use the inductance pickup coil as a magnetic probe. Thus the mag-ear can be used to locate the exact positions of wires by listening to the 60 Hz hum radiated by the lines. You will be able to locate wires to within a quarter of an inch, and you can even tell through a wall cover which side of the switch a wire is connected to. Simply move the probe around until the 60 Hz hum is a maximum. In most cases, it will not even be necessary to have the wall switch on, as a great amount of hum voltage is induced in all the electrical wiring in buildings.

This strong hum will aid you in locating house wiring and also in locating conduit buried in concrete walls, sidewalks, or patios. By knowing on which side of a wall stud the electrical wiring is running, you can drill or drive nails without fear of making accidental contact with a hot wire.

The Barkhausen Effect

The Barkhausen effect is a means of observing magnetic domains. A *magnetic domain* is the region in which the magnetic fields of atoms are grouped together and aligned. Favored domains are those domains with a high propensity for alignment or magnetic domains. Barkhausen discovered that a change in current in an electrical coil encircling a ferromagnetic sample produced a noise in the sample that could be heard when amplified sufficiently and fed to a speaker. This noise could be heard even though the magnetizing force applied to the sample was changed smoothly. From these experiments, Barkhausen concluded that the magnetism in the sample does not increase in a strictly continuous manner, but instead by small, abrupt, discontinuous steps or increments. These steps are now called *Barkhausen jumps*. They are caused by the discontinuous movement of mobile magnetic boundaries between magnetic domains.

Research

Barkhausen reported on his work with magnetic domain movement in the technical literature in 1919. Since that time, limited research has been conducted as to the practicality and usefulness of Barkhausen noise jumps. More recently Southwest Research Institute of San Antonio, Texas, has been doing pioneering work in the area of developing an easy-to-use completely nondestructive method for measuring residual stress. The method, which is a proprietary development, uses the Barkhausen noise analysis method of residual stress measurement. It is applicable to most ferromagnetic materials. Briefly, with this method, a controlled magnetic field is applied to the region of the part being examined. A small probe is used to sense the Barkhausen noise caused by abrupt movement of magnetic domain walls. Analysis of the Barkhausen noise during controlled experiments has established that high-amplitude signatures (patterns) are associated with tensile stresses, low-amplitude signatures are associated with compression stresses, and intermediate signatures are obtained from unstressed regions.

Applications of Barkhausen Noise

Many industrial and private uses of Barkhausen noise detection are possible. Some of the commercial uses of Barkhausen noise analysis include listening to turbine engine blades, disks, and compressors. Various ball bearings can be checked in this manner, and their future failure predicted. Checking for metal fatigue on a submarine is another possible application, with detector probes being placed about the submarine

hull. In another scenario, metal beams can be attached to dam structures to sense the generation of Barkhausen noise and to warn of any structure shifting. Many other applications will also come to mind.

The amount of signal generated is sufficient so that a sensor pickup probe can be placed 2 to 3 inches from the metal sample and still pick up enough signal for sampling. The pickup can be noncontacting because it is picking up a radiated magnetic field.

Listening to Barkhausen Noise

Barkhausen noise is as a result of the sudden irregular motion of the domain boundaries as the favored domains grow at the expense of their neighbors. The Barkhausen noise we can hear is a hiss or white noise. Using the Barkhausen effect, we can "hear" the magnetic domains shift under the influence of a changing magnetic field. Barkhausen noise can be heard accompanying the motion of a single domain boundary.

Locating Buried Ferrous Metals

You can use the Barkhausen noise technique to look for buried ferrous metals such as nails or screws in wood or concrete. Figure 7-6 shows how a magnet held close to the inductive pickup coil induces a magnetic field in a nail or screw below surface level. The nail or screw then radiates Barkhausen noise, which the pickup coil senses. The output is amplified by the mag-ear, and a nail can be closely pinpointed with this technique.

Two-Inch Diameter Pickup Coil and Applications

You can utilize the two-inch magnetic pickup coil and the mag-ear to detect and locate buried waste pipes, locate wiring and rods in walls and concrete, and even listen to moving subway trains. You easily can wind your own 2-inch diameter pickup sensing coil by winding 50 to 100 turns of a thin, enamel-covered wire around a 2- to 3-inch-diameter cardboard tube, as shown in Figure 7-7. Use *American wire gauge* (AWG) No. 28 to 32, enamel-covered wire and wind the turns on top of each other so the loop is about ½ inch thick. Figure 7-6 illustrates the coil connections

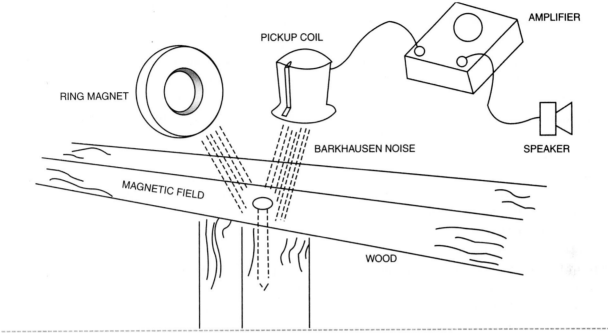

Figure 7-6   *Locating ferrous metal objects*

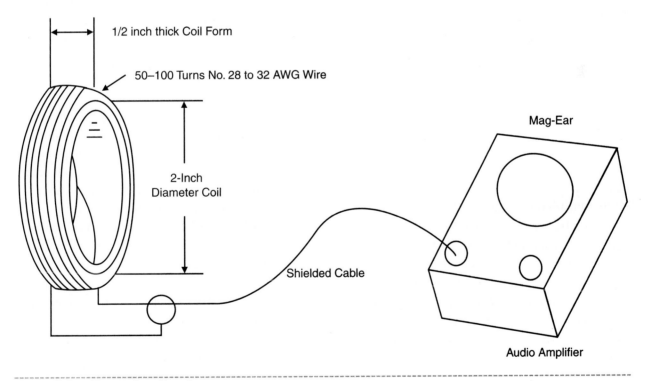

Figure 7-7   *Two-inch coil probe*

to the mag-ear audio amplifier. Take the output of the 2-inch coil and connect it to the mag-ear circuit and you will be able to explore a whole new avenue of magnetic fields.

Locating Buried Water Pipes

Around any city or town, a maze of electrical wires spreads across the countryside. This maze of wires includes the giant transmission lines going cross-country and carrying a voltage of 138 KV, the 4,300-volt line going down your alley, and the 240- or 120-volt line coming into your house. Your city or town also has a maze of water mains serving whole sections of the town, and l-inch pipes coming into individual yards and serving individual homes or apartments. Now an interesting thing happens; the power maze of wires introduces, or induces, a 60 Hz component of the power-line energy into the water distribution system serving your house. If you want to listen in or to find the water pipes, the signal you will be looking for is a strong component of the 60 Hz power line. The strongest signal you will hear is a 120 Hz hum, but harmonic energy will extend up to at least 1,200 Hz.

Locating Residential Water Pipes

In order to locate buried copper or galvanized-iron water pipes in your front or back yard, start at some location where you can see a water faucet or water meter. If your question is where to dig or not to dig, you must learn where the pipes run. When many houses are built, the water cutoff valve is promptly covered up by soil and grass. Trying to locate a buried water cutoff valve during an emergency is hectic. This technique will help you locate the line to your house.

If the water mains run down your street, you will want to start looking in the front yard. Walk along the front of your property line near the house or build-

ing, with the 2-inch loop dangling 2 to 6 inches above the ground or grass. The plane of the loop should point in the direction you are walking. You will eventually find an area where the 120 Hz hum is rather strong. At this point, turn in the direction of the street and swing the pickup back and forth, seeking the maximum hum as you walk slowly toward the street. As you do, you will soon run across the water meter. The cutoff is usually located a few feet from the meter. You also can check with neighbors to see how far their cutoffs are from the meter. Because you have determined the direction the pipe takes, you should now be able to locate the cutoff valve.

Plastic water pipes are not at all cooperative, or conductive. Consequently you will not be able to sense any yard installation with a plastic pipe, even if the meter and cutoff valve are metal.

Locating the Water Main

Once you have learned to locate your small 1- to 2-inch residential water pipe, you can go on to larger things and look for the 5- to 10-inch water main that runs down your street. The sound you will listen for is the same as the one you listened for with the small pipe. But one thing may surprise you; you will be able to locate a water main from over 100 feet away. All it takes is skill and understanding what the loop antenna is telling you.

There are two ways to locate a water main using the 2-inch loop probe: (1) by identifying increasing hum intensity and (2) by changing the loop plane angle. The increasing hum intensity is the same technique you used to find the small water pipe. After you have found your cutoff valve and meter, proceed with caution into the street. Swing the plane of the loop back and forth across the path you are taking, keeping the hum signal the loudest you can as you proceed. If you lose the hum signal, back up a few steps and look for the signal in the other three directions a pipe could go. Soon you will find a strong hum at a right angle to your path, and this is the water main. At this time, look above you to see if you are not following some overhead power lines. You will find, surprisingly, that the loudest hum signal comes from the buried water main and not the usual neighborhood power line.

The changing-loop-plane-angle method of finding the water main allows you to recognize the signal when you have tilted the loop so you are receiving a *minimum* hum signal. Minimum hum at any particular location is received when the broadside of the loop is at a right angle to the line from the loop center to the water main. As you move closer to the main, the tilt angle becomes less until you are finally over the pipe. When you pass the pipe, the tilt will be in the other direction.

The tilt of the loop to obtain a minimum signal from the water main changed as you moved closer to the pipe. You observed also that the line pointed to the pipe and not the ground. Consequently, when you are over the pipe, the line will point directly straight down. If the pipe is buried 5 feet deep, and you move back 10 feet from directly over the pipe, the loop will make a 45-degree angle with the ground. If you move 10 feet to the other side of the center line, the loop will again make an angle of 45 degrees with the ground.

Locating Wiring and Rods in Walls and Concrete

It is dangerous to drill, nail, or screw into a wall without knowing where the electrical wiring or conduit runs. With the 2-inch loop as a probe, you will be able to trace the direction the wiring takes by looking for a maximum hum in the mag-ear. An electrical circuit that is turned on while you are searching for wiring will help make the signal stronger. In addition, if you have a lamp dimmer you can turn on, set it at minimum brightness as this causes the lamp dimmer to emit the most interference, or *hash*. The signal will then be conducted throughout most of the house or apartment and this will help you in searching out all possible wiring directions.

When you probe with the 2-inch loop and swing it back and forth as you did for the buried water pipe,

you will be able to trace wiring that goes underground and passes under sidewalks. You should be able to place its location to within $1/2$ inch in concrete that is 6 inches thick. With practice, you will become adept at locating buried wiring and electrical conduit. While you may not be able to locate plastic or PVC water pipe because it will not conduct electrical current, you will be able to hear hum conducted by buried reinforcement rods, especially in sidewalks or driveways. Using this technique is almost like taking an X-ray picture of the sidewalk or wall, locating electrical wiring and reinforcement rods.

Listening to Subway Trains

An electric subway train deep under ground generates a cloud of electrical noise as it travels swiftly along on its appointed rounds. You can hear this electrical noise with the 2-inch loop antenna and the mag-ear. The noise of a subway train is unlike any other sound and has the appearance of a cracking noise, as if many wires carrying current were being scraped together. And, indeed, this is how power is supplied to the train as each car has its own power takeoff from the third rail. As the train travels along, crossing different power supply grids, the front of the train may not have power but the back part will, until all cars have moved to the new power supply grid. As a result of all this power switching, much electrical noise is generated by the train that can be received or detected many hundreds of feet away on the surface of the ground.

ELF Monitor

There is a growing concern over possible health hazards from low-frequency electromagnetic fields. Some new evidence suggests this really isn't the case. However, to some it appears that the *extremely low-frequency* (ELF) magnetic fields given off by many household appliances and computer monitors can be of sufficient strength to be potentially hazardous. If you are concerned by the possible radiation safety issues of ELF fields, you can construct a simple ELF monitor to check and modify your environment. The ELF monitor described in this chapter is simple to build and should cost less than $25.

You might be wondering if ELF radiation is such a health hazard, why has it taken so long for anyone to mention its dangers? To answer that question, we must look at how scientists first sized up the potential harm of low-frequency magnetic fields. It was originally believed that weak low-frequency fields could not have a significant impact on living systems. This belief was based on the amount of thermal energy the ELF fields could produce in biological tissue. The energy produced by ELF radiation is much smaller than the normal thermal energy internally generated by a cell's metabolic processes. In addition, the quantum energy of the fields is far too low to break any chemical or nuclear bonds in the tissue. Therefore, scientists felt DNA would be safe from mutating if exposed to ELF radiation. Additionally, the electric field of the body is much greater than the induced field from ELF radiation. Looking at all these factors, it's easy to understand why the scientific community quickly dismissed epidemiological studies that indicated a statistically significant hazard associated with ELF radiation.

As a result, the scientific community has of late been portrayed in the press as a bunch of hacks or bureaucratic puppets genuflecting for grants from government agencies and power companies. However, the reason for the quick dismissal was one of genuine disbelief, not one of a mass cover up. Although in truth, a few scientists have stepped over the line and maligned good researchers based upon the profit and loss statement of their employers, these scientists are few in number and the entire scientific community should not be condemned based upon their isolated unethical endeavors.

Concerns over the radiation from televisions and computer monitors (which are closely related in operation and technology) is nothing new. A number of years ago, concern arose as to whether radiation given off by color televisions would have a negative impact on health. This concern was based primarily on ionizing radiation (or low-level X-rays) whose intensity fell off dramatically a few inches away from the TV screen and turned out to be incidental. But more insidious than this overt threat is one that has passed unnoticed until quite recently: the low-frequency magnetic fields generated by the coils on the TV *cathode ray tube* (CRT).

Computer monitors, or *video display terminals* (VDTs), generate these low-frequency magnetic fields, which emanate in all directions. They present a greater hazard than TV's because a user must sit closer to them during use. In fact, numerous reports have been made from female computer operators of cluster miscarriages. The word *cluster* here refers to a greater-than-average incident of miscarriages among a group of women.

A recent study involves 1,583 pregnant women and cluster miscarriages. The study was performed by doctors Marilyn Goldhaber, Michael Polen, and Robert Hiat of the Kaiser Permanente Health Group in Oakland, California, and showed that female workers who used computers more than 20 hours a week had double the miscarriage rate as female workers who did similar work without computers. What the study didn't take into account was the incidence of malformations and cancers in the children born to women who use computers as compared to the children of women who don't use computers. Excessive ELF fields emitted by computer monitors is an industrywide problem, virtually all CRT computer monitors emit excessive ELF radiation unless specifically stated otherwise. In a recent ELF study of 10 popular monitors, all of the monitors tested emitted excessive ELF radiation at close range. The only recommendation that can be offered at this time is to increase the distance between you and the monitor. A working distance of 2 feet is recommended.

The ELF field propagates from all points around the monitor, not just from the front of the screen. That fact becomes important in offices where computer terminals are in close proximity to one another, because workers can be exposed not only from their own monitor but also from a coworker's monitor. It's important to realize that the ELF field given off will vary somewhat from monitor to monitor.

Shielding

It would be nice if we could purchase a radiation shield for our monitors, something similar to the antiglare shield on the market, but none exist. Be very careful, there are some antiglare screens on the market that make a claim of blocking the electric and magnetic fields given off from the monitor. First, electric fields, as far as I know, have not been reported to have any negative impact upon health. Second, the magnetic field the screens claim to block are in fact the visible-light-frequency fields generated by the CRT. Those high-frequency magnetic fields have not been shown to have a negative impact on health. These antiglare screens have no impact on the low-frequency (60 Hz) magnetic fields emitted that we have discussed here.

There is no easy way to shield the monitor to reduce the propagating ELF field. If you are concerned, one possibility is to use an alternative type computer monitor. *Liquid crystal display* (LCD) and plasma-display monitors do not emit ELF fields. However their drawbacks are higher cost and lower resolution.

Precautions Around the Home

Other sources of ELF exist around the typical home. An appliance in the home may generate a very strong ELF field, but if the appliance is used only a short time, its risk factor is probably low. Electric razors fall into this category. Line-operated (plugged into a wall socket rather than battery-powered) razors produce extremely strong ELF fields and are held in very close proximity to the body, but because they are used for only a short time, the total exposure or dose is small and they are probably safe. In contrast to the electric razor is the electric blanket, which emits a much lower ELF field strength, but for a much longer exposure time.

Dr. Nancy Wertheimer, who first published the epidemiological study showing a correlation between 60 Hz power lines and the increased incidence of childhood cancer in this country, has also performed similar research on users of electric blankets. She has found a higher incidence of miscarriages among pregnant women who use electric blankets as compared to pregnant women who did not. For users of electric blankets, the following recommendations can be made. One, switch to ordinary blankets. If you like electric blankets, use one to heat your bed before going to sleep, but unplug the blanket before you actually get into bed. It is not sufficient to just turn off the blanket because many blankets still produce an ELF field as long as they're plugged into a socket.

Fluorescent lights are much more efficient (more light per electrical watt) than ordinary incandescent bulbs. Because of that, they have become the standard lighting system used in most offices, industry, and now around the home. However fluorescent lights require a ballast transformer that generates an ELF field. If you're using a small fluorescent lamp as a desk light, you may want to consider switching to an incandescent lamp, which generates virtually no ELF. That also applies to the new energy-saver fluorescent lamps that replace standard incandescent bulbs. They are okay for overhead lighting, but you may want to reconsider using them for close-up work or desk lighting.

Television sets fall into the same category as computer monitors. Like our monitors, they produce a field that propagates from around the entire set. The ELF field will propagate through standard building materials such as wood and plaster. So if a TV set is placed against a wall, the ELF will propagate through the wall into the adjoining room. So it becomes important not to place a bed against a wall with a TV set on the other side. Again, an LCD monitor is a good choice if you're concerned with ELF radiation.

The small motors in AC-powered clocks produce ELF fields. If the clock is an alarm clock that lies close to the sleeper's head, it could be giving that person a significant dose of ELF radiation during the night. The recommendation would be to move the clock a significant distance away. Or purchase either a battery-powered clock or a digital clock that produces a negligible field.

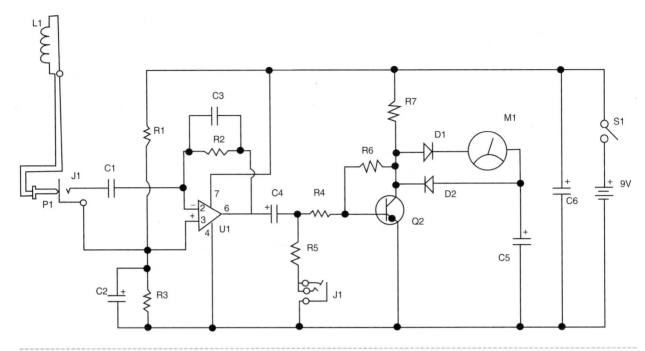

Figure 7-8 *Electromagnetic field detector circuit*

Hair dryers fall in the same category as electric shavers. Because using them means short-term, high field-strength exposure, they are probably safe for most people. Notable exceptions are people who use them in their occupation, such as hair stylists and hair dressers. Electric baseboard heaters are another potential problem appliance. A minimum of 4 feet of distance is recommended from such heaters.

It's impossible to state at this time exactly what a safe, long-term dose is because it hasn't been established. Effects have been reported at dose rates as low as 1.2 to 3 milligauss. And the controversy still rages as to the impact and extent of ELF fields on human health. Try to limit long-term exposure to one milligauss or less. Of course, it is difficult to know what your ELF exposure level is without a milligauss meter. Using an ELF meter around your home, apartment, or work space will enable you to identify potentially hazardous ELF fields and their sources so you can implement corrective action.

The diagram shown in Figure 7-8 illustrates a low-cost ELF monitor that you can build. This circuit is extremely sensitive and provides both an analog meter and provisions for an audio output, so you can listen to the magnetic field at the source. The circuit is based on the 1 mH choke acting as the sensor coil, and the op-amp at U1, an LF351, acting as a low-noise op-amp. The network of C3 and R2 form a gain path and filter between pins 2 and 6 of the op-amp. The output of U1 at pin 6 is fed to capacitor C4. An audio output is provided at J1. The output from capacitor C4 is also fed to transistor Q1, which drives the indicating meter M1. Diodes D1 and D2 provide protection to the meter. The ELF monitor is powered by a single 9-volt transistor radio battery, through switch S1.

The ELF monitor can detect any magnetic field with frequencies between 50 Hz and 100 kHz, and by using headphones you can estimate the frequency of the magnetic field. This low-cost ELF meter is a useful addition to any curious experimenter or science enthusiast.

The ELF monitor can be constructed on a perfboard or on a printed circuit board, as desired. Keep wiring between component leads as short as possible. Remember, when building the ELF monitor circuit to pay careful attention to the correct polarity of the diodes and capacitors when installing them on the circuit board, in order to avoid damaging the circuit upon power-up. IC sockets are highly recommended, to prevent major headaches if the circuit fails at a later date and you have to remove the IC. When installing the IC, be sure to observe the correct orientation. There are usually two different types of identifiers for integrated circuits: Often there is either a cutout on the top of the plastic IC case, or sometimes there is a round indented circle. Pin 1 will always be to the left of the plastic cutout or indented circle. When finished constructing the circuit board, carefully check it over for cold solder joints and any solder bridges that might have formed while you were building the circuit. Once you are satisfied with your inspection, you can move on to mounting the circuit board and hardware inside an enclosure.

The sensor input jack at J1, the headphone jack at J2, as well as power switch S1 can all be mounted on the top front of the chassis box. Once you have selected your meter, it too can be mounted on the front panel of the chassis box. Mounting the meter in the chassis box may prove to be the most challenging aspect of this project. Unless you have the exact size chassis punch, you will have to mark where the meter is to be mounted, and then drill a number of small holes around the inside circumference of the meter template, very close together. Then, you will need to cut out the metal between the small holes, and finally file the crude circle into a smooth circle so you can insert the meter. Next, mount a 9-volt battery holder on the bottom of the chassis, and secure a location for the circuit board within the chassis box. Once all the switches, meters, and jacks have been installed, you can mount the circuit board into a metal chassis box on plastic standoffs using ½-inch 4-40 machine screws.

Finally, you will need to make up the sensor-probe cable assembly. Locate a plastic film canister and drill a small hole in the nonopening end of the container. Place the sensor coil (L1) inside the plastic film canister; you can glue the coil or wrap the coil with foam and then insert the foam and coil into the film canister. Solder the ends of the coil sensor to the lead of a mini two-conductor shielded audio cable. At the opposite end of the 4- to 6-foot sensor probe cable, solder an ⅛-inch mini phone plug, which can be inserted into the input jack to the ELF circuit on the chassis box. After the circuit board and all the components have been mounted, you can test the ELF monitor to make sure it is working correctly.

Using the ELF Monitor

To test the unit, turn on a television. Starting from approximately 2 feet away, slowly move the sensor closer to the set. Once you are close enough to the TV, you will begin to see the monitor's meter needle begin to move across the meter face. As you walk around checking various appliances, you'll probably find that you can lower your ELF exposure. For instance, one of my computers has an external power supply that emitted strong radiation; by simply moving the power supply farther away from my work space to another location, the level of radiation decreased significantly. Other simple things, such as changing from a fluorescent desk lamp to an incandescent one, will further reduce your exposure. The ELF monitor can also detect a static magnetic field, when it is moved into or out of the field.

Electromagnetic Field Detector Parts List

Required Parts

R1, R3, R7 10K ohm, 1/4-watt resistor

R2 2.2 megohm, 1/4-watt resistor

R4 2.2K ohm, 1/4-watt resistor

R5 10-ohm, 1/4-watt resistor

R6 1-megohm, 1/4-watt resistor

C1 100 nF, 35-volt capacitor

C2 10 uF, 35-volt electrolytic capacitor

C3 150 pF, 35-volt capacitor

C4, C5 220 uF, 35-volt electrolytic capacitor

C6 100 uF, 35-volt

electrolytic capacitor

D1, D2 1N4148 silicon diode

L1 1 mH coil (see text)

U1 LF351 op-amp

Q1 2N3904 NPN transistor

M1 analog meter

S1 SPST on-off switch

J1, J2 1/8-inch mini switched phone jack

P1 1/8-inch mini phone plug

B1 9-volt transistor radio battery

Miscellaneous PC board, battery clip, wire, sensor cable, chassis, sockets, etc.

Electronic Compass

Most of us have used a common magnetic compass, which often consists of a light-weight balanced magnet suspended on a pivot. The magnet, free to rotate, is affected by Earth's magnetic field and assumes a position in which its north-seeking arrow points to Earth's magnetic North Pole. The geographical North Pole of Earth is of course offset from the magnetic North Pole by about 10 or 15 degrees in most areas of the United States.

Most low-cost compasses leave something to be desired and are usually affected by any tilt of the case or friction in the pivot. With the development of solid-state magnetic detecting devices such as Hall-effect sensors, it is possible to construct a low-cost, reliable magnetic compass that has no moving parts and eliminates the disadvantages of inexpensive mechanical types. Because this project contains no moving or mechanically sensitive parts, it is an extremely rugged device that can tolerate all potential stresses encountered when hiking or traveling through rough terrain. Taking a reading on the compass is quick, easy, and very reliable.

This solid-state compass uses a unique detection system that produces two sharply defined points centered on the direction of magnetic north, as indicated by an LED. That permits a quick, accurate reading. The project is housed in a plastic enclosure, is small and lightweight, and is powered by a common 9-volt battery. Because the compass circuit is energized only when it is used to make a reading, the battery's useful life approaches that of its shelf life.

Development of a magnetically sensitive solid-state compass is made possible through a phenomenon called the *Hall effect*, discovered in 1879 by Edwin Hall. He observed that a small voltage was developed at the edges of a current-carrying gold foil when the foil was exposed to a magnetic field. Solid-state technology now provides small, low-cost Hall-effect devices that are very sensitive and able to detect Earth's extremely weak magnetic field.

The Hall-effect sensor is a small sheet of semiconductor material in which a bias current flows. The Hall-effect output of the sensor takes the form of a voltage measured across the width of the conducting material and will be negligible in the absence of a magnetic field. If the biased Hall sensor is placed in a magnetic field with the flux at right angles to the flow of current, a voltage output directly proportional to the intensity of the magnetic field is produced. Additionally, the voltage will be a function of the angle between the lines of force and the plane of the sensor. Maximum Hall-effect output voltage occurs when the face of the sensor is at a right angle to the lines of force, and zero voltage is produced when the lines of force are parallel to the face of the sensor. The Hall-effect sensor is further enhanced by using integrated-circuit technology to add a stable high-quality DC amplifier to the device. It then provides a usable linear output voltage, which is sensitive enough to react to Earth's magnetic field (about $1/2$ gauss).

The diagram in Figure 7-9 illustrates the schematic of the Hall-effect electronic compass. The Hall-effect devices shown at U3 and U4 are three-terminal linear devices, which are driven by a regulated 5-volt supply provided by fixed-voltage regulator U1. The output of each of the sensors is a DC voltage that varies linearly from a quiescent value of 2.5 volts as their position with respect to the lines of force of the magnetic field changes. A typical sensor has an output voltage sensitivity of about 1.3 millivolts per gauss.

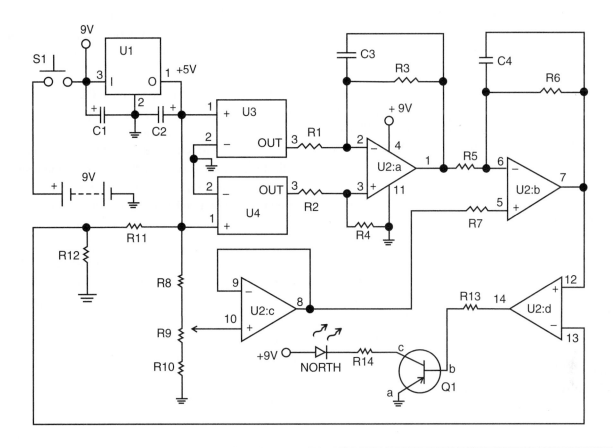

Figure 7-9 *Electronic compass circuit*

Two Hall-effect generators are used in the circuit to provide twice the sensitivity of a single sensor. The two devices are oriented in opposite directions so that the change in output voltage of one sensor will be positive, whereas the change in the other will be negative as the compass is rotated. The voltage differential between the two output terminals of the sensors is a representation of the magnetic field intensity and direction. The voltage differential produced by the Hall generators is fed to a differential amplifier, U2:a. As a result, the output of U2:a (pin 1) will be a minimum (null) when the compass is facing the magnetic North Pole and a maximum when it faces the South Pole.

The change in output voltage of U2:a is too small to allow a simple method of determining the null voltage as the compass is rotated. Therefore, U2:b is used as an inverting amplifier with a gain of 100 to further increase the change in voltage. A DC offset, provided by sensitivity-adjuster potentiometer R9 and voltage follower U2:c, permits the DC output

voltage of U2-b to be set to a usable level to drive the next stage.

Op-amp U2:d is used as a voltage comparator, with a fixed reference of about 3.4 volts fed to its negative input. Thus, when the one output of U2:b fed into the positive input of the comparator exceeds the 3.4-volt reference level, the output of U2:d (pin 14) goes high, applying forward bias to Q1. That in turn illuminates LED1 to indicate that a voltage exceeding the reference exists at U2:b (pin 7). The use of a voltage comparator to detect the change in output voltage of U2:b (pin 7) produces two sharply defined points and allows a more accurate determination of the magnetic North Pole.

The indicator LED will be illuminated over a small arc as the compass is rotated full circle and will remain off over the rest of the 360-degree span. The sensitivity control (R9) allows adjustment of the width of the arc. Once the two LED switching points are determined, true magnetic north is then the position at the center of the arc. Power is provided by a common 9-volt battery. The circuit draws about 25 mA, and, because the compass usually is powered for only a few seconds at a time, battery life is extremely long; several hours of continuous compass operation is also possible. Circuit stability with a falling battery voltage is ensured by the 5-volt regulator, U5. When the battery is exhausted and cannot deliver sufficient current to operate the circuit, the LED will appear dim or will not illuminate at all.

Construction

The compass circuit can be built on the printed circuit board, as the circuit is very compact. The prototype is housed in a 2½-inch square by 1-inch high plastic enclosure that has sufficient room to accommodate both the board and the 9-volt battery. A metal enclosure must not be used for this project, as it can attenuate or distort Earth's weak magnetic field. The power switch and sensitivity control are mounted on the side of the enclosure to allow easy operation of the compass.

The operation of the electronic compass depends upon the Hall-effect sensors being placed in opposite directions and exactly parallel, as indicated in Figure 7-10. Note that the orientation of the sensors is determined by the marked face of the device, with pin 1 being on the left side when looking at the markings. The sensors must be positioned so that they are aligned squarely with the rectangular shape of the printed circuit board. That way the compass direction will be accurate when the project is assembled into the enclosure.

Many of the resistors specified in the parts list are metal-film types. The use of such components ensures maximum stability of the circuit in varying ambient temperatures and reduces the need to periodically adjust the sensitivity control. Ordinary carbon resistors are not temperature stable and should not be used in place of metal-film types. Also, it's a good idea to use a socket for U2.

It is recommended that you use a miniature momentary pushbutton switch for Sl. This will ensure that battery power will never be inadvertently left on when the device is not in use. The sensitivity control, R9, may be placed on the side of the enclosure to allow circuit adjustment when necessary. You should use a battery clip for B1. Be very careful to wire the battery clip with the correct polarity.

When the circuit board is completed, examine it very carefully for shorts and cold solder joints. It is much easier to correct problems at this stage rather than later on if you discover that your project does not operate.

Use a photocopy of the artwork in Figure 7-11 for the top of the compass. You can simply glue it in place. Indicator LED1 is placed at the north indication of the compass by drilling a suitable sized hole in the plastic top where the letter *N* would be. Be very careful when drilling; some plastics will shatter with

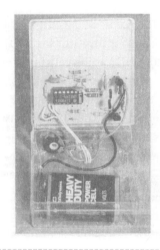

Figure 7-10   *Placement of Hall-effect sensors*

Figure 7-11   *Compass dial master*

excessive stress. Be sure to properly orient the top of the enclosure in accordance with the final position of the PC board.

Checkout Procedure

When you are satisfied that all the wiring is correct, the checkout procedure must be performed. Be sure to use a fresh 9-volt battery. Checkout requires a DC voltmeter connected to ground and the output terminal of U1. Apply power to the circuit, check for +4.75 to +5.25 volts. Measure the resistance between the 5-volt bus and ground: A normal reading is about 600 ohms. Measure the terminal voltage of the battery to be sure that it is delivering at least 7 volts under load to U1. Replace a weak battery if necessary.

Next, measure the output voltage of U2 pin 1, and verify the voltage range of potentiometer R9. (Compass orientation is not important at this time.) The voltage should be about 2 to 3 volts DC. Measure and record the DC voltage that you observe at U2:a pin 1.

Measure the voltage change at U2:c pin 8 as the sensitivity control is rotated over its entire range. The difference between the highest and lowest readings should be about 0.45 volts. Ideally the center of the measured voltage range should be close to the voltage recorded earlier at U2 pin 1.

If necessary, change the values of R8 and R10 so that the voltage range obtained at U2:c pin 8 is somewhat centered about the voltage reading at U2:a pin 1. This ensures proper adjustment range of the sensitivity control for the particular pair of Hall generators that are used in your compass project. Once the sensitivity range is correct, rotate R9 over its range while observing the LED. At one end of the setting, the LED should be extinguished, and at the other end it should be illuminated. If not, check the polarity of LED1 and the orientation of Q1. Check pin 14 of U2:d to be certain it swings from about zero to battery voltage as R9 is rotated over its range. Check pin 13 of U2:d for a voltage of about 3.4 volts as set by R11 and R12. Problems in this area may warrant replacing U2. If everything else checks out alright, check the soldering before changing the IC.

When the LED operates as described, the project is ready to be tested under actual operating conditions. Before you start, make sure that there are no magnetic fields nearby and the project is not shielded by a large mass of iron or steel. While holding the unit horizontally in any direction, apply power and carefully adjust R9 so that the LED is at the switchover point between on and off; while doing this, allow at least 10 seconds for the circuit to stabilize. Flickering of the LED is normal as the circuit switches back and forth. Once R9 is set, rotate the compass over a 360-degree arc or full circle, and note that the LED will be on over part of the arc and off over the rest. If necessary, readjust potentiometer R9 very slightly to obtain this result. The optimum setting for R9 will be at the point where the arc of illumination is as small as possible.

As the compass is rotated over the illuminated arc, note the two on/off points. When the compass is positioned halfway between those points, it is facing the magnetic North Pole, and the scale indications on its face indicate all other direction.

Using the Compass

Always be sure that the battery is reasonably fresh, and take along an extra one before starting out on an excursion with the compass. Note that a weak battery will be indicated by a dim or totally unlit LED. Avoid taking a compass reading in any area where there may be a magnetic field from a nearby device or where the earth's magnetic field is shielded by a large mass of metal.

Hold the compass in a horizontal position and rotate it full circle while observing the LED. Adjustment of the sensitivity control is indicated if the LED is totally on or totally off as the compass is rotated. Always allow at least 10 seconds operating time for the circuit to stabilize. Once the sensitivity control is adjusted, it should not require readjustment unless the project is subjected to an extreme change in temperature.

Don't forget that the electronic compass circuit can be used for things other than a simple direction finding. It provides an electronic means of finding north, so it should be easy to interface the compass to other devices that may need to know where north is, such as a robot, for example.

Electronic Compass Parts List

Required Parts

R1, R2   4,750-ohm, 1%, 1/4-watt resistor (metal film)

R3, R4, R12   100,000-ohm, 1%, 1/4-watt resistor (metal film)

R5, R7, R11   47,500-ohm, 1%, 1/4-watt resistor (metal film)

R6   475,000-ohm, 1%, 1/4-watt resistor (metal film)

R8, R10   249,000-ohm, 1%, 1/4-watt resistor (metal film)

R9   50,000-ohm potentiometer

R13   47,000-ohm, 1/4-watt resistor

R14   560-ohm, 1/4-watt resistor

C1, C2, C3   0.1 uF, 50-volt ceramic capacitor

C4   0.01 uF, 50-volt ceramic capacitor

U1   AN7805 5-volt regulator IC

U2, U3   UGN3503U Hall-effect sensor (Allegro)

U4   LM324 quad op-amp

LED1   red LED

Q1   2N3904 NPN transistor

B1   9-volt transistor radio battery

S1   SPST switch

Miscellaneous   PC board, wire, IC sockets, battery clip, etc.

Sudden Ionospheric Disturbance Receiver

A *sudden ionospheric disturbance* (SID) is an abnormally high plasma density in the ionosphere caused by an occasional sudden solar flare, which often interrupts or interferes with telecommunications systems. The SID results in a sudden increase in radio-wave absorption that is most severe in the upper *medium-frequency* (MF) and lower *high-frequency* (HF) ranges.

When a solar flare occurs on the sun, a blast of ultraviolet and X-ray radiation hits the day side of the earth after 8 minutes. This high-energy radiation is absorbed by atmospheric particles raising them to excited states and knocking electrons free in the process of photoionization. The low-altitude ionospheric layers (D region and E region) immediately increase in density over the entire day side of the earth. Radio evil geniuses may already know that the D ionosphere layer is the lowest layer and is absorbed during the day by sunlight. The E layer is the next layer up, followed by the F layer, which is used for shortwave propagation.

Earth's ionosphere reacts to the intense X-ray and ultraviolet radiation released during a solar flare and produces shortwave fadeout on the day side of the earth as a result.

Shortwave radio waves (in the HF range) are absorbed by the increased particles in the low-altitude ionosphere causing a complete blackout of radio communications. This is where we get the term *shortwave fadeout*. These fadeouts last for a few minutes to a few hours and are most severe in the equatorial regions where the sun is most directly overhead.

The ionospheric disturbance enhances *long-wave* (VLF) radio propagation. SIDs are observed and recorded by monitoring the signal strength of a distant VLF transmitter.

You can investigate the phenomena of SIDs by building a special receiver and a low-cost data-logger setup. You can not only observe when solar flares are occurring, you also can collect and analyze the data and display it on your computer. The SIDs receiver is a great opportunity to observe first hand when a solar event occurs, and you can use it to predict when radio blackouts will effect radio propagation, which is extremely useful for amateur radio operators.

VLF Signal Propagation

Why do VLF signals strengthen at night instead of getting weaker? If propagation is basically via the waveguide effect, why doesn't the signal drop down at night when the waveguide disappears with the D layer? Is there some kind of reduced absorption at night? If so, where is it taking place and why? Also, what accounts for the big fluctuations in signal strength at night, apparently more or less at random?

The strength of the received signal depends on the effective reflection coefficient of the region from which the radio wave reflects in its multihop path between Earth and the ionosphere. In daytime the reflecting region is lower, the air density is higher, and the free electron density is controlled strongly by the solar radiation. At nighttime the reflecting region is higher, the air density is lower, and the free electron density is controlled by variable ambient conditions as well as variable influences from electron precipitation from above. At noon the electron density is about 10 electrons/cm^3 at an altitude of 40 kilometers, 100 electrons/cm^3 at an altitude of 60 kilometers, 1,000 electrons/cm^3 at an altitude of 80 kilometers, and 10,000 electrons/cm^3 at an altitude of 85 kilometers.

At night these figures become 10 electrons electrons/cm^3 at 85 kilometers, 100 electrons/cm^3 at 88 kilometers, 1,000 electrons/cm^3 at 95 kilometers, and then it remain somewhat the same up to at least 140 kilometers. At night the electron density in the lower part of the D region pretty much disappears. At 40 kilometers, the electron collision frequency is about one billion collisions per second, whereas at 80 kilometers, the collision frequency drops to one million collisions per second. The reflection coefficient depends on (among other things) the number density of free electrons, the collision frequency, and the frequency of the radio signal. It is found by a mathematical integration throughout the entire D region, and of course the result depends on what time of the 24-hour day one performs the integration. A paper titled *VLF Signal Propagation: A Discussion by SID Observers* by the American Association of Variable Star Observers (www.aavso.org) mentions reflection coefficients of the order of 0.6 at night and on the order of 0.4 at noon. The paper calculates what the sunrise VLF signal strength signature should look like for certain transmission paths and compares them with actual signatures. The results were very good.

We can think of the E layer propagating the signal at night. Then the prominent sunrise pattern we see is a shift from E layer propagation back to D layer as the sun rises and forms the daytime D layer. The sunset pattern is the reverse. An interesting feature of waveguide mode propagation is that the signal is split into two components, which can form an interference pattern.

You can build your own SIDs receiver and begin your own investigation of solar flares and their effects on radio propagation. The SIDs receiver is a simple VLF receiver designed to be used with a loop antenna, which can be placed either inside or outside. The receiver monitors the strong VLF signal from the U.S. Navy's station NAA 24 KHz transmitter in Cutler, Maine.

The diagram shown in Figure 7-12 illustrates a SIDs receiver consisting of two integrated circuits, a Texas Instruments RM4136 and a National LF353 operation amplifier. The SIDs receiver begins with the loop antenna, which is connected to a miniature 600-to-600 ohm matching transformer at T1. The output from the secondary coil of T1 is fed to two protection diodes at D1 and D2 at the front end of the receiver. The output of the diode network is then coupled to a 100 pF capacitor at C1. Capacitor C2 is connected from the output of C1 to ground at the input of the first op-amp stage U2:b. IC sections U2:a and U2:b form a amplifier/filter and tuning. The *tuning* is controlled via R5, which is placed in a shielded enclosure to prevent circuit oscillation. The potentiometer is kept separate in a small shielded box

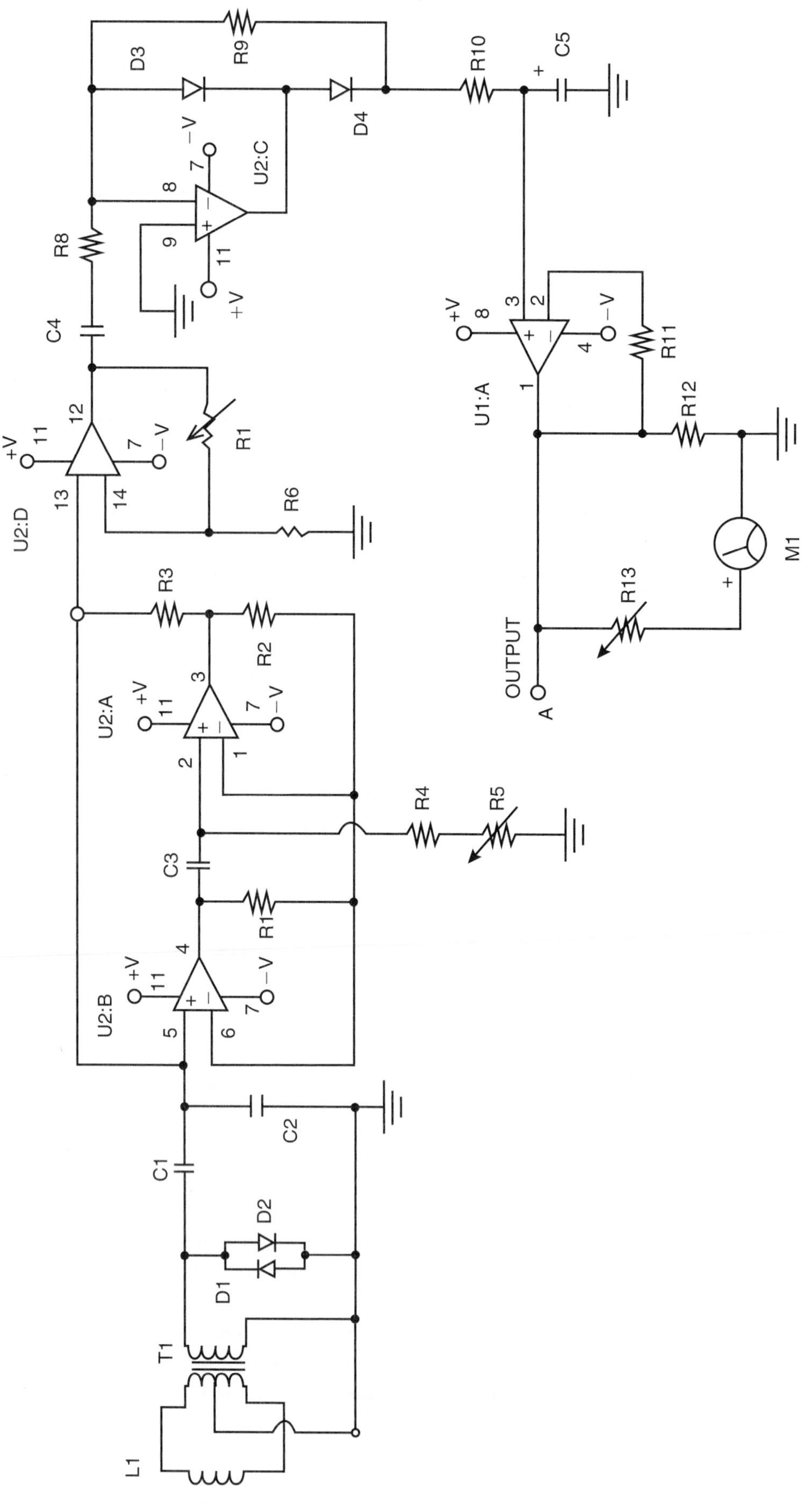

Figure 7-12   *SIDs receiver*

formed by some scrap pieces of thin circuit board soldered together. The output from U2:a is next fed to IC:d and U2:c, which together form an amplifier/integrator section. A final buffer amplifier section at U1 is used to drive the 0 to 1 mA meter at the output. The output at (A) can be used to feed a low-cost data logger.

The SIDs receiver is best built on a printed circuit board, although other RF building techniques could be used. The receiver is relatively simple to build, but the tuning potentiometer R5 should be shielded. This can be done with the use of some scrap circuit board material soldered together to form a small box. You will have to use a soldering gun or a higher temperature soldering iron to solder large areas of circuit board material. The SIDs receiver contains silicon diodes that must be installed with respect to polarity markings on the diodes. The white or black band on a diode's markings usually denote the cathode of the diode. Electrolytic capacitors also must be installed with respect to polarity marking. Capacitors are generally marked with a plus or minus marking. Be sure to take the extra time to install these correctly. The SIDs receiver utilizes two ICs, which must be installed correctly to avoid damaging the circuit. ICs are usually marked in one of two ways. Often a plastic cutout can be seen at the top of the IC, with pin 1

of the IC to the left of the cutout. Some ICs have an indented circle near pin 1 of the IC. The use of IC sockets is highly recommended, in order to avoid circuit board meltdown if a component fails at a later date for some reason.

The SIDs receiver utilizes ICs that require the use of a dual power supply providing both a plus and minus voltage to the circuit. Because the circuit is meant to be left on for long periods of time, the use of a dual-voltage wall-powered power supply is recommended. The diagram shown in Figure 7-13 illustrates a simple dual plus-and-minus voltage power supply that can be used to power the SIDs receiver. A 9-volt 500 mA center-tapped transformer is used to drive a bridge rectifier that provides both a plus and minus voltage, where the plus leg is sent to a 5-volt plus regulator (7805 regulator) and the minus leg of the bridge is sent to a minus voltage regulator (a 7905 minus 5-volt regulator). The power supply could be built on perf-board or on a printed circuit board if desired.

The SIDs receiver and power supply can be both installed in a metal chassis box enclosure. The prototype receiver is mounted in a sloping cabinet, as seen in Figure 7-14. The power supply is mounted along side the SIDs receiver. The tuning control as well as the power switch and meter are all mounted on the

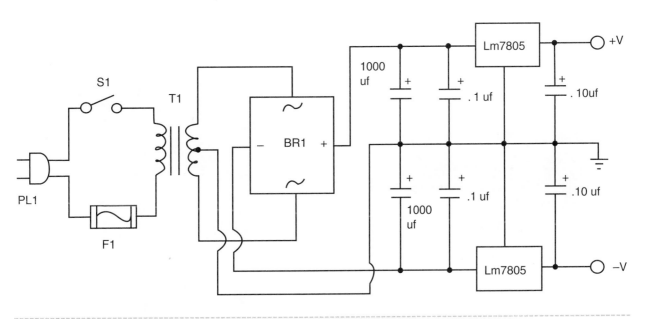

Figure 7-13   *SIDs dual power supply*

top front panel of the enclosure. The power cord is lead out of the rear panel through a strain relief. A power fuse is installed inside a chassis-mounted fuse holder on the rear of the panel. A dual-binding post (banana chassis jack) is mounted on the rear panel to connect the antenna to the SIDs receiver. The SIDs receiver requires a rather large loop antenna, which can be placed inside your attic or outside away from large metal structures or aluminum siding. The loop antenna is shown in Figure 7-15.

The loop is a diamond-shaped loop consisting of 50 turns of solid #24 enamel-coated or plastic-coated wire. The specific wire requirements are not critical. The loop has an enclosed area of about 9 square feet. The lead-in from the loop antenna should form a loose twisted pair. This balanced lead-in is routed as far away from any metal as possible to the 600-to-600 ohm matching transformer at T1. This transformer arrangement reduces 60 Hz hum interference from entering the SIDs receiver.

The output of the SIDs receiver can be coupled to a solid state data logger from the output of the receiver at point (A) at the receiver output terminals. In order to save and later view and correlate your recorded data, you will need to acquire some form of data logger. Three good options are available for saving your data. The first option is the ONSET Computer HOBO series of data loggers (see appendix for contact information). ONSET offers a number of different models from 8-bit to 12-bit models, all reasonably priced from around $60.00. The data loggers are powered by a small button battery, which will last for a long time. ONSET also offers low-cost software for

uploading and downloading information to the data logger. Starting times and dates as well as voltage parameters and timing between samples can be preset. These are a great alternative. Another data logger option is the new introductory starter package DI-194RS 10-bit resolution PC data logger kit from DATAQ for $24.95. This data logger option is a real bargain, providing both hardware and software, and it can get you started recording data in just a short time. The company offers many other data recorder options as well. Check the DATAQ web site for details at www.dataq.com.

SIDs Research Opportunities

You can join the foremost group involved with SIDs research. The *American Association of Variable Start Observers* (AAVSO) SID Program is made up of solar observers who monitor VLF radio stations for sudden enhancements of their signals. As we now know, Earth's ionosphere reacts to the intense X-ray and ultraviolet radiation released during a solar flare. We also know that the ionospheric disturbance enhances VLF radio propagation. By monitoring the signal strength of a distant VLF transmitter, SIDs are recorded and indicate a recent solar flare event. All SID monitoring stations are home built by the observers. A SID station operates unattended until the end of each month. Recordings are then analyzed for the beginning, end, and duration of SID events.

Figure 7-14 *Prototype receiver mounted on sloping cabinet*

Figure 7-15 *Loop antenna*

A simple analog-to-digital converter design for specific use with the VLF receivers is available by contacting the chairman of AAVSO. SID observers submit strip charts or computer plots to the SID coordinator for visual inspection at the end of each month. Many observers analyze their own strip charts and computer plots. Analyzed results are submitted via e-mail to the AAVSO SID analyst for correlation with other observers' results. The final SID report combines individual observers' reports with the AAVSO SID coordinator's visual analysis. SID event results are sent monthly to the *National Geophysical Data Center* (NGDC) for publication in the *Solar-Geophysical Data Report* where they are accessed by researchers worldwide. The reduced SIDs data and particularly interesting plots are reproduced in the monthly *AAVSO Solar Bulletin* mailed to all contributing members.

SIDs Receiver Parts List

Required Parts

R1, R2, R3, R4 3.3K ohm, 1/4-watt resistor

R5 10K ohm potentiometer (chassis mount)

R6 1K ohm, 1/4-watt resistor

R7 100K potentiometer (PC mount)

R8 100-ohm, 1/4-watt resistor

R9 10K ohm, 1/4-watt resistor

R10 470K ohm, 1/4-watt resistor

R11 56K ohm, 1/4-watt resistor

R12 22K ohm, 1/4-watt resistor

R13 5K ohm potentiometer (PC mount)

C1 100 pF, 35-volt ceramic capacitor

C2 1,500 pF, 35-volt Mylar capacitor

C3 0.001 uF, 35-volt ceramic capacitor

C4 1 uF, 35-volt tantalum capacitor

C5 10 uF, 35-volt tantalum capacitor

C6, C7 1 uF, 35-volt electrolytic capacitor

C8, C9 0.1 uF, 35-volt ceramic capacitor

D1, D2 1N914 silicon diode

D3, D4 1N34 germanium diode

U1 LM353 op-amp (National Semiconductor)

U2 RM4136 op-amp (Texas Instruments)

U3 79L05 5-volt regulator (+ volts)

U4 78L05 5-volt regulator (- volts)

T1 600-to-600 ohm interstage/matching transformer

L1 loop antenna (see text)

S1 DPST toggle switch (on-off)

B1, B2 9-volt transistor radio battery

Miscellaneous PC board, IC sockets, wire, solder, solder lugs, chassis hardware, screws, standoffs, etc.

Earth Field Magnetometer

Magnetic fields are all around us. The earth itself produces a magnetic field, which is why compasses work. Anytime an electrical current flows in a conductor, a magnetic field is generated. That is why transformers, inductors, and radio antennas work. Several different devices could be used to sense a magnetic field. One of the ones most familiar to electronics hobbyists is the Hall-effect device. However, in this section we'll take a look at a magnetic sensor that is as easy to use but is more sensitive, more linear, and more temperature stable than the typical Hall-effect devices. And just like Hall-effect devices, it can be used to make a variety of instruments, including magnetometers and gradiometers.

For those unfamiliar with them, *magnetometers* are used in a variety of applications in science and engineering. One high-tech magnetometer is used by Navy aircrafts to locate submarines. Radio scientists use magnetometers to monitor solar activity. Archeologists use magnetometers to locate buried artifacts. And marine archeologists and treasure hunters use them to locate sunken wrecks and sunken treasure.

The earth field magnetometer that we will be exploring uses a flux-gate magnetic sensor. This device, in essence, is an over-driven magnetic-core transformer where the "transducible" event is the saturation of the magnetic material. These devices can be made very small and compact, yet will still provide reasonable accuracy.

The most simple form of flux-gate magnetic sensor is shown in Figure 7-16. It consists of a nickel–iron rod used as a core, wound with two coils. One coil is used as the excitation coil, and the other is used as the output or sensing coil. The excitation coil is driven with a square-wave signal with an amplitude high enough to saturate the core. The current in the output coil will increase in a linear manner so long as the core is not saturated. But when the saturation point is reached, the inductance of the coil drops and the current rises to a level limited only by the coil's other circuit resistances.

If the simple flux-gate sensor was in a magnetically pure environment, then the field produced by the excitation coil would be the end of the story. But magnetic fields are all around us, and these either add to or subtract from the magnetic field in the core of the flux-gate sensor. Magnetic field lines along the

axis of the core have the most effect on the total magnetic field inside the core. As a result of the external magnetic fields, the saturation condition occurs either earlier or later than would occur if we were dealing only with the magnetic field of interest. Whether the saturation occurs early or later depends on whether the external field opposes or reinforces the intended field.

A better solution, in this version of the flux-gate sensor, is to have two independent cores, each of which has its own excitation winding. A common pickup winding serves both cores. The excitation coils are wound in a series-opposing manner so that the induction generated in the cores precisely cancels each other if the external field is zero. However, in the presence of an external field, pulses are produced in the pickup coil. These can be integrated in a low-pass filter to produce a slowly varying DC signal that is proportional to the applied external magnetic field.

Toroidal-Core Flux-Gate Sensor

The flux-gate sensors that use a linear or straight core suffer from two main problems. First, the desired signal is small compared with the signal on which it rides, so it is difficult to discriminate properly. Second, there must be a very good match between the cores and the excitation winding segments on each winding. Although those problems can be overcome, it makes the sensor more expensive, limiting the design's popularity.

A better solution is to use a toroidal doughnut-shaped magnetic core. This type of core relieves the problem of picking small signals in the presence of large offset components. It also reduces the drive levels required from the excitation source. In the toroidal-core flux-gate sensor, we can get away with using a single excitation coil wound over the entire circumference of the toroidal core (see Figure 7-17). The pickup coil is wound over the outside diameter of the core. Another advantage of the toroidal-core version of the flux-gate sensor is that a pair of

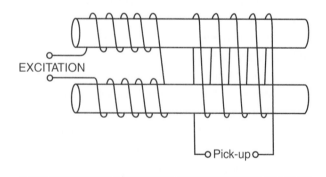

EXCITATION

Pick-up

Figure 7-16   *Flux-gate magnetic sensor*

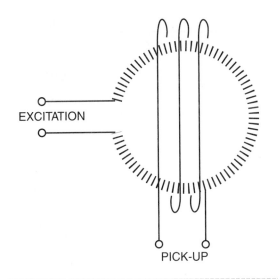

Figure 7-17 *Toroid flux-gate sensor*

orthogonal (i.e., right angle) pickup coils can be installed that will allow null measurements to be made. The maximum sensitivity occurs when the magnetic H-field (or the strength of the magnetic field) is orthogonal to the pickup coil, whereas minimum sensitivity occurs when the pickup coil and H-field are aligned with each other. As you can see, when two pickup coils are at right angles to each other, one will be at maximum sensitivity when the other is at a null (minimum sensitivity).

The Flux-Gate Sensor

A compact and reasonably low-cost line of flux-gate sensors, designated FGM-*X*, is made by Speake & Co. Ltd. and distributed in the United States by Fat Quarters Software (see appendix for contact information). The FGM-3 device is the one used in this project. It is a 62-millimeter-long by 16-millimeter-diameter (2.44 inches by 0.63 inch) device. Like all the devices in the line, it converts the magnetic field strength to a signal with a proportional frequency. The FGM-3 sensor has only three leads: Red is +5 VDC (power); Black is 0 volts (ground); and White is the output signal (a square-wave whose frequency varies with the applied field). The output signal is a train of *transistor-transistor logic* (TTL) compatible pulses with a period that ranges from 8 to 25 mS, or a

frequency range of 40 to 125 KHz. The detection sensitivity of the FGM-3 device is +/−0.5 oersted (+50 μTesla). That range covers the earth's magnetic field, making it possible to use the sensor in Earth-field magnetometers. Using two or three sensors together provides functions such as compass orientation, three-dimensional orientation measurement systems, and three-dimensional gimbaled devices such as virtual-reality helmet display devices. They can also be used in applications such as ferrous metal detectors, as underwater shipwreck finders, and in factories as conveyer-belt sensors or counters. A host of other applications exist where a small change in a magnetic field needs to be detected.

The series also includes two other devices, the FGM-2 and the FGM-3h. The FGM-2 is an orthogonal sensor with two FGM-1 devices on a circular platform at right angles to one another. That orthogonal arrangement permits easier implementation of orientation measurement, compass functioning, and other applications. The FGM-3h is the same size and shape as the FGM-3, but is about 2.5 times more sensitive. Its output frequency changes approximately 2 to 3 Hz per gamma of field change, with a dynamic range of +/−0.15 oersted (about one-third of Earth's magnetic field strength).

The output signal in all the devices in the FGM series is a +5-volt (TTL-compatible) pulse whose period is directly proportional to the applied magnetic field strength. This relationship makes the frequency of the output signal directly proportional to the magnetic field strength. The period varies typically from 8.5 μs to 25 μs or a frequency of about 120 KHz to 50 KHz. For the FGM-3, the linearity is about 5.5 percent over its +/−0.5 oersted range.

The response pattern of the FGM-*X* series sensors is shown in Figure 7-18. It is a figure-eight pattern that has major lobes (maxima) along the axis of the sensor and nulls (minima) at right angles to the sensor axis. This pattern suggests that for any given situation, there is a preferred direction for sensor alignment. The long axis of the sensor should be pointed toward the target source. When calibrating or aligning sensor circuits, it is common practice to align the sensor along the east-west direction in order to minimize the effects of Earth's magnetic field.

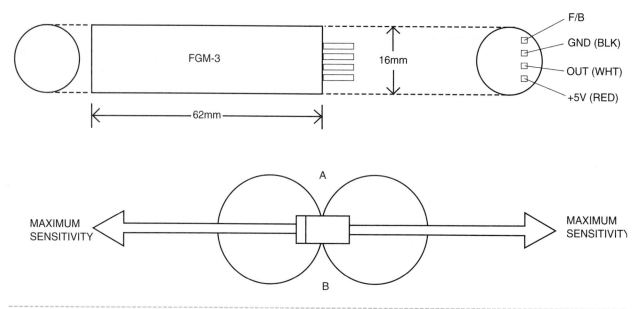

Figure 7-18 *FGM-3 magnetometer sensor*

Flux-Gate Magnetometer

The earth field magnetometer project depicted in Figures 7-19 and 7-20 illustrates the circuit for a flux-gate magnetometer based on the FGM-3 flux-gate sensor.

The heart of the circuit flux-gate magnetometer circuit shown in Figure 7-21 is the special interface chip at U1, a Speake's SCL006A IC. It provides the circuitry needed to perform magnetometry including earth field magnetometry. It integrates field fluctuations in 1-second intervals, producing very sensitive output variations in response to small field variations. The magnetometer is of special interest to people

doing radio-propagation studies and those who want to monitor solar flares. It also works as a laboratory magnetometer for various purposes. The SCL006A is housed in an 18-pin DIP IC package and is shown at U1. The FGM-3 flux-gate sensor is coupled to the input pin at pin 17 of U1. The reference oscillator of the SCL0006A is controlled by components X1, C1, C2, and R2. A sensitivity switch, S1, provides four positions, each with a different overall sensitivity range. Switch S2 is used to reset the SCL0006A, when pressed for about 2 seconds.

The second op-amp at U2, is an Analog Devices AD557 digital-to-analog converter. The magnetometer circuit is designed so that it could be run from a 9- to 15-volt battery for use in the field. Power supplied from the battery is controlled by power switch S3,

Figure 7-19 *Magnetometer control box*

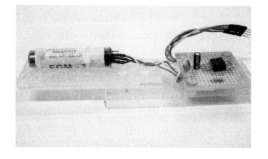

Figure 7-20 *Magnetometer sensor and regulator assembly*

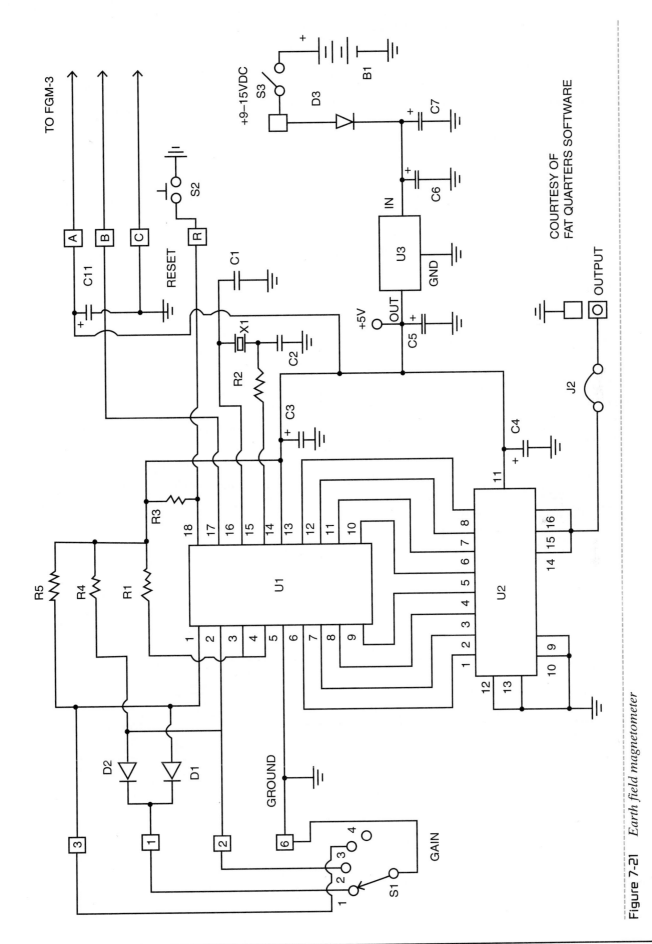

Figure 7-21   *Earth field magnetometer*

which supplies 9-volts to the regulator. For stationary research, the magnetometer should be powered from an 110 VAC to the 9- to 15-volt DC power supply. The regulator U3 drops the 9-volt input supply power down to 5 volts to power the magnetometer circuit. The output signal is a DC voltage that can be monitored by a strip-chart or X-Y paper recorder; a voltmeter or the signal can be fed into a computer using an analog-to-digital converter. In this project we use the output to drive a portable mini data logger such as the ONSET corporation HOBO H08-002-02. You will want to select the 2.5-volt model logger. You could also elect to use a digital multimeter with RS-232 output, which can be fed to a computer's serial port for data logging.

Construction

The earth field magnetometer project should be built on a two small printed circuit boards. The main circuit shown in Figure 7-21 is constructed on a 3½-inch × 41/2-inch PC board, and the FGM-3 sensor is paired with a second 5-volt regulator shown in Figure 7-22. Regulator U4 is mounted on a 2 × 3 inch PC board, attached to a sheet of flexible plastic using plastic screws. The FGM-3 sensor probe is then mounted on the 2 × 4½ inch plastic sheet. The plastic sheet is used to mount the probe, ahead of the regulator in order to give the flux-gate probe some isolation from sur-

rounding metal. The flux-gate probe is mounted length-wise and parallel to the length of the PVC pipe. The FGM-3 sensor and remote regulator were then housed in a 2½-inch diameter by 12-inch long PVC pipe, which is separated by a length of three-conductor cable.

When constructing the main magnetometer circuit board, you should install IC sockets. In the event of a circuit failure at a later date, they will make parts replacement much easier. When installing the two ICs, you will need to make sure that you correctly orientated them in order to avoid damage to the circuit. Most ICs will have either a small indented circle on the left side of the plastic package or a small rectangular cutout in the top center of the plastic package. Pin 1 of the IC is always to the left of either the indented circle or the cutout.

The magnetometer circuit contains both electrolytic capacitors and diodes. Electrolytic capacitors are polarity sensitive and must be installed correctly to avoid damage. Diodes are usually marked with a white or black band at one end; this band denotes the cathode end of the diode. After the circuit board has been completed you should inspect the circuit board for cold solder joints as well as for solder bridges between PC circuit board lands. Also look for stray component leads, which may have stuck to the underside of the board during the assembly process. These steps help ensure that the ICs will not be damaged upon power-up. Note that you could also elect to pur-

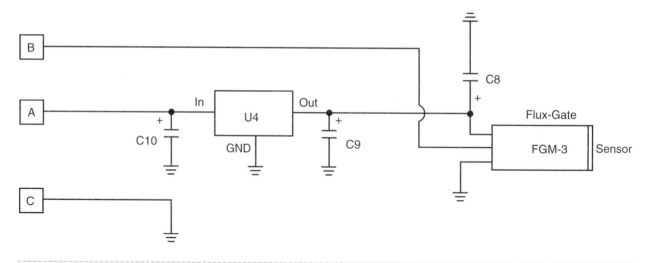

Figure 7-22   *Flux-gate sensor probe assembly*

chase a flux-magnetometer kit including PC board, the FGM-3 sensor, ICs, and most other parts except the gain switch, which can be obtained from Fat Quarters Software for about $75.00. The Fat Quarters circuit board pin layout is shown in Figure 7-23.

Now that the magnetometer circuit board has been completed, you can locate a suitable metal enclosure for your magnetometer project. The earth field magnetometer prototype main circuit board is mounted in a 6 × 3 × 3 inch metal chassis box as shown in Figure 7-24. The gain, power, and reset power switches, and the power indicator LED are also mounted on the front panel of the chassis box. The magnetometer's output connections are brought to a red and black binding post on the front panel of the chassis. A screw-terminal block is mounted at the side of the chassis box to allow connections between the main electronics and the probe assembly.

The flux-gate probe as described earlier is mounted in a 2½ × 12 inch length of PVC pipe. Circuit board guides are epoxied to the inside of the PVC pipe to allow the circuit board to be slid into the PVC pipe. One 2½-inch end cap is then cleaned and cemented to

the outside probe end of the PVC pipe. A ½-inch hole is drilled and tapped in the center of the second PVC end cap. A brass nipple fitting is cemented in the tapped hole to allow the three-conductor cable to exit the probe assembly in the PVC pipe. Before cementing the final end cap to the PVC pipe you will want to test the earth field magnetometer.

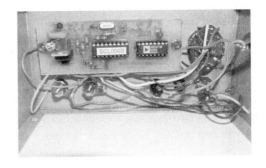

Figure 7-24   *Earth field magnetometer circuit board mounted in a metal chassis box*

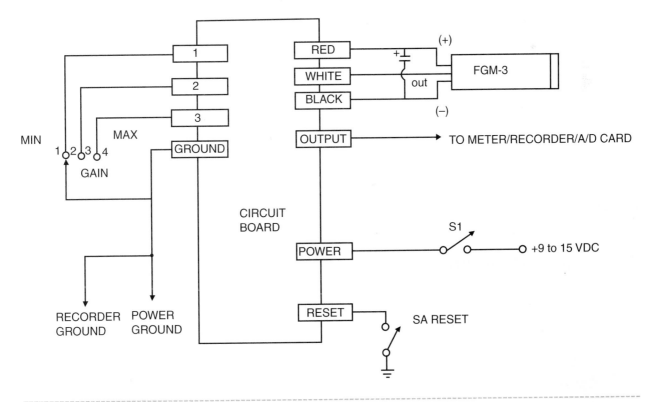

Figure 7-23   *Magnetometer PC board*

Testing

Testing the earth field magnetometer is straightforward. Connect a three-wire shielded cable between the probe assembly and the main circuit board. Connect a digital multimeter to the output of the magnetometer circuit and select the 2-volt setting on the multimeter. Be sure to observe polarity connections to the meter. Connect a power supply to the magnetometer circuits and switch the power to the circuits. Put the probe assembly on a nonmetal surface and very slowly rotate the probe assembly in a circle. As the probe assembly is rotated slowly, the meter should show a significant difference in the reading when the probe is rotated through north. Note, that the circuitry in U1 takes about 7 to 8 seconds for a reading to be updated, so the results will not show immediately. Once the probe assembly has been tested, you can cement the last end cap to the PVC pipe.

For solar storm observations, the magnetometer probe should be buried outside away from your house or out-buildings to ensure no metal influence. Since the flux-gate probe is not temperature compensated, the probe assembly should be mounted at least 3 feet underground for best results. A permanent AC to DC power supply should be used to power the magnetometer circuits for best results. As mentioned previously, you could use an inexpensive HOBO data logger, a serial port multimeter, or an analog-to-digital card inside your computer for data logging purposes.

Your earth field magnetometer can now be put into service collecting data on solar storms, diurnal variations, and more. Have fun!

Earth Field Magnetometer Parts List

Required Parts

R1, R3, R4, R5 4.7K ohm, 1/4-watt resistor

R2 100 ohms, 1/4-watt resistor

C1, C2 15 pF, 35-volt capacitor

C3, C4 47 nF, 35-volt electrolytic capacitor

C5, C10 2.2 uF, 35-volt capacitor

C6, C9 0.1 uF, 35-volt capacitor

C7 100 uF, 35-volt electrolytic capacitor

C8, C11 10 uF, 35-volt electrolytic capacitor

D1, D2 1N914 silicon diode

D3 1N4001 silicon diode

X1 10 MHz crystal

U1 SCL006A magnetometer output IC

U2 AD-557 digital-to-analog converter chip

U3, U4 LM7805, 5-volt regulator

J1, J2 wire jumpers

S1 four-position rotary gain switch

S2 normally open push-button switch

S3 SPST toggle power switch

Logger HOBO H08-002-02 (ONSET)

Miscellaneous PC board, wire, IC sockets, chassis, etc.

Chapter Eight

Sensing Electric Fields

The earliest study of nature's wonders began with the observations relating to static electricity or electrostatics. Many early observations of static electricity involved animal skins and hair and glass or stone objects. These early observations lead to the discovery of electrical attraction principles, electrical waves, and later the basis for electronics as we know it today. In this chapter you will discover and learn about electrostatic fundamentals, electric fields, and electromagnetic fields. You will learn that all electromagnetic energy, regardless of its frequency, has certain common properties. Two things magically bind electric and magnetic energy together into electromagnetic radiation. These are the traveling electric and magnetic fields, which are always at right angles to each other, always at right angles to the direction of propagation, and always becoming weaker with the distance they travel.

You will discover applications for the classical electroscope, the Leyden Jar, the static tube, as well as how to build and use a cloud chamber. Practical projects in this chapter include an ion detector, an electronic electroscope, and an atmospheric electricity monitor. Advanced projects presented for junior scientists include an advanced electronic electroscope and a cloud charge monitor that can detect and display the charges from clouds traveling overhead. The chapter's final project is an electric field disturbance monitor, which can be used to detect human bodies in an electrical field and then sound an alarm. The electric field disturbance monitor could be used for research or could form the basis for a home or camping alarm system.

Electrostatic Fundamentals

For centuries people have pondered lightning, lodestones, and static electricity phenomenon without developing even rudimentary explanations or models. In our modern technical world, it is hard to imagine anything as dramatic as lightning going unexplained, but electricity is usually invisible, and, without electronic gadgets, it is rather elusive. Fortunately, modern day electronics experimenters have no shortage of electronic equipment from which to study. Even a poorly financed experimenter can acquire precision equipment that would have served proudly in a modern laboratory only a few decades ago. And junk yards have mountains of scrap electronic components that would have been worth a fortune to that same lab. The concepts are simple and materials are easy to obtain.

Electrical interactions are surprisingly simple and a few basic laws of electricity model the real world with stunning precision. As far as anyone knows there is an absolute minimum quantity of charge—an amount of charge that cannot be subdivided. This is the amount of charge that an electron and proton possess. This packet of charge has a magnitude denoted by e and comes in two *polarities*, denoted by the plus and minus signs (a positive and a negative charge). All electrical charges are made up of these tiny packets. Charges with the same polarity physically repel each other, and opposite charges attract each other.

The charge of an electron is generally defined as the negative charge. A little more that 6,240,000,000,000,000,000 electrons are needed to make the common unit of charge called the *coulomb*. The unit of current, the *ampere*, is defined as the flow of one coulomb per second. Clearly, the charge on one electron is rather small. If the electric charge on the electron were represented by a tiny drop of water, one coulomb would fill a lake more than 30 miles across and over 100 feet deep. The current flowing in a 1-amp flashlight bulb is equal to one such lake per second! Despite this incredible flow rate, an electron entering a short wire at the battery might take several minutes to reach the bulb. Obviously, a tremendous number of free electrons are in the wire! Most electronic circuits deal with huge numbers of electrons, and the discrete nature of the charge carriers is insignificant.

Because there are 6.24×10^{18} electrons in a coulomb, the charge on a single electron is $1/(6.24 \times 10^{18})$ coulombs, or 1.6×10^{-19} coulombs. R.A. Millikan and associates are credited with being the first to accurately measure this charge using an ingenious apparatus in what is now known as the Millikan oil-drop experiment. Tiny drops of oil from an atomizer were injected between two metal plates with an adjustable voltage between them. The resulting electric field would attract one polarity of charge to the top plate, and, if the voltage was just right, the force of gravity could be perfectly balanced, freezing the particle in midair. At this point, gravity multiplied times the oil drop's mass equals the charge multiplied times the electric field. Millikan discovered that the random charge on the droplets was always an integer multiple of 1.6×10^{-19} coulombs.

Try this experiment from elementary school. Obtain a clear Lucite rod or a black plastic pocket comb. Rub the rod on your hair, a piece of fur, or silk cloth with rapid strokes. Hold the rod near some tiny pieces of paper or sawdust and observe the results. Now watch closely!

You should have observed the expected attraction of the neutral particles to the charged plastic. The careful observer will also have noticed that many of the attracted particles will suddenly fly away from the rod after a minute or two—as if repelled. The reaction is nearly immediate if the little pieces of paper are a slightly damp. Try dropping a damp piece of paper or a tiny piece of aluminum foil past the charged plastic. It will be attracted, but as soon as it touches, it will fly away. The rapid, erratic motion is hard to follow, so watch carefully! The charge on the rod attracts the neutral particle, but as soon as contact is made the particle picks up some of the charge and is repelled because like charges repel each other. Dry particles do not pick up the charge as quickly because they are less *conductive* and remain attracted to the rod for a longer period.

If you have never experimented with static electricity, take a few moments and try a few projects described in this chapter. Millikan's amazing experiment was the result of sound scientific reasoning combined with an engineering common sense that comes only from tinkering. You have charged rubber balloons to stick them to the wall, no doubt! But did you ever determine if the charge on the balloon is the same polarity as the charge on your pocket comb? How about the charge on a glass rod rubbed with silk? See if you can find out through experimentation. Here's a useful hint: Like charges repel; opposites attract. Another hint: Glass and silk are positively attractive, whereas cheap plastic and fake fur have negative connotations!

The force between two tiny particles with charges q_1 and q_2 is given by *Coulomb's law*: The force that one particle exerts on another is directly proportional to the product of their charges and inversely proportional to the square of their separation.

$$F_{elec} = K \times Q_1 Q_2/r_2$$

where q_1 is the net charge of particle one, q_2 is the net charge of particle two, r is the separation of the particles, and e_0 is equal to 8.85×10^{12} Nm³/C. C is called the coulomb, which is the unit of electric charge.

This force is a vector that points along the line between the two charges. This simple equation can give one food for thought! If the force is proportional to the product of two charges then why does the charged plastic rod pick up neutral particles? Because one of the charge terms is zero, shouldn't the force be zero? The answer is not obvious and may leave one wondering if the equation has any real world relevance.

When the charged rod is held near a small object, a charge is induced in the object. If the object is a conductor, the like charges will be repelled to the far side of the object and the opposite charges will be attracted to the side nearest the charged rod. Because the charges are segregated at different distances, their contribution to the total force will be different (note the $1/r^2$ term). The charges in an insulating material are not free to move about as they are in a metal, but the charges can redistribute on a microscopic level resulting in a somewhat weaker attraction to the charged rod. If a conductive object is touching a conductive surface, it can accumulate a net charge because like charges can leave the object entirely and opposite charges can accumulate. The resulting attraction can be quite strong.

To find out if Coulomb's law might have any real world application you may wish to do another experiment. This experiment will require the construction of a unique differential electroscope (see Figure 8-1). The traditional electroscope consists of two metal leaves hanging freely from a wire like a sheet draped over a clothesline. The leaves are usually placed inside a glass jar to block the wind. The wire protrudes through an insulated top to allow for the deposition of charge. A differential electroscope has two isolated supports for the metal leaves, allowing different charges to be applied to each leaf. For the purposes of this experiment, construction can be quite simple, requiring only a small block of styrofoam, two large nails, and a couple of strips of aluminum foil about 4 inches long and $1/2$ inch wide. Push the two nails through the foam about $1^1/2$ inches apart, forming two hangers for the foil strips. Bend the ends of the foil strips around the nails so that the strips hang freely.

Collect some charge on your plastic rod or pocket comb and then transfer the charge to one of the leaves by dragging the charged rod across the head of the nail on the back side of the foam. Do not touch the other nail. Notice that only a little attraction or perhaps a slight repulsion is noted between the leaves! Now touch the head of the nail supporting the uncharged strip. The two strips slam together! Because the strips are very thin compared to the distance between them, the amount of force due to the induced charge is small. Whether the charge is on one face of the leaf or the other, it is almost the same distance from the other leaf so $1/r^2$ is about the same. Because the net charge on the neutral leaf is zero, the force between the leaves is always nearly zero. But when you touch the neutral leaf, the induced charge can skyrocket because there is now a place for the like charge to go and a plentiful source of opposite charge (your body).

Coulomb's law appears to be looking pretty good after all, and the physics of what happens when a comb picks up little particles involves induced charge, not simply different levels of charge. Attraction occurs because the charged rod induces a significant opposite charge, or *polarization*, in the otherwise neutral objects.

A *coulomb* can be stated as one amp for one second, a seemingly modest amount of charge but actually an extremely large electrostatic charge. According to Coulomb's law, two plates 1 yard apart, each with one coulomb of charge, will attract or repel

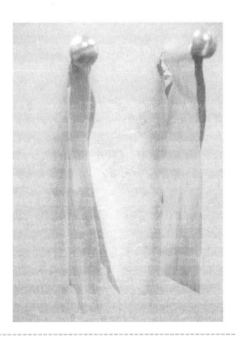

Figure 8-1   *Differential electroscope*

each other with billions of pounds of force! Fortunately, charged bodies usually contain much less than a coulomb!

Electric Fields

The concept of a *force field* is quite familiar to us earth-bound humans. We spend our entire lives in the earth's gravitational field, which seems perfectly uniform and constant. The lines of force are clearly *down* —the direction the fine china heads when it slips out of your hands. A *line of force* is simply an imaginary line that traces the path an object would take due to the influence of a field. For diagramming purposes, it might be useful to associate a line with a certain amount of force so that more lines indicate more force. A person standing on the moon would be "skewered" by fewer lines of force than a person standing on the earth because the moon's gravitational field is weaker than Earth's. The force on an object is proportional to the line density. Therefore when the lines are far apart, the force is weak, and when they are closely spaced, the force is high. Remember that the line density can be hard to judge because only a few lines are usually drawn.

Lines of force are useful also for showing the general shape of a field, only hinting at the field strength at any particular point. One could draw so many lines that they blur together in shades of gray, darkening in areas of greater force, but it would be difficult to determine the direction of the force if the individual lines could not be seen. The following figures show typical lines of force for an attracting body and a simulation of how the lines of force might look if thousands were drawn. Remember, as an object moves along a line of force, it usually encounters a changing level of force. Lines of force are readily seen in Figure 8-2, and field strength is more easily seen in the Figure 8-3, which simulates a very large number of lines.

The lines of electric force are defined to point in the direction a positive charge would move; so electrons have lines of force pointing at them, and protons have similar lines of force pointing away. When

several charges are in close proximity, the line of force at a particular point will be the vector sum of the contributions from all of the charges.

Unlike the gravitational field, electric fields have two polarities and can work together or oppose each other as the diagrams depict. Notice how a positive particle precisely between the two positive charges would experience no force because the two fields cancel each other. But if the charge is slightly above or below the center line, it is squeezed out like a watermelon pit between wet fingers! Just like gravity, the electric field is a force field—where the field intensity at any point is equal to the amount of mechanical force that would be applied to a charge of one coulomb positioned at the point. The units are newtons per coulomb. The electric field surrounding a charged Lucite rod can be detected by holding the

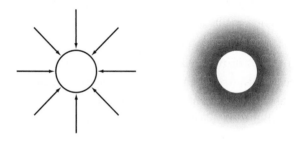

Figure 8-2   *Lines of force*

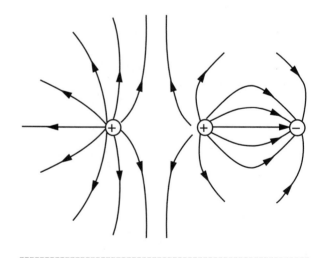

Figure 8-3   *Field of strength*

rod near the electrode of a sensitive electroscope. Without the rod making contact, the leaves will spread due to the induced charge caused by the field.

Electromagnetic Waves

After the discovery of infrared light by Herschel, a Danish physicist named Hans Christian Oersted found an electrical current that could change the direction of a compass's needle. André-Marie Ampère also found that different electrical currents could attract or repel one another. It was in 1865 when James Clerk Maxwell proposed the idea of electromagnetism. He showed that electricity and magnetism were closely related and, if made to surge back and forth, could produce alternating electromagnetic waves, which would move at the speed of light. Maxwell then concluded that light itself must be an electromagnetic wave.

Electric and magnetic fields interact in a variety of ways. A moving electric field creates a magnetic field. That is, as an electric field moves through space, it gives up its energy to a companion magnetic field. The electric field loses energy as the magnetic field gains energy. Thus, we see a gradual transfer of energy from one form of energy to another, but no loss or gain in the total energy of the wave. Conversely, a moving magnetic field creates an electric field, thus a magnetic field moving through space will transfer its energy gradually to a companion electric field.

A static or unvarying electric field will produce a static magnetic field. A varying electric field produces a varying magnetic field. The reverse is also true. A static magnetic field produces a static electric field. If the magnetic field varies so too will the electric field.

When an electric and magnetic field vary in strength over time, they form electromagnetic waves. A flow of electric current creates a corresponding magnetic field. If the current flow is unvarying, the magnetic field will not vary and no electromagnetic waves are produced. A varying, or oscillating, field will produce a changing magnetic field. Together, changing electromagnetic fields produce electromagnetic radiation that travels in waves. This characteris-tic of electricity and magnetism is used in radios to send information to distant receivers.

Infrared, visible, and ultraviolet light are not the only types of waves. Further down the spectrum past infrared lay microwaves, radar waves, television waves, and radio waves. Microwaves are thought to have been released during the Big Bang, which could have been the start of the universe. These slow waves are used in microwave ovens to heat food by changing the alignment of water molecules. Even slower are radar waves. Radar is an acronym for *radio detection and ranging*. Radar scanners send out short radio waves and detect the echoes from objects in their path. Even further down the spectrum are television waves. These waves send out sound and pictures to our TV sets. And the slowest known waves of all are radio waves. Radio waves are produced not only by radio stations but also by stars and galaxies.

At the ultraviolet end of the spectrum are three other types of waves, all with higher frequencies than ultraviolet. Directly above (higher in frequency) ultraviolet are X-rays. While X-rays can pass through the soft parts of our bodies, they can not penetrate bone. This is why they are used today to take pictures of broken bones. Gamma rays are radioactive and are released by certain atomic nuclei. With their large amounts of energy, they are able to penetrate metal and concrete and kill living cells. This type of wave is released by nuclear bombs, causing multitudes of destruction.

The highest-energy and most frequent waves are cosmic rays. They are made up of particles of atomic nuclei, electrons, and gamma rays. The earth's atmosphere protects us from these waves, which come from outer space (see the electromagnetic spectrum chart in Figure 8-4). Electromagnetic radiation exists all around us and throughout space. It is produced through the interaction of electric and magnetic fields. These two fields always exist together.

Waves are measured by their length (wavelength) and by the frequency with which they pass a point in space (frequency). The length of a wave is usually measured in meters. The frequency is measured by the number of waves or cycles that pass a given point in one second. One cycle per second is called a *hertz*

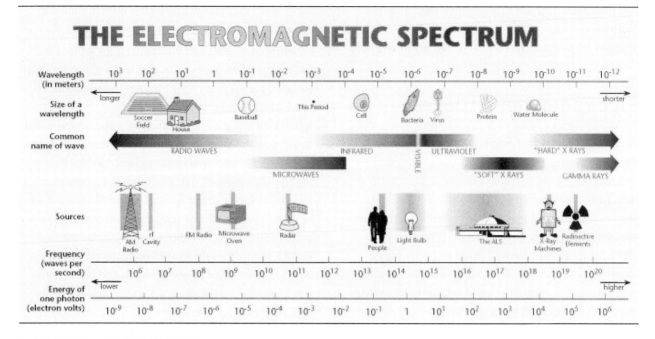

Figure 8-4 *Electromagnetic spectrum chart*

(Hz). The shorter the wavelength the higher the frequency of the radiation. The higher the frequency, the more energy the wave has.

Electromagnetic radiation always travels at the speed of light in a vacuum, or nearly the speed of light if it is traveling through a medium of some kind. The speed of light is approximately 300,000,000 meters per second (300,000 kilometers per second). For practical purposes we can say that radio waves always travel at the speed of light.

While the wavelength and the frequency of radio waves will vary, the speed of light does not. These three aspects of electromagnetic radiation are related to one another and can be described using mathematical formulas. The following formula can be used to determine either the frequency or the wavelength if you know one of the two:

Wavelength = 300 / Frequency in MHz

Building a Classic Electroscope

Even though the electroscope is a bit obsolete, it still possesses charm and elegant simplicity. And in fact it

works! The classic electroscope is constructed in a large clear glass jar or bottle with an attractive brass wire hanger for the foil leaves and a polished brass knob at the top for the electrode (as seen in Figure 8-5). Laboratory-grade electroscopes use extremely thin gold foil to minimize the weight and therefore maximize the deflection. Similar voltage sensitivities may be achieved with aluminum foil if very long leaves are used. These larger leaves will make the electroscope more sensitive. And because the *capacity* of the electroscope will be higher, more charge will be necessary, but the sensitivity to voltage can still be quite good.

Figure 8-5 *Electroscope*

Various chemistry beakers make attractive electroscopes with necks suitable for stoppers and wide chambers for the leaves. It is usually desirable to choose a wider jar than the wine bottle so that the leaves do not become attracted to the glass when highly charged. Hanging short leaves can be a bit tricky because they must swing independently. Give one leaf arms for hanging like a person hanging from a bar. The other leaf has a single narrow arm in the middle like a monkey hanging by its tail! The two leaves can move away from each other without the pivot points interfering with each other. The long leaves in the picture were simply attached to the screw on the top because they are long enough to flex. Smooth the leaves by gently rubbing them with a finger against a flat surface.

If you wish to make an electroscope with a scale, replace one of the leaves with a stiff strip of metal or copper-clad circuit board. Fasten the flexible leaf to the stiff leaf at the top and place a paper scale behind the leaves. If the flexible leaf is at least ½-inch wide, it will be stiff enough and therefore will not twist and stick to the scale. Enclose the scope in a large, clear plastic box with square sides (the type sold for decorative purposes will do nicely). A colored Lucite rod mounted in a wooden handle makes a nice accessory for this unique conversation piece. Actually, ordinary glass bottles are inferior to clear plastic bottles due to surface conductivity. So, if you want the very best performance, consider using a clear plastic jug.

Building a Leyden Jar

After experimenting with a comb and an electroscope for a few minutes, the experimenter will notice that the electroscope leaves can be charged in steps: Each touch of the comb causes more deflection. The leaves accumulate the bits of charge and the voltage builds. Bigger leaves will require more charge to achieve the same voltage because the electrons are spread over a larger area. If we coat the entire inside of the jar with aluminum, the capacity of the jar to hold charge will be much higher.

Named for its discovery by the physicist Pieter van Musschenbroek at the University of Leyden in the mid-1700s, a *Leyden jar* is a device that early experimenters used to help build and store electric energy. It was also referred to as a *condenser* because many people thought of electricity as fluid or matter that could be condensed. Today someone familiar with electrical terminology would call it a *capacitor*.

Basically, the Leyden jar is a cylindrical container made of a dielectric (that's an insulator, like plastic or glass) with a layer of metal foil on the inside and on the outside. With the outside surface grounded, a charge is given to the inside surface. This gives the outside an equal but opposite charge, see Figures 8-6 and 8-7. When the outside and inside surfaces are connected by a conductor . . . SNAP! You get a spark and everything returns to normal. The amount of charge one of these devices can store is related to the voltage applied to it multiplied times its capacitance. In simple terms, capacitance depends on (1) the *area* of the foil or metal, (2) the *type of material* between the two layers of foil, and (3) the *thickness* (generally the thinner the better) of that material.

A more modern version is made by lining the inside and coating the outside of the jar with aluminum foil, thereby forming two capacitor plates. The attraction of opposite charges between the two plates gives the foil Leyden jar the capacity to store a significant charge. The outer foil is usually the *ground plate*, and the inner foil is the *hot plate*. A connection is made between an electrode on the top of the jar and the foil with a wire or metal chain.

The Leyden jar in the Figure 8-7 was made from an empty spaghetti sauce jar, a rubber stopper, a long screw, aluminum foil, and black electrical tape. Spray

Figure 8-6   *Leyden jar*

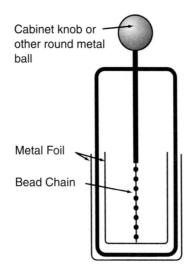

Cabinet knob or other round metal ball

Metal Foil

Bead Chain

Figure 8-7   *Leyden jar interior*

adhesive was used to affix the foil to the inside surface and the electrode connection was made by tangling a long piece of bare wire with some of the foil on the inside and fastening the other end to the screw with electrical tape. A ground wire was fastened to the outer foil, and the foil was wrapped with a coating of fiber tape for protection. A 1-inch gap was left at the top of the jar for insulation. This gap should be cleaned with alcohol and wrapped with electrical tape to enhance the insulation.

The Leyden jar may be tested by connecting it to an electroscope with a stiff wire that touches only the electrodes. A few discharges from a charged pocket comb or plastic rod should give an indication on the electroscope. If the electroscope discharges quickly, there may be too much leakage—probably due to the glass being dirty or too much humidity in the air. As with the electroscope, you will have an easier time with a clear plastic jar (try a peanut butter jar). Electrostatics experiments are best performed on dry days! If everything is working well, charge the Leyden jar with repeated comb charges and then bring your finger near the electrode while holding the ground wire in the other hand. You should get a spark, and the electroscope will indicate a sudden drop in charge.

Building a Static Tube

A static tube is a simple way to charge a Leyden jar (see Figures 8-8 and 8-9). A static tube consists of a piece of PVC tubing with some fur, wool, paper towel, chamois, or other material wrapped around it. You can construct your own static tube, first obtain a piece of ¾-inch diameter PVC pipe about 4 feet long and a piece of cotton fabric. To operate, simply hold the fabric or paper with the left hand wrapped around the pipe. Then push and pull the pipe through the fabric, using large strokes, while holding the pipe near some fine wire tip pickups to collect the charge. The wire tip pickups can be connected to a Leyden jar or electrostatic motor. It is also a good idea to hold in the left hand, along with the fabric, a piece of metal that is connected to ground. This gives the generator a good voltage reference point and keeps the body from building up a big charge that will zap you the next time you touch a grounded object. With the spark gap set at about ¼ inch, a spark can be produced across it with almost every stroke of the pipe.

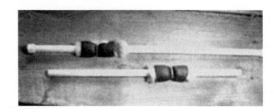

Figure 8-8   *Static tube*

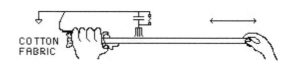

COTTON FABRIC

Electrostatic Generator

Figure 8-9   *Electrostatic generator*

How to Charge a Leyden Jar with a Static Tube

Set the jar somewhere so that the outside surface is grounded (not on a glass table or on a plastic container, for example) or have some one hold it (even better). Get your static-generating tube and hold it so that the tube is right above the ball on top of the Leyden jar. The closer the better. The hand holding the sliding pad remains still. I find it helpful to anchor my elbow by resting it on the table. The other hand slides the tube back and forth. This part requires some finesse: You have to draw the tube across the ball on the jar so that the charge on the tube "goes into" the jar. If you listen carefully, you can hear this happen, and, in a very dark room, you can see it (give your eyes time to adjust). There is no problem if the tube touches the ball except that you might knock the jar over. After several strokes, the charge that builds up is sufficient to create a visible spark when the Leyden jar is discharged. Figure 8-10 illustrates a static tube charging a Leyden jar.

Other Experiments that Use the Static Tube

This tube works great for demonstrating the attractive and repulsive powers of electricity. If you charge it up and hold it over bits of paper or other light material,

it will pick those things right up. If you hold it over an arm, it will make the hair stand on end. A dramatic way to demonstrate repulsion is to hang a balloon (rubber not Mylar) by a string, then charge the balloon by rubbing it with fur. Move the charged tube near the balloon and you will see very clearly that like charges repel. You can do similar things with ping pong balls on a foam plate, or perhaps a plastic soda bottle. For more information on static electricity experiments see the following websites:

www.amasci.com/emotor/statelec.html

www.alaska.net/~natnkell/staticgen.htm

www.stevespanglerscience.com/category/10/

Simple Electronic Electroscope

The circuit shown in Figure 8-11 is a simple electronic version of the classic electroscope, which is quite useful for detecting electrostatic charges. A small metal sphere of whip antenna is being used as the sensing antenna and is coupled to the simple electrometer via an RCA jack. The input RCA connector is fed directly to a high-value 1-megohm resistor. Capacitor

Figure 8-10   *Static tube charging a Leyden jar*

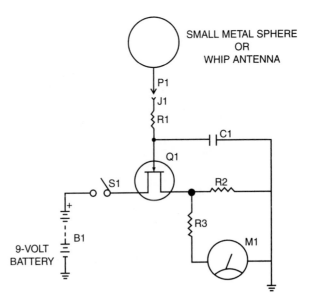

Figure 8-11   *Electronic electroscope circuit*

C1 is used to reduce AC noise, but lowers the sensitivity a tiny bit. The MPF102 FET transistor and R1 form a voltage divider. When the FET's gate is earth grounded, the divider's output will be about 4.5 volts, giving a half-scale reading on the meter M1. The sensitive meter is a 200 uA meter. A positively charged object (like a cotton-rubbed glass) will give a positive deflection from half-scale, and a negatively charged object (a plastic comb) will give a negative meter deflection. The entire circuit can be built in a few minutes using a perf-board or circuit board. The whole circuit including the 9-volt transistor radio battery will fit in a small metal chassis box. For best results the electrometer should be housed in a metal box and connected to an earth ground if possible.

Simple Electronic Electrometer Parts List

Required Parts

R1   1-megohm, 1/4-watt, 5% resistor

R2   150K ohm, 1/4-watt, 5% resistor

R3   33K ohm, 1/4-watt, 5% resistor

C1   220 pF, 100-volt capacitor

Q1   MPF102 FET transistor

M1   200 uA current meter

S1   SPST toggle switch

B1   9-volt transistor radio battery

P1   RCA plug

J1   RCA chassis jack

Antenna   small metal sphere or whip antenna

Miscellaneous   wire, battery clip, perfboard

Ion Detector

An ion detector detects static charges and free ions in the air (see the ion detector circuit shown in Figure 8-12). It can be used to indicate the presence of ion emissions. Health enthusiasts might use the ion detector to detect beneficial negative ions near waterfalls or after a rain storm. The ion detector can be used to detect leakage of high voltage in power supplies and radiation circuits. The ion detector can also be used to detect static electricity and electrostatic fields around your home or workshop. The antenna or charge collector can be made from a short piece of bare #12 or #14 copper wire. The antenna is then fed to a 100-megohm resistor, which is coupled to the first transistor, a 2N2907, at Q1. The emitter of Q1 is next fed to

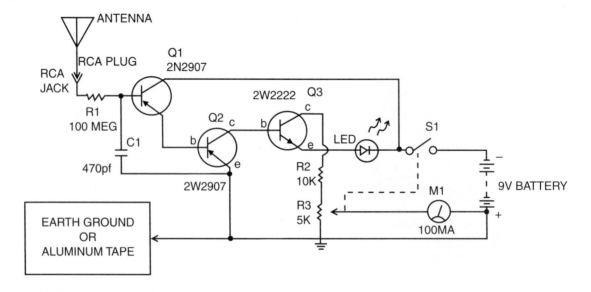

Figure 8-12   *Ion detector circuit*

the base of a second transistor at Q2. The output of the second-stage amplifier at the collector of Q2 is finally sent to the base of transistor at Q3, a 2N2222. The output of the emitter at Q3 is then coupled to the LED indicator. The overall sensitivity of the circuit is controlled by potentiometer R3. A 100 uA meter at M1 is used to provide an indication of the presence of ions nearby.

The ion detector circuit is powered by a 9-volt transistor radio battery. Switch S1 is used to apply power to the circuit. The ion detector circuit can be readily built in less than an hour on a perf-board or circuit board. For best results, a small aluminum chassis box should be used to house the ion detector circuit. The short antenna has an RCA plug soldered to one end, which mates to an RCA jack on the chassis box, which is then coupled to the ion detector circuit. The power switch, LED, and meter can all be mounted on the top front of the aluminum enclosure. This circuit should have some sort of a ground connected to it. The ground connection can be made by either touching the aluminum foil electrode with your hand or to an earth ground.

Ion Detector Parts List

Required Parts

R1 1-megohm, 1/4-watt, 5% resistor

R2 10K ohm, 1/4-watt, 5% resistor

R3 5K ohm potentiometer (chassis mount)

C1 470 pF, 35-volt, capacitor

Q1, Q2 2N2907 transistor

Q3 2N2222 transistor

D1 red LED

M1 100 uA panel meter

S1 SPST toggle switch

B1 9-volt transistor radio battery

P1 RCA plug

J1 RCA chassis jack

Antenna 6-inch whip antenna

Ground earth ground or aluminum plate

Miscellaneous PC board, wire, battery clip, etc.

Atmospheric Electricity Monitor

Nearly everyone is familiar with lightning, a very dynamic example of electricity in the atmosphere. This and smaller electrical discharges in clouds are the primary cause of radio static. The atmospheric electricity monitor, illustrated in Figure 8-13, can be used to measure atmospheric electricity.

If a needle is fastened to an insulated wire at the top of a 10-meter pole, electricity will flow from the earth to the atmosphere or vice versa. Under fair-weather skies, little if any current flow can be detected with this device because several thousand volts are needed before an ordinary needle can "go into corona." However, if an air-ionizing source such as a cartridge of radioactive polonium is mounted at the top of the pole, it produces a localized zone of ions that serves as an effective electrical connection to the atmosphere. Using this device in conjunction with an ultrasensitive microammeter will allow you to easily measure currents as low as 0.01 microampere. This is the type of fair-weather current that produces an electrical gradient near the earth of +60 to +100 volts per meter.

As the atmosphere becomes cloudy, the corona current increases until values as high as 5 to 10 microamperes may be observed. indicating the growing proximity of a thunderstorm. The electrical gradient under an active thunderstorm often shows values of 5,000 to 10,000 volts (plus or minus) per meter.

A number of devices are available for measuring and studying the electricity of the atmosphere, some of which are highly sophisticated and very expensive.

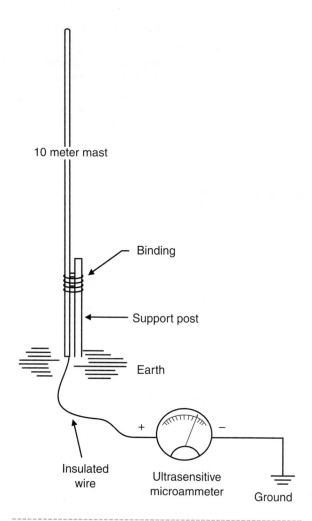

Figure 8-13   *Atmospheric electricity monitor*

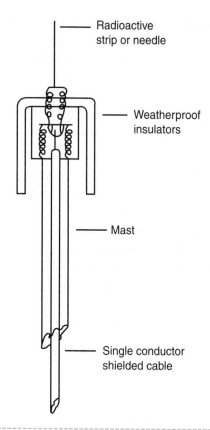

Figure 8-14   *Antenna detector assembly*

This project offers a simple, portable, and highly informative device that will allow you to do your own atmospheric electricity research. At the heart of the atmospheric electricity monitor is the antenna system. A four-section telescoping aluminum radio mast (four 2-meter sections) can be used as the detecting pole or antenna. The insulators can be mounted on the top end of the mast with a single-conductor polystyrene insulated cable carried down inside the pipe to the sensing instrument (as shown in Figure 8-14). In order for the system to operate effectively, the atmospheric electricity monitor must be referenced to ground as shown. A 4- to 6-foot-long copper ground rod should be driven into the ground as the reference electrode. When the weather is severe, microammeter readings should be made from an

outdoor sheltered place. Note that placing an antenna outdoors, high on a pole, acts as a lightning rod, therefore be extremely careful when using this system during an electrical storm. When the system is not in use, the antenna should be shorted to a good earth ground. The meter display is an ultrasensitive microammeter, which can be readily located through Newark Electronics. The air ionizing radioactive cartridge is available from Nuclear Products Co. (see the appendix for contact info).

Advanced Electrometer

The field of electrostatics concerns itself with charges, potentials, and forces and is often considered to have first been studied in the late eighteenth and early nineteenth centuries. In today's modern world, electrodynamics is the dominant study in electrical engineering. Electrostatics is making somewhat of a come back, and the forces and charges involved can be used for a number of novel purposes in our day-to-day lives. *Xerography*, or the process of copying images by the action of light on an electrically charged photoconductive surface, is a prime example.

The experimenter can arm him- or herself with an old classic electroscope and investigate a number of rather large electrostatic effects. In order to gain real insight and do more subtle and difficult experiments, an electrometer is often needed. The electronic electrometer is nothing more than an ordinary *voltmeter* (VOM). The difference between your digital VOM and the electrometer is one of input impedance, which relates to the amount of load the instrument places on the circuit under test. Most good electrometers have an input impedance of 200 terraohms. This is tens of thousands of times greater than the average electronic VOM (1 to 10 megohms). It will immediately amaze the experimenter just how electrical the world really is, once armed with an electrometer.

Modern electronic electrometers are very expensive. Five-thousand dollars will buy a fairly nice instrument, while $10,000 would be needed for a superlative one. Keithley Instruments and Victoreen are major suppliers of these specialized instruments. This project will supply the experimenter with schematics and a broad overview for assembling a first class electrometer with a very modest outlay of cash. The advanced electrometer circuit depicted in Figure 8-15 is a unity-gain impedance translator or electrometer amplifier, in the strictest sense, as the experimenter will have to supply the readout system (usually a VOM or oscilloscope). It will be assumed that the experimenter has a modicum of experience in assembling simple electronic circuits.

The advanced electrometer has a frequency response around 10 Hz, which is limited by the low-pass filter circuitry so no special PC boards are needed. Point-to-point or wire-wrap-type connections are all that is required. You could also assemble this circuit on the simple pad per hole, 2 × 3 inch circuit boards found in your local Radio Shack store. The key point is shielding and insulation. The front end of the system uses a special precision CMOS FET IC by National Semiconductor (LMC 6081), which can be purchased from Digikey electronics. The only other IC is an

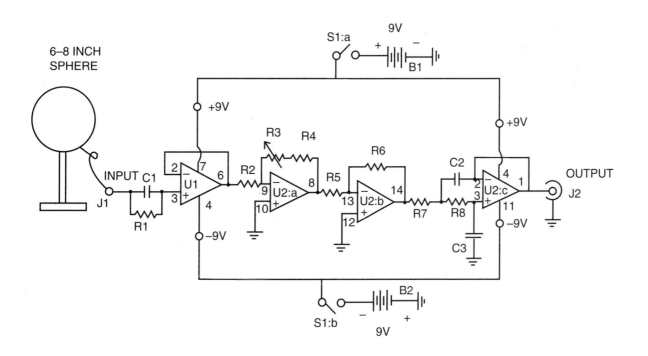

Figure 8-15   *Advanced electrometer circuit*

LM-324N, found at most Radio Shack stores. IC sockets are highly recommended. Radio Shack also has aluminum chassis cases that will be needed for shielding. The input connector must be a Teflon insulated, chassis mount, female, UHF connector PL-259.

The specifications of the output connector are unimportant and can be an RCA-type phone jack. A BNC chassis jack is used for the output connector. The system is powered by two common 9-volt transistor radio batteries. Parts layout is critical only in that the part of the PC board containing the input IC (LMC 6081) must be located immediately at the input connector. The critical input pin to this IC (pin 3) is carefully lifted and not inserted in its socket at all. The two components from this IC lead to the input connector center terminal and are connected floating in midair! All batteries and wiring should avoid this area and be on the opposite side of the board and shielded container box. Beyond this, the builder is totally at liberty to do as he or she pleases.

Construction

Begin by soldering the IC sockets to the PC board. Next make all the connections to all the various tie points. When the board is wired, add the battery clip leads. The box should be a fully enclosed all-metal one. Place the input connecter at one end and the output connector at the other end of the box. For convenience in some models, I put the input connector on the top of the box at one end. Holes must be drilled for standoffs or long machine screws to support the board near the input connector. The batteries are best placed in the opposite side of the box near the output connector. Make sure at each step of the drilling process that all components will fit in the box and not short out or touch one another. The power switch should also be positioned near the output end of the box. If you still haven't picked up on it, the input end of the box is special! Only the input IC and the ultrashort input connection, and air, are allowed.

Testing and Calibrating

With the shielded box still apart, connect a 1.5-volt battery to the input. Turn the unit on. Measure with a VOM the voltage on the battery. Write it down. Now go to the output with your VOM and adjust the gain potentiometer until the reading exactly equals the input voltage of the battery. This sets the instrument for unity gain. Your electrometer is now calibrated!

Next, check the finished system. Assemble the box and case for final use. I prefer to use a banana plug with a stiff 24-gauge rod soldered inside the plug to make a vertical whip antenna if the input connector is on top of the instrument. If not, simply use a short flexible wire soldered to a banana plug and connect it to a stainless steel salad bowl that is setting on a clean drinking glass. It is crucial that no path of any resistance exist to ground from the isolated and insulated metal object. Ground the chassis of the instrument (a good solid earth ground is needed.) Connect the output to an oscilloscope or VOM. Set the scope's sweep for a long-period sweep of 10 seconds or more across the screen. Set your meter for the 10-volt range if you have no scope. Turn on the instrument. Now move about around a bit and notice the varying voltage you are impressing on the antenna! The impedance is so high (terraohms) that the instrument can resolve voltages that would be overloaded and swamped back to zero by other meters of lower impedance. If you have encountered any problems in getting the system to work, you have made a simple mistake somewhere. Recheck your batteries. Are they viable? Are they hooked up correctly? Are the ICs in their sockets correctly? Recheck all wiring. You can blow the input IC very easily by touching the input lead while your body is electrostatically charged.

It seems bizarre that we use static-sensitive components in an instrument and then use them in static experiments. This instrument gets its ultrahigh impedance from the use of feedback and the delicate microthin layer of insulation separating the FET gate lead from the main substrate in the device. This is a layer that is easily punctured by static. Therefore such puncture means the destruction of the device and the IC. Wise application of the electrometer device is the best precautionary measure you can take. Never leave anything hooked to the input connector when the instrument is not in use.

The size of the isolated capacity (item hung up as an antenna and hooked to the input) relates to the amount of charge collected, and thus to the voltage collected. This instrument will respond to voltages in the range of $+/-7$ volts or so. Large collectors around high-voltage equipment such as Van DeGraf generators, Wimhurst machines, and Tesla coils will collect hundreds of volts and thus destroy the input IC. The voltage collected also relates to proximity of the source of charge for any given collector size. If in doubt, use small collectors or a whip antenna hooked to the input. Never walk across the room and simply touch the input collector or connector. You may carry thousands of volts on your body! Always ground yourself on the grounded instrument box. Remember, the box must always have a solid connection to a real earth ground to function properly.

Your electrometer can be used to determine the polarity of various insulators. Plastics are all negative. Polished, clean glass is always positive. Tapping one's toe in time with a piece of music while sitting on a modern carpet can induce a $+/-10$-volt potential charge on a can of Spam 5 feet away! The study of atmospheric electricity can be investigated. A people-proximity detector could be devised. No matter what your intended use, you will see in short order that every object is continuously exchanging charges with other objects in our everyday world. We exist in a sea of charge. For more information on electrostatics, an excellent book by A.D. Moore titled *Electrostatics* is available at a very modest price, and many experiments are suggested in it to introduce the amateur scientist to the fascinating world of electrostatics.

Advanced Electrometer Parts List

Required Parts

R1   10-megohm, 1/4-watt film resistor 5%

R2, R6   10K ohm, 1/4-watt film resistors 5%

R3   10K ohm potentiometer linear taper (PC type)

R4   3.9K ohm, 1/4-watt film resistor 5%

R7, R8   100K ohm, 1/4-watt film resistors 5%

C1   10 pF disc capacitor, 600 volts

C2, C3   0.22 uF, 50-volt, 10% Mylar capacitor

UI   LMC 6081 integrated circuit (Digikey)

U2   LM 324 integrated circuit (Digikey)

SP-1   8-inch metal sphere

S1   DPST toggle switch

J1   UHF chassis jack

J2   BNC chassis jack

B1, B2   9-volt transistor radio battery

Miscellaneous perfboard, IC sockets, enclosure, battery clips

Cloud Charge Monitor

Did you ever wonder how the charges were changing in the clouds directly overhead during a thunderstorm? Does an early-warning voltage exist right before a strike or do the charges jump from cloud to cloud too quickly? Is the polarity of the charge always the same? Build the cloud charge monitor and watch the charges ebb and flow for yourself.

The cloud charge monitor is an extremely sensitive device capable of detecting subtle changes in the accumulated overhead charge. The device consists of a charge-sensing antenna, a 60 Hz notch filter, a self-zeroing integrator, a signal limiter, and a leakage zero adjustment (see Figure 8-16). The monitor is a fairly sophisticated device with a few special construction requirements. Beginners may wish to solicit some help with construction and testing. For the more advanced experimenter each circuit function is described along with possible variations and performance enhancement ideas.

An overhead cloud charge is sensed by an antenna fashioned from a large aluminum-foil pizza pan or similar metal sheet. An insulated wire connects the pan antenna to the electronic circuit, using the following method. First, strip back several inches of the wire's insulation. Next, unroll a few inches of the pan's rim. Now, roll the rim back wrapping the bare wire inside. Then, crimp the rim in a few places with pliers to ensure a good connection. Finally, cut a hole in the center of the pan just large enough for ½-inch PVC. The pan is mounted at the end of a short piece of ½-inch electrical conduit PVC pipe. It may be secured by cutting small washers or donuts about ¼-inch tall from a ½-inch PVC coupling. Glue one washer about ⅓-inch from the end of the pipe, slip on the pan (with a hole punched in the center just large enough for the pipe), then glue another washer on top. Press the two washers together as the glue sets to firmly secure the pan. Next, thread the insulated wire down through the pipe. The pan may now be insulated by gluing two round pieces of trash-bag plastic, cut to be slightly larger in diameter than the pan. The bottom piece will need a hole for the conduit. (Satisfactory operation is possible without insulation.) The other end of the pipe may be secured to a metal box in a similar manner. An ordinary electrical outlet box is a good choice because it has a center punch-out just the correct size for the conduit, it has mounting ears that can be used to nail or fasten the box to a board, and it has plenty of room for the amplifier. A bottom cover is not necessary.

Construction

The cloud charge monitor circuit prototype is constructed on a piece of copper-clad circuit board using the "dead bug" technique. So named because the amplifier IC is mounted upside-down (leads pointing up) and will look a bit like a dead cockroach. Bend the ground pin (pin 4) back to the board and solder it directly to the foil. The other leads may

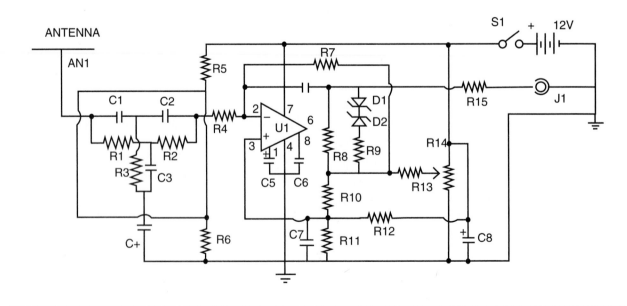

Figure 8-16   *Cloud charge monitor circuit*

be air wired, or little islands may be made by cutting small pieces of copper-clad board. The "bites" that come out of a typical nibbling tool are about the correct size! They may be soldered in place if the board has copper on both sides, or they may be glued. Make sure to inspect the edges to ensure that no metal sliver is shorting the island to the ground plane.

Other construction techniques are also fine, but a couple of points in the circuit should be wired in a special way. R6, R7, and pin 2 should be air wired and should not be allowed to touch anything. The base of the IC must be clean and free of solder flux. Use alcohol or lacquer thinner to clean the package if necessary. (The dead bug technique makes inspection easy!) The input filter is built on a separate piece of board, which may be mounted on the main board, but the circuit foil is biased to 6 volts instead of ground. R4 and R5 connect directly to the foil, which is indicated by a dotted line in the schematic. This bias helps to reduce leakage current from the input circuit to ground. If the builder has access to Teflon-insulated terminals then the input circuit may be air wired between terminals without concern for leakage. The insulated wire from the antenna should not rest against the metal box, so cut it short enough that it is stretched from the conduit to the circuit board. It is a good idea to solder a ground wire from the circuit ground foil to the metal box and to strain-relieve the power and signal cable.

The output of the circuit may be connected to a variety of readouts including a simple 1 mA current meter with a 5.6K series resistor or multimeter. The output voltage will read about 6 volts when no charge is present. R14 is adjusted to zero any offset due to leakage, and it may be remotely located with the meter and power supply, if desired. Leave R13 in the metal box, however. Remember, this circuit has a *very*-long time constant on the auto-zero circuit so the effect of any adjustment will take several minutes.

R7 may be made from a single resistor or several lower value resistors connected in series. The purpose of R7 is to automatically drive the output toward zero charge indication (6 volts out) so that the user isn't required to constantly adjust an offset control. The rate of auto-zero is determined by the value of

R7, the integrating capacitor CS, and the antenna capacity. The effective value of R7 is much higher than 220 megohms thanks to the bootstrapping circuit formed by R8 and R10. If R10 is decreased, the amount of bootstrapping will increase and the feedback resistor will seem even larger. A lower limit of 10K is recommended unless the experimenter is quite experienced! The effect of a larger feedback resistor is to increase the auto-zero time constant. A similar effect is realized by increasing C5, but a proportional loss in sensitivity will be evident. Lowering C5 will increase the sensitivity, if desired, but the sensitivity is sufficiently high for storm monitoring, as shown.

The bootstrapping technique also amplifies the op-amp's offset voltage and drift, but the chopper-stabilized ICL7650 has virtually no offset. Other CMOS op-amps may be used and might be more desirable because they have better overload recovery behavior. If the experimenter has extremely high-value resistors in the junk box, R10 could be increased or even removed and the diodes eliminated, creating a circuit with less sensitivity to op-amp drift and excellent overload recovery properties. Try a CA3160 op-amp or one of the numerous new CMOS amplifiers.

It can be quite difficult for the beginner to work with circuits that have several minute time constants and extremely high impedance, but the experience can be quite educational. To slew the circuit back to the center point after an overload, simply touch the ground or the +12 volts with one finger (depending on which way the circuit must slew), and touch pin 2 with the end of a 10-megohm resistor held in the other hand. To verify that the circuit is working, rub a CD or other piece of plastic on your hair and bring it close to the antenna. The meter should move up or down and then slowly drift back toward center. A 12-volt battery was used to supply power to the circuit.

Antenna experimentation is encouraged, but be very careful in placing an antenna outside unprotected from lightning. Place the device near a large window or hang it from the ceiling of a porch away from the rain. A different antenna can be made by taping aluminum foil to a clean window and covering it with plastic wrap. The edge of the foil should be several inches away from the window frame at all points to avoid leakage. Ordinary roofs are usually too conductive for an attic installation, but installing on a plastic or glass greenhouse roof or large skylight might work well. Those experimenters who are experts at handling the threat of lightning strikes might wish to build a Plexiglas umbrella for an outside antenna, but make sure that no discharge path to ground exists when the plexiglass becomes wet. The outdoor antenna should be mounted at ground level rather than high in the air, for safety reasons. The photo shown in Figure 8-17 illustrates a cloud charge monitor mounted on a short tripod and placed under a plexiglass roof enclosure.

Cloud Charge Monitor Parts List

Required Parts

R1, R2, R4, R8   10-megohm, 1/4-watt resistor

R3   5-megohm, 1/4-watt resistor

R5, R6, R11, R12   100K ohm, 1/4-watt resistor

R7   220-megohm, 1/4-watt resistor

R9, R13   1-megohm, 1/4-watt resistor

R10   47K ohm, 1/4-watt resistor

R14   100K ohm potentiometer

R15   470-ohm, 1/4-watt resistor

C1, C2   250 pF, 35-volt Mylar capacitor

C3   500 pF, 35-volt Mylar capacitor

C4   10 nF, 35-volt Mylar capacitor

C5, C6   220 nF, 35-volt Mylar capacitor

C7   100 nF, 35-volt Mylar capacitor

C8   10 uF, 35-volt electrolytic capacitor

D1, D2   1N750 zener diodes

U1   ICL7650 chopper-stabilized (or CA3160)

AN-1   pizza pan antenna

J1   RCA phone jack

S1   SPST toggle switch

B1   12-volt lantern battery

Miscellaneous perfboard, IC socket, wire, terminals, enclosure, etc.

Figure 8-17   *Cloud charge monitor mounted on a short tripod*

Electrical Field Disturbance Monitor

The electrical field disturbance monitor project will instruct you how to build a unique monitoring device to measure the voltage fluctuations or disturbances of the earth's natural electric field that are caused by conductive objects moving near the device. The magnitude and frequency of the fluctuations can be used to determine the actual nature of the object. As an example, large moving objects, such as trucks, produce large voltage swings with low-frequency components whereas small fast moving objects, such as birds, produce smaller field changes with higher frequencies. Moving humans, on the other hand, produce a variety of frequencies associated with arm and leg motions.

Unlike other projects that have very specific purposes in mind, this construction project provides the experimenter with a device that can be used to conduct a wide variety of object-motion experiments, including simple motion alarm applications. For example, with sufficient signal processing and analysis, it may even be possible for the experimenter not only to detect the motion of a human but to determine *which* human was moving. The objective of this construction project is to provide the experimenter with the equipment necessary to conduct research in the nature of the field changes.

I'm hoping also that this device will excite a new generation of experimenters into exploring this little known phenomenon. Before we get into the details of the circuit, let's cover some of the basics first—the phenomenon of the earth's electric field gradient. In his *Lectures on Physics*, Richard Feynman stated that as you go up from the surface of the earth, the electrical potential increases by about 100 volts per meter. Thus a vertical electric field gradient of 100 volts per meter exists in the air.

As a means of explanation, imagine the existence of a very sensitive voltmeter that could measure the voltages present in the open air. If you pushed the negative terminal of the instrument's probe into the earth's surface, and you positioned the positive lead 1 meter above the surface, about 100 volts would be detected. If you then moved the probe vertically by another meter above the surface, the voltmeter would measure 200 volts. This voltage difference would continue to increase as you moved the positive probe upward until it reached the top of the atmosphere, some 150,000 feet (46,000 meters) up. At that point, the instrument would finally measure an average potential difference of about 4 million volts.

This naturally occurring 100 volts per meter electric field gradient exists everywhere in the earth's atmosphere and can even penetrate inside most buildings. You might ask: If such voltages exist in open air, then why isn't the average 2-meter tall human shocked by the 200 volts that should be present between his feet and the top of his head? The reason you don't feel anything is because the air is too poor a conductor of electricity to allow enough current to be delivered by the voltage. Also, because the human body is filled with salt water, which is a good conduc-

tor, the body actually distorts the voltages as it moves through the field, reducing the actual potential difference across the body.

To illustrate this effect, suppose an air potential voltmeter was positioned to measure the atmospheric voltage between ground level and a position 2 meters above the ground surface. Without any conducting objects nearby, the device would measure a potential of 200 volts. But, when a conducting human walked next to the probe, the voltage is shunted to a near ground level potential and the instrument's reading would drop toward zero volts. As the human walked away from the meter, the field would again gradually be restored to the 200-volt reading. In addition to the average voltage changes as a person walks near the detector, the shifting contact of the human's foot with the ground and his or her arm motion cause small higher frequency fluctuations in the measured voltage. Such signal changes are used as the basis for detecting human motion near the detection circuit.

The field disturbance monitor is illustrated in Figures 8-18 and 8-19. It uses a telescoping whip antenna that is mounted on top of a metal box to probe the air for field changes. By raising and lowering the antenna, you can increase or decrease the field change sensitivity. The design I chose uses an off the shelf metal box to house the electronics. The circuit is powered by a standard 9-volt battery. Three LED indicator lights provide system status. One of the LEDs indicates a positive field change, and another indicates a negative field change. A third power indicator light doubles as a battery-voltage indicator. If the light fails to turn on, it is an indication that the 9-volt battery needs to be replaced. An alarm sensitivity dial can set the minimum disturbance level to trigger an alarm. A loud piezoelectric-type beeper sounds whenever the alarm level is exceeded. A toggle switch allows the alarm feature to be turned off. An output jack at the rear of the monitor can be used to connect the monitor to a remote alarm device if desired. I have also included an output jack that can be used to send the monitor's processed disturbance signal to some remote recording device. Finally, to ensure consistent operation, an earth ground jack is also included at the rear of the monitor's enclosure. Connecting the ground jack to a true earth ground improves disturbance sensitivity.

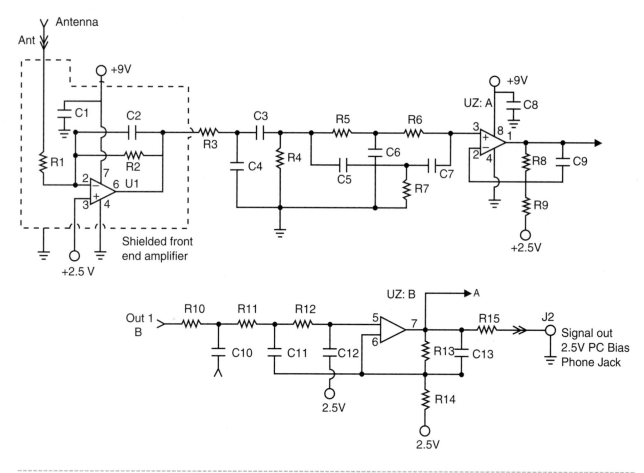

Figure 8-18 *Electronic field disturbance monitor circuit, part 1*

Circuit Description

Most moving objects, including humans, produce electric field disturbances with frequencies that range between 0.1 Hz and 15 Hz. However, when the disturbance monitor is used indoors, those signals must compete with rather large fields produced by nearby 50 Hz and 60 Hz power lines and appliances. Because the motion signals of interest can be as much as 1,000 times smaller than the signals produced by power lines, much of the monitor's electronic circuit is dedicated to removing the unwanted power line frequencies. If the monitor is to be used only outside or away from 60 Hz power-line interference, a less aggressive filter design can be used.

The circuit can be broken into several sections. The most important part of the design is the front-end section. Many different front-end circuits were tried over the years. The circuit included in this project is

both simple and effective. It uses an operational amplifier U1 that is wired as an impedance amplifier. The circuit has a very high-input impedance and a low-output impedance. The small voltage signals collected by a telescoping whip antenna, caused by an object moving near the monitor, are routed to the amplifier circuit. The 1-gigaohm feedback resistor R2 provides the amplifier with a DC feedback path, while the capacitor C1, in parallel with resistor R2, reduces the gain of the amplifier at high frequencies. Without the capacitor C1, the amplifier would easily be swamped by AC fields associated with any 50 to 60 Hz power lines or line-powered devices nearby. The 100K resistor R1 is placed between the antenna and the amplifier to protect the amplifier from being damaged by any high-voltage static charge that could be picked up by the antenna. The 1-gigaohm resistor value is important for studying human motion signals. Higher resistance values, going up to 100 gigaohms, have been tried and cause the monitor's frequency

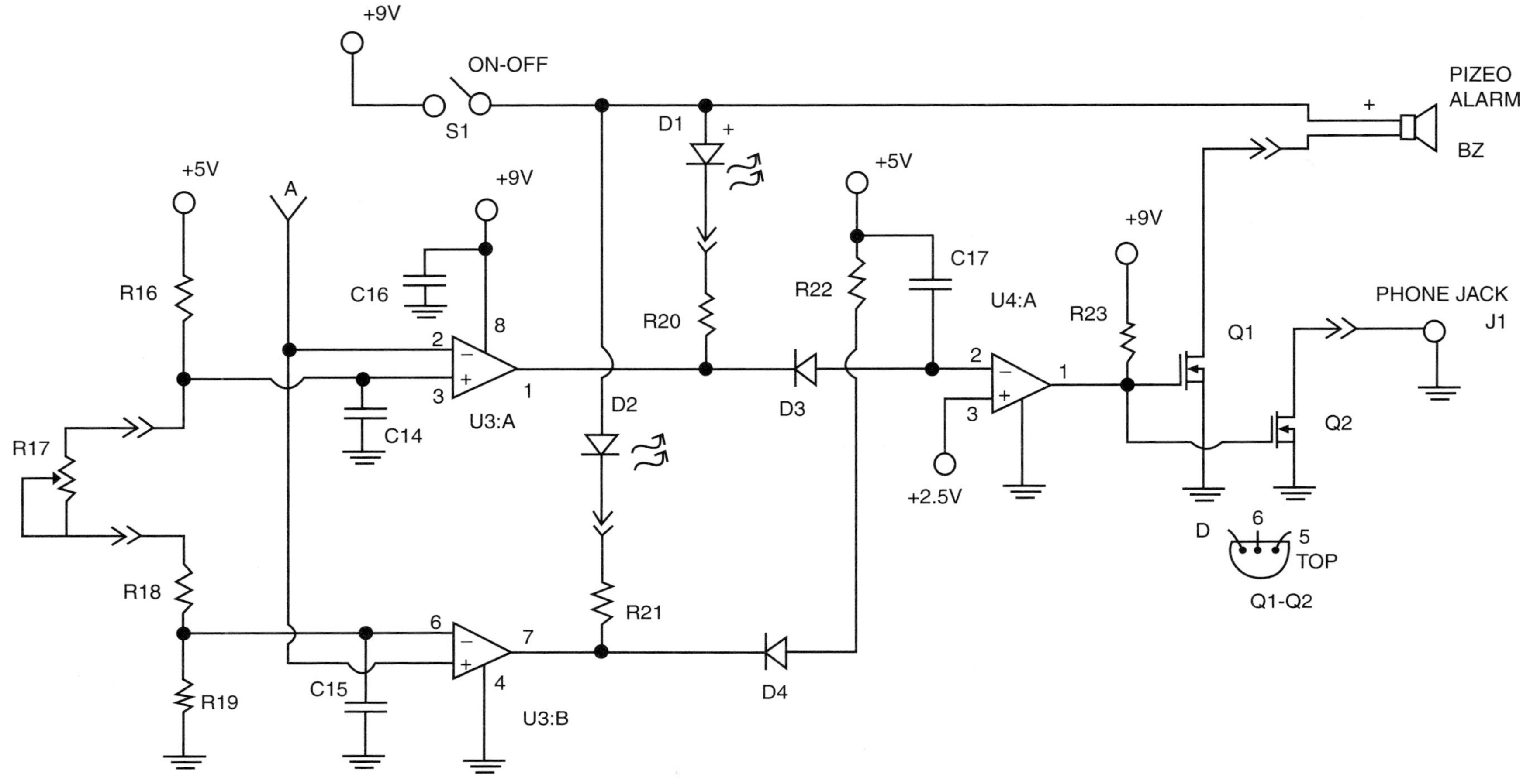

Figure 8-19 *Electronic field disturbance monitor circuit, part 2*

response to be excessively low. Electric field changes from nearby rain storms can be measured with high resistance values. A resistance value lower than 1 gigaohm makes the monitor more sensitive to higher frequencies.

The signal that emerges from the front-end circuit will contain a large amount of power-line noise. Even when the monitor is used outside away from visible power lines, there will still be some unwanted power-line signals collected. The passive filter section after the front-end stage contains three networks: (1) One low-pass filter network begins the process of attenuating the unwanted high frequencies. (2) One high-pass filter network is designed to block the slow DC shift that will occur at the front-end circuit. The values selected start rejecting frequencies below 0.1 Hz. (3) To remove much of the fundamental 50 to 60 Hz noise signals, a third notch filter network is used. As shown on the schematic, components were selected for a 55 Hz notch-filter center frequency. The selection is a compromise between the 50 Hz and 60 Hz international power-line frequency standards in use around the world. The notch filter should reduce the power-line frequency noise by a factor of $1/50$ (-34 db).

The output of the notch filter is connected to a noninverting operational amplifier (U2:a) at the first buffer stage. The values chosen give the signals of interest a gain of about ×6 while rejecting some of the unwanted higher frequencies.

The output of the first buffer stage is connected to a three-pole active low-pass filter, and to a second buffer stage. The combination of the passive components and the operational amplifier (A2:b) boosts the signals of interest with a gain of ×6 while filtering the high-frequency signals that may still remain. The overall gain for the two amplifiers is about ×36 (+31 db). Note that all three of the operational amplifier stages are biased at 2.5 volts. The signals of interest will therefore swing above and below 2.5 volts.

The output of the second buffer stage is routed to a phone jack at the rear of the monitor enclosure. Using a shielded cable connected to the phone jack, the signal can be fed to a strip chart recorder or to a digital recorder that is connected to a computer. Using some digital signal processing schemes, a lot of information can be squeezed from the raw signals generated by the monitor. As stated previously, identification of specific individual humans is possible by carefully monitoring the frequency signature produced by a person's arm and leg motions during walking. In a similar manner, certain animals and insects can be identified.

For some experiments, you may want to know if the signals exceed a certain level. This feature is especially useful if the monitor is to be used in a motion alarm application. The signals that emerge from the signal-processing circuits are routed to two comparators (U3:a and U3:b). The two comparators determine if the signal has sufficient amplitude to be considered an alarm condition. Comparator U3:a is referenced above the 2.5-volt bias point, while the low U3:b stage is referenced below the 2.5-volt bias voltage. The upper comparator is triggered when the signal swings above the upper threshold (positive voltage change), and the lower comparator is triggered when the signal swings below the lower threshold (negative voltage change). The variable resistor R17 symmetrically controls both thresholds and allows a single knob to set the trigger sensitivity levels. With the values chosen, the comparators can be triggered by voltage changes as small as $+/-0.05$ volts or as high as $+/-1.5$ volts. An LED connected to the output of each comparator provides an indication of either a positive or a negative disturbance. Both LEDs are mounted on the front side of the monitor's metal panel. Diodes D3 and D4 sum the two comparator outputs and, with the aid of another comparator A4a, drive the transistors Q1 and Q2. The two transistors Q1 and Q2 are used to drive the piezoelectric alarm connected to the monitor or an external alarm that is connect by a shielded cable to the remote alarm output jack at the rear of the monitor. The external alarm output can sound a remote beeper or close a relay. If desired, the local alarm feature can also be turned off when the alarm selector switch is placed in the off position. In the off position, the switch disconnects the 9-volt source to the local beeper alarm and to the two indicator LEDs. When the system is operated in the alarm off mode, the overall current will be much less and will extend the operating time from the battery.

To provide the system with two well-regulated voltages from the 9-volt battery, two voltage regulators are used, as shown in Figure 8-20. The regulator U5 generates 5 volts, and U6 produces 2.5 volts. The 2.5-volt supply generates the midsupply bias voltage used in the signal processing circuits, whereas the 5-volt supply is used in the alarm and the voltage monitor circuits. Because the current demands of both supplies are very low, low-power regulators are used.

Another voltage comparator, U4:b, is wired as a battery-voltage monitor. Should the battery voltage drop below about 6.8 volts, the power indicating LED will not turn on, telling the user that the battery needs to be replaced.

Circuit Assembly

A metal box should be selected to house the monitor electronics. A telescoping whip antenna makes an excellent probe for collecting the electric field changes near the monitor. The antenna can easily be raised or lowered to increase or decrease sensitivity. A banana plug soldered onto the end of the whip antenna can be plugged into a matching insulated banana jack, mounted on top of the box. The metal chassis forms an electric shield around the electronic circuits inside and forms a reference capacitor plate with a large cross-sectional area. As the previous discussion suggests, the metal chassis and the whip antenna form the two vertical points in space needed to detect the electric field change.

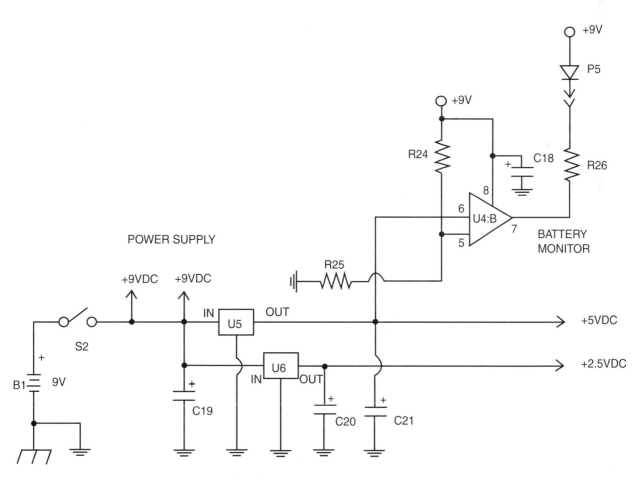

Figure 8-20 *Field disturbance monitor power supply*

The metal box selected has a top section that slides over the bottom section in two U-shaped pieces. All of the chassis-mounted parts can then be mounted onto the top section. The circuit board can be mounted, upside down, to the inside top of the box. Four 1-inch-long metal standoff legs should be used to suspend the circuit board.

The front-end amplifier circuit should be enclosed within a metal can that is soldered to the circuit board and connected to circuit ground. The shield can be easily constructed using sections cut from tin-plated steel sheets purchased from any hobby store. The can will help prevent other signals generated inside the enclosure from being picked up by the very sensitive front-end circuit. The open top of the can should be made about 1 inch high so it nearly touches the inside of the metal enclosure. It should be positioned so it surrounds the banana jack that the whip antenna plugs into. You can connect the banana jack terminal to the front-end circuit by feeding a wire that is soldered to the jack through a hole in the circuit board.

The signal output jack and the earth ground terminal should be installed on the rear of the box. A 9-volt battery clip is used to secure the 9-volt battery inside the box. Although having the battery inside the box is less convenient, a fresh battery should provide many days of experiments. The three LED indicator lights, the piezoelectric beeper, the alarm-level control knob, and the two toggle switches should be mounted on the front of the panel. Connections from the circuit board to the alarm LEDs, beeper, rear output signal jack, and rear alarm jack should all be made using shielded cables. When completed, the disturbance monitor will resemble a portable radio with a single, large, whip antenna protruding out from its top. By gluing a $^1/_4$-20 nut on the bottom of the box, the assembly can be attached to a standard metal camera tripod.

Operation

The system works best when the metal box is attached to a metal tripod. If a plastic or wood tripod is used, an earth ground reference should be established by connecting a wire from the unit's earth ground terminal to a metal rod, which can be driven into the ground. An old screwdriver with the wire attached makes a convenient grounding tool. If the unit is used indoors, a good earth ground can be obtained by connecting the monitor's ground terminal to a water pipe or to the ground terminal of a power outlet. The monitor will still work, even without an earth ground, but it will lack overall sensitivity, especially to the low-field-change frequencies.

I should also mention that the monitor works much better in dry environments. Moist air tends to be slightly more conductive than dry air, and therefore less static electricity is produced by walking humans. Because the monitor's frequency response extends down to one cycle in 5 seconds, you should expect the monitor to go into an alarm condition for several seconds after the power is first turned on. Likewise, when the monitor is subjected to a large field change, which may saturate the front-end circuit, the alarm may sound for several seconds, even after the field has stabilized.

A standard phone jack mounted on the rear panel can be used to route the disturbance signals to a data collection device. When using the signal output jack, a shielded audio type cable should be used to route the signal to the recording device. The output signal is designed to have a 2.5-volt center-bias point, so the signals of interest will swing above and below 2.5 volts. This voltage range is ideal for many digital recording systems that are connected to a computer.

When you are satisfied that the circuit board and all the components are wired correctly, attach a fresh 9-volt battery to the battery clip. Make sure the unit is turned off when you connect the new battery. Extend the whip antenna to a mid-24-inch length. Lay the monitor on a high wooden table or on top of a wooden shelf. Position the sensitivity dial to the midpoint. Switch the alarm switch to the on position. Switch the power switch to the on position, and back away about 6 feet from the box. Stay perfectly still and do not move. The alarm should sound for several seconds. After the alarm stops, you should be able to move one of your feet and see the two LED indicator lights turn on and off while the alarm beeper sounds. Try walking past the unit and notice how the two indicator lights turn on and off matching your foot steps.

Electrical Field Disturbance Monitor Parts List

Required Parts

R1 100K ohms, 1/4-watt resistor

R2 1-gigaohm, 1/4-watt resistor

R3 10K ohm, 1/4-watt resistor

R4 22-megohm, 1/4-watt resistor

R5, R6, R11, R12 620K ohm, 1/4-watt resistor

R7 300K ohm, 1/4-watt resistor

R8, R13 56K ohm, 1/4-watt resistor

R9, R14 10K ohm, 1/4-watt resistor

R10, R18 22K ohm, 1/4-watt resistor

R15 470-ohm, 1/4-watt resistor

R16, R25 1-megohm, 1/4-watt resistor

R17, R19, R23 1-megohm potentiometer

R20, R21, R26 4.7K ohm, 1/4-watt resistor

R22 4.7-megohm, 1/4-watt resistor

R24 360K ohm, 1/4-watt resistor

C1, C8, C9, C10, C13 0.1 uF, 35-volt disc capacitor

C14, C15, C16, C18 0.1 uF, 35-volt disc capacitor

C2 22 pF, 35-volt Mylar capacitor

C3, C4 0.47 uF, 35-volt Mylar capacitor

C5, C7 0.0047 uF, 35-volt Mylar capacitor

C6, C12 0.01 uF, 35-volt disc capacitor

C11 0.022 uF, 35-volt disc capacitor

C17 0.22 uF, 35-volt disc capacitor

C19 47 uF, 35-volt electrolytic capacitor

C20, C21 10 uF, 35-volt electrolytic capacitor

D1, D2 red LED

D2, D3 1N4148 silicon diode

Q1, Q2 BS170 transistor

U1 LP661 integrated circuit

U2 LP662 integrated circuit

U3, U4 LMC6762 integrated circuit

U5 ZMR500C (Zetec)

U6 ZMR250C (Zetec)

BZ piezo buzzer

S1 SPST toggle switch

J1, J2 RCA phone jacks

B1 9-volt transistor radio battery

ANT telescoping whip antenna

Miscellaneous PC board, IC sockets, wire, solder, chassis, battery clips

Radio Projects

Electromagnetic energy encompasses an extremely wide frequency range. Radio frequency energy, both natural radio energy created by lightning and planetary storms as well as radio frequencies generated by humans for communications, entertainment, radar, and television are the topic of this chapter. This RF, or *radio frequency* energy, covers the frequency range from the low end of the radio spectrum (10 to 25 KHz), which is used by high-power navy stations that communicate with submerged nuclear submarines, all the way through the radar frequency band of 1,000 to 1,500 MHz. Beyond that, the spectrum extends through approximately 300 GHz. The radio frequency spectrum actually extends almost up to the lower limit of visible light frequencies.

In this chapter we will explore different types of radio receivers that you can construct and use. Our first project is an electronic lightning detector, used to warn of oncoming electrical storms—a great project for weather enthusiasts and researchers. Our next project is the ELF natural radio, which can be used to listen to those mysterious sounds produced by Mother Nature, such as tweeks, pops, whistlers, and the dawn chorus. These ascending and descending frequency sweeps are caused by electrical storms on the other side of the earth. Why not build your own shortwave receiver and explore the world of radio from foreign broadcasters on the other side of the globe. Hear exciting music and news from European and African radio stations. Our final advanced project in this chapter is the Jupiter radio telescope project, which will permit you to listen to the strange sounds of storms on the planet of Jupiter. This radio receiver project is a great starting point for some great amateur research on radio astronomy.

Radio History

One of the more fascinating applications of electricity is in the generation of invisible ripples of energy called radio waves. Following Hans Oersted's accidental discovery of electromagnetism, it was realized that electricity and magetism were related to each other. When an electric current was passed through a conductor, a magnetic field was generated perpendicular to the axis of flow. Likewise, if a conductor was exposed to a change in magnetic flux perpendicular to the conductor, a voltage was produced along the length of that conductor.

Joseph Henry, a Princeton University professor, and Michael Faraday, a British physicist, experimented separately with electromagnets in the early 1800s. They each arrived at the same observation: the theory that a current in one wire can produce a current in another wire, even at a distance. This phenomenon is called *electromagnetic induction*, or just *induction*. That is, one wire carrying a current induces a current in a second wire. Up until this time, scientists knew that electricity and magnetism always seemed to affect each other at right angles. However, a major discovery lay hidden just beneath this seemingly simple concept of related perpendicularity, and its unveiling was one of the pivotal moments in modern science.

The man responsible for the next conceptual revolution was the Scottish physicist James Clerk Maxwell (1831–1879), who unified the study of electricity and magnetism in four relatively tidy equations. In essence, what he discovered was that electric and magnetic fields are intrinsically related to one another, *with or without* the presence of a conductive path for electrons to flow. Stated more formally, Maxwell's discovery was this:

A changing electric field produces a perpendicular magnetic field, and a changing magnetic field produces a perpendicular electric field. All of this can take place in open space, the alternating electric and magnetic fields supporting each other as they travel through space at the speed of light. This dynamic structure of electric and magnetic fields propagating through space is better known as an *electromagnetic wave*.

Later between the years 1886 and 1888 Heinrich Hertz, the German physicist who is honored by our replacing the expression *cycles per second* with *hertz* (Hz), proved Maxwell's theory. Shortly after that in 1892, Edouard Branly, a French physicist, invented a device that could receive radio waves (as we know them today) and cause them to ring an electric bell. Note that at the time, all the research being conducted in what was to become radio and later radioelectronics was done by physicists.

Then finally in 1895, the father of modern radio, Guglielmo Marconi of Italy, put all this together and developed the first wireless telegraph. The wire telegraph had been in commercial use in Europe for a number a years.

Types of Radio Waves

There are many kinds of natural radioactive energy composed of electromagnetic waves. Even light is electromagnetic in nature. So are shortwave, X-ray, and gamma-ray radiation. The only difference

between these kinds of electromagnetic radiations is the frequency of their oscillation (alternation of the electric and magnetic fields back and forth in polarity). By using a source of AC voltage and a special device called an *antenna*, we can create electromagnetic waves (of a much lower frequency than that of light) with ease.

It was discovered that high-frequency electromagnetic currents in a wire (antenna), which in turn result in a high-frequency electromagnetic field around the antenna, will result in electromagnetic radiation that will move away from the antenna into free space at the velocity of light (approximately 300,000,000 meters per second).

In radio broadcasting, a radiating antenna is used to convert a time-varying electric current into an electromagnetic wave, which freely propagates through a nonconducting medium, such as air or space. An antenna is nothing more than a device built to produce a dispersing electric or magnetic field. An electromagnetic wave, with its electric and magnetic components, is shown in Figure 9-1.

When attached to a source of radio-frequency signals (a generator or a transmitter), an antenna acts as a transmitting device, converting AC voltage and current to electromagnetic-wave energy. Antennas also have the ability to intercept electromagnetic waves and convert their energy into AC voltage and current. In this mode, an antenna acts as a receiving device.

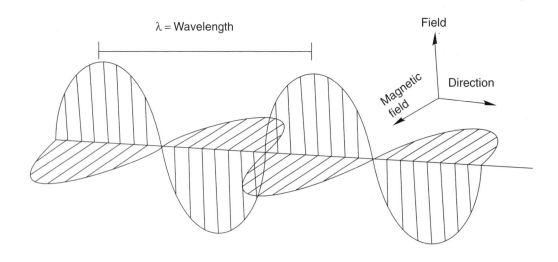

Figure 9-1 *Electric and magnetic components of an electromagnet wave*

Radio-Frequency Spectrum

Radio-frequency energy is generated by humans for communications, entertainment, radar, television, and so on. This *radio frequency* (RF) energy covers the range from the low end of the radio spectrum (10 to 25 KHz), which is the domain occupied by the high-power navy stations that communicate with submerged nuclear submarines, through the familiar AM broadcast band (550 to 1,600 KHz), on through the shortwave band (2,000 to 30,000 KHz), which makes use of multiple reflections from the ionosphere that surrounds the earth. The next band of frequencies are used by the very high-frequency television channels (54 to 216 MHz), followed by the very popular *frequency modulation* (or FM) band (from 88 to 108 MHz). Following the FM broadcast band are aircraft frequencies, UHF television channels, and the radar frequency band of 1,000 to 1,500 MHz. Beyond that, the spectrum extends through approximately 300 GHz. The radio frequency spectrum actually extends almost up to the lower limit of visible light frequencies with just the infrared frequencies lying in between it and visible light. Table 9-1 illustrates the division of radio frequencies. A radio frequency spectrum chart is shown in Figure 9-2.

Table 9-1

Radio Frequency Bands

Hertz		Specific Band
3 to 30	kilohertz	Very low frequencies (VLF)
30 to 300	kilohertz	Long-wave (LW) band
300 to 3,000	kilohertz	Medium-wave (MW) band
3 to 30	megahertz	Shortwave (SW) or high-frequency (HF) band
30 to 300	megahertz	Very-high-frequency (VHF) band
300 to 3,000	megahertz	Ultra-high-frequency (UHF) band
3 to 30	gigahertz	Super-high-frequency (SHF) band
30 to 300	gigahertz	Microwave frequencies
Above 300	gigahertz	Infrared, visible light, ultraviolet, X-ray, and gamma rays

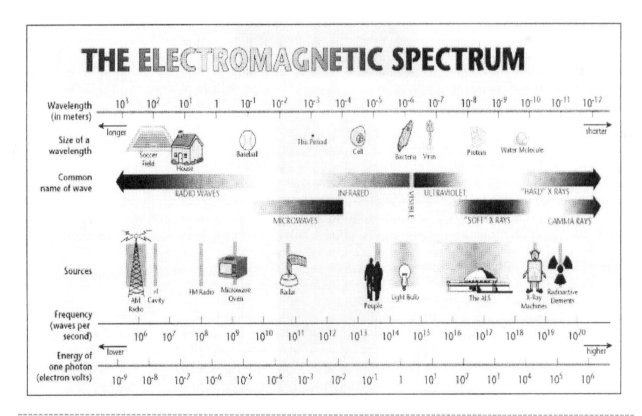

Figure 9-2 *Radio frequency spectrum chart*

In this chapter we will learn how to detect radio energy caused by lightning. The *very-low-frequency* (VLF) radio receiver will permit you to listen to low-frequency natural radio signals, such as whistlers and the dawn chorus. The shortwave radio will open up a whole new world of listening to DX radio stations from all around the globe. Science-minded enthusiasts will be interested in building a radio receiver that will permit you to listen and record natural radio energy emitted from Jupiter.

Detecting Lightening

Intense electrical storms created by unstable weather conditions occur frequently all around the world. At any given point in time, there is likely an electrical storm occurring somewhere in the world. Most people, especially hikers, boaters, and even backyard party goers, would be grateful to have an advance warning of an approaching electrical storm. Now you can construct your own lightning monitor to obtain an advance warning of an approaching electrical storm before you see or hear the lightning. The sensitive lighting detector project will alert you to an oncoming storm from over 50 miles away, giving you time to take cover or go inside to safety.

Many have wondered: During a lightning strike, is the earth considered positive or negative? In an electrical storm, the storm cloud is charged like a giant capacitor. The upper portion of the cloud is positive and the lower portion is negative. Like all capacitors, an electrical field gradient exists between the upper positive and lower negative regions. The strength or intensity of the electric field is directly related to the amount of charge buildup in the cloud. The charge is created by colliding water droplets.

As the collisions continue and the charges at the top and bottom of the cloud increase, the electric field becomes more intense—so intense, in fact, that the electrons at the earth's surface are repelled deeper into the earth by the strong negative charge at the lower portion of the cloud. This repulsion of electrons causes the earth's surface to acquire a strong, positive charge.

The strong electric field also causes the air around the cloud to break down and become ionized (a plasma). A point is reached (usually when the gradient exceeds tens of thousands of volts per inch) when the ionized air begins to act like a conductor. At this point, the ground sends out feelers to the cloud, searching for a path of least resistance. Once that path is established, the cloud-to-earth capacitor discharges in a bright flash of lightning.

Because an enormous amount of current exists in a lightning strike, there's also an enormous amount of heat. (In fact, a bolt of lightning is hotter than the surface of the sun.) The air around the strike becomes superheated, so hot that the air immediate to the strike actually explodes. The explosion creates a sound wave that we call *thunder*.

Cloud-to-ground strikes are not the only form of lightning though. There are also ground-to-cloud strikes (usually originating from a tall structure) and cloud-to-cloud strikes. These strikes are further defined into normal lightning, sheet lightning, heat lightning, ball lightning, red sprite lightening, blue jet lightening, and others. For more information on lightning, check out these two great Web sites:

http://science.howstuffworks.com/lightning.htm

www.lightningstorm.com/tux/jsp/gpg/lex1/map display_free.jsp

Satellites are often used to follow lightning strikes around the world and haven't advanced to the point where they can accurately map local areas. Two major types of sensors are commonly used: magnetic direction finders and VHF interferometery. The *National Lightning Detection Network* (NLDN), which is operated by *Global Atmospherics, Inc.* (GAI) in Tucson, Arizona, is a network of more than 130 magnetic direction finders that covers the entire U.S.A.—more than twice the coverage of existing weather radar networks. Each direction finder determines the location of a lightning discharge using triangulation and is capable of detecting cloud-to-ground lightning flashes at distances of up to 250 miles and more. Processed information is transmitted to the Network Control Center, where it is then displayed in the form of a grid map showing lightning across the United States.

Recently, the *National Space and Aeronautics Administration* (NASA) has improved the resolution of the system by adding acoustical measurements to the mix. Although the flash and resulting thunder occur at essentially the same time, light travels at 186,000 miles per second, whereas sound travels at the relative snail's pace of one-fifth of a mile per second. Thus, the flash—if not obscured by clouds—is seen before the thunder is heard. By counting the seconds between the flash and the thunder and dividing by 5, an estimate of the distance to the strike (in miles) can be made.

In the NASA lightning sensors, low-frequency receivers detect the lightning strike. The leading edge of the electric-field pulse is used to start a timer, and the leading edge of the thunder pulse stops the timer. A microcontroller in each receiver transmits the time measured to a processing station, where the times are converted to distances that are used to compute the location of the lightning strike to within 12 inches. The only drawback is that the NASA sensors have to be located within a 30 mile radius of the strike to be accurate.

Lightning Detector

Lightning flashes generate a broad spectrum of radio frequencies with especially intense emissions in the

VLF band. You can build your own VLF radio receiver that is capable of monitoring lightning flashes from an on-coming electrical storm. The lightning receiver circuit is shown in Figure 9-3. This receiver is designed to pick up a band of frequencies near 300 KHz, a range that is a fairly empty except for lightning static. These radio crackles are picked up by an antenna with the help of the 10 *millihenry* (mH) choke. Short antennas behave as though a very tiny capacitor is connected in series, and this choke resonates with this capacitor allowing current to flow into the receiver. The 330 *microhenry* (uH) and 680 pF capacitor form a tuned circuit at 300 KHz, and the 0.01 uF capacitor couples this tank circuit (or a coil/capacitor combination used as a resonant circuit) into the base of the first transistor amplifier.

The amplified radio signal on the collector of Q1 is coupled into the base Q2 through capacitor C3. The circuit gain is controlled by the potentiometer at R5. The collector of transistor Q3 is coupled through diode D2 and R7 to the base of transistor Q4, which acts as a meter driver circuit. The output of Q4 provides an output to the averaging meter circuit, which shows a fairly steady reading that is proportional to the lighting activity.

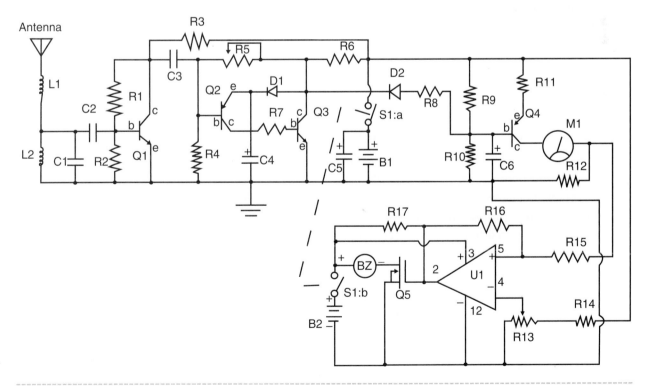

Figure 9-3   *Lightening detector circuit (Courtesy Charles Wenzel)*

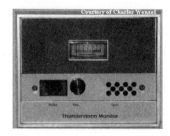

The DC output at the meter circuit also provides an output, which is used to drive an additional comparator circuit that can be used to drive higher current loads such as for alarms and motors. Note that a separate power supply is used to power the external load that, in this diagram, is a buzzer but could be a higher current load if desired. The meter circuit drives the comparator via the input resistor. A 10K ohm resistor at R11 drives the positive input of the LM339 comparator at pin 5. A 3-volt bias is applied to the minus input of the op-amp at pin 4 and the threshold is adjusted via the potentiometer at R13. The comparator output at pin 2 drives the VMOS n-channel power transistor at Q5.

Construction

The lighting detector circuit is built on a glass epoxy circuit board measuring 4 × 6 inches, but you could elect to build the circuit on a perf-board if desired. The circuit is simple to construct and has no special wiring considerations because it operates at low frequencies around 300 KHz. When constructing the lighting detector, be careful to observe the correct polarity when inserting the electrolytic capacitors. Also note that you must observe the proper orientation when installing the diodes and the transistor semiconductors. The lightning detector utilizes a single IC at U1. Locate an IC socket and install it prior to installing the IC. When installing the IC be sure to correctly insert the IC into its socket. Most ICs have either a cutout or notch at the top of the IC or an indented circle near pin 1 of the IC. Pin 1 is usually just to the left of either the cutout or the indented circle.

The prototype lightning detector is powered from two C cells. As mentioned earlier, to drive a higher load, the comparator circuit is powered by a second battery determined by the load that you choose. If, for example, you select a 6-volt load or buzzer at the output of the comparator circuit, you would have to provide a 6-volt source by using four C cells.

You will need to locate two battery holders, one for the main circuit and one for the load circuit. The main circuit can use a two-C-cell battery holder, and the battery holder chosen for the load circuit is dependant upon the load that you have chose. A 6-volt load

Figure 9-4 *On-off switch, gain control, and buzzer on from panel of chassis*

would require a 4-cell battery holder, and so on. After completing the circuit board, be sure to check and recheck your circuit board for errors and shorts.

The lighting monitor circuit was enclosed in a 6 × 8 × 5 metal chassis box. The on-off switch, a DPST switch was used for power in the circuit and was placed on the front panel of the chassis along with the meter, gain control, and the buzzer, as shown in Figure 9-4. The prototype uses two chassis-mounted potentiometers, and they are both mounted on the front panel of the enclosure as well.

A binding post or banana-type jack is placed on the top of the chassis for securing the antenna. Depending upon your choice of antenna, you could use a telescoping antenna, a wire antenna attached to the binding post, or a wire whip antenna attached to a banana plug.

Operation

Operation of the lightning monitor is quite simple. Once the circuit has been built and placed in the chassis, your next task will be to power up the circuit and test it. Place the batteries into the holders. Insert the antenna into the input jack and power up the circuit via switch S1.

Next, adjust the sensitivity control via potentiometer R5 for maximum sensitivity, or fully clockwise. Finally, adjust the comparator potentiometer at R13. Turn the R13 clockwise until the alarm sounds and then back down the control until the sound stops, and the comparator control is set for normal operation. Your lighting monitor is now ready for storm detection! You'll be glad you built it once the next storm rumbles into your town.

Lightning Detector Parts List

Required Parts

R1, R2, R4, R9, R14
1-megohm, 1/4-watt resistor

R3, R15 10K ohm, 1/4-watt resistor

R5 200K ohm potentiometer

R6 4.7K ohm, 1/4-watt resistor

R7 3.9K ohm, 1/4-watt resistor

R8 100K ohm, 1/4-watt resistor

R10 4.7-megohm, 1/4-watt resistor

R11 5K ohm, 1/4-watt resistor

R12, R17 1K ohm, 1/4-watt resistor

R13 50K ohm potentiometer

R16 10-megohm, 1/4-watt resistor

C1 680 pF, 35-volt disc capacitor

C2 0.001 uF, 35-volt disc capacitor

C3 0.001 uF, 35-volt disc capacitor

C4 1 uf, 35-volt electrolytic capacitor

C5 10 uF, 35-volt electrolytic capacitor

C6 100 uF, 35-volt electrolytic capacitor

D1, D2 1N914 silicon diode

L1 10 mH choke coil

L2 330 uH choke coil

Q1, Q3 2N4401 transistor

Q2, Q4 2N4403 transistor

Q5 VN10KM VMOS power transistor

U1 LM339 quad comparator IC

M1 100 uA mini panel meter

S1 DPST toggle switch

BZ 6-volt buzzer or other load (see text)

B1 D or C cells (2 cells to equal 3 volts)

B2 see text

Miscellaneous PC board, wire, binding post, antenna, etc.

ELF/VLF Radio or Nature's Radio

Few people know of the beautiful radio music produced naturally by several processes of nature, processes including lightning storms and the aurora, that are aided by events occurring on the sun. The majority of Earth's natural radio emissions occur in the *extremely-low-frequency* and *very-low-frequency* (ELF/VLF) radio wave spectrum.

These natural radio signals have caught the interest and fascination of a small but growing number of hobby listeners and professional researchers for the past four decades. Whistlers, one of the more frequent natural radio emissions to be heard, are just one of many natural radio sounds the earth produces at all times in one form or another. Whistlers are magnificent sounding bursts of ELF/VLF radio energy initiated by lightning strikes that fall in pitch. A whistler, as heard in the audio output from a VLF whistler receiver, generally falls lower in pitch, from as high as the middle-to-upper frequency range of our hearing downward to a low pitch of a couple hundred *cycles per second* (or hertz). Measured in frequency terms, a whistler can begin at over 10,000 Hz and fall to less than 200 Hz, though the majority are heard from 6,000 down to 500 Hz. Whistlers can tell scientists a great deal about the space environment between the sun and the earth and also about Earth's magnetosphere.

The causes of whistlers are generally well known today though not yet completely understood. What is clear is that whistlers owe their existence to lightning storms. Lightning strike energy happens at all electromagnetic frequencies simultaneously. The earth is literally bathed in lightning-stroke radio energy from an estimated 1,500 to 2,000 lightning storms in progress at any given time, triggering over a million lightning strikes daily. The total energy output of lightning storms far exceeds the combined power output of all manmade radio signals and electric power generating plants.

Whistlers also owe their existence to the sun and to the earth's magnetic field (magnetosphere), which surrounds the planet like an enormous glove. Streaming from the sun is the *solar wind*, which consists of energy and charged particles, called *ions*. And so, the combination of the sun's solar wind, the earth's magnetosphere surrounding the entire planet, and lightning storms all interact to create the intriguing sounds and great varieties of whistlers.

How whistlers happen from this combination of natural solar-terrestrial forces is (briefly) as follows: Some of the radio energy bursts from lightning strikes travel into space beyond Earth's ionosphere layers and into the magnetosphere, where they follow approximately the lines of force of the earth's magnetic field to the opposite polar hemisphere. They travel along ducts formed by ions streaming toward Earth from the sun's solar wind. Solar-wind ions get trapped in and aligned with Earth's magnetic field. As the lightning energy travels along a field-aligned duct, its radio frequencies become spread out (dispersed) in a similar fashion to light shining into a glass prism. The higher radio frequencies arrive before the lower frequencies, resulting in a downward falling tone of varying purity.

A whistler will often be heard many thousands of miles from its initiating lightning strike, and in the opposite polar hemisphere! Lightning storms in British Columbia and Alaska may produce whistlers that are heard in New Zealand. Likewise, lightning storms in eastern North America may produce whistlers that are heard in southern Argentina or even Antarctica. Even more remarkably, whistler energy can also be bounced back through the magnetosphere near (or not-so-near) the lightning storm from which it was born!

Whistlers are descending tones. Their duration can range from a fraction of a second to several seconds, with the rate of frequency shift steadily decreasing as the frequency decreases. A whistler's note may be pure, sounding almost as if it was produced using a laboratory audio signal generator. Other whistlers are more diffuse, sounding like a breathy *swoosh* or like they are composed of multiple tones. On occasion, some whistlers produce echoes or long progressions of echoes, known as *echo trains*, that can continue for many minutes.

Whistlers can occur in any season. Some types are more likely to be heard in summer than winter and vice versa. Statistically, the odds are especially high for good whistler activity between mid-March and mid-April.

Whistlers and the sounds of the dawn chorus can not be heard equally well everywhere in the world. Reception of these is poor in equatorial regions and best at geomagnetic latitudes above 50 degrees. Fortunately the continental United States and Canada are well positioned for reception of whistlers and other signals of natural radio.

Considered by many listeners to be the Music of Earth, whistlers are among the accidental discoveries of science. In the late nineteenth century, European long-distance telegraph and telephone operators were the first people to hear whistlers. The long telegraph wires often picked up the snapping and crackling of lightning storms mixed with the Morse code buzzes or voice audio from the sending stations. Sometimes, the telephone operators also heard strange whistling tones in the background. These tones were attributed to problems in the wires and connections of the telegraph system and then disregarded.

The first written report of this phenomenon dates back to 1886 in Austria when whistlers were heard on a 22 kilometer (14 mile) telephone wire without amplification. Then during World War I, while the German scientist H. Barkhausen was eavesdropping on Allied telephone conversations, he heard the whistlers. In order to pick up the telephone conversations, he inserted two metal probes in the ground some distance apart and connected them to the input of a sensitive audio amplifier. He was surprised to hear whistling tones that lasted for 1 or 2 seconds and glided from a high frequency in the audio range to a lower frequency where they disappeared into the amplifier or background noise level. On occasion, the whistlers were so numerous and loud that he could not detect any type of telephone.

Later, in 1928 three researchers from the Marconi Company in England reported on the work they had done in whistler research. Eckersley, Smith, and Tremellen established a positive correlation between whistler occurrence and solar activity, and they found that whistlers frequently occur in groups preceded by a loud click. The time between the click and the whistlers was about 3 seconds. They concluded that there must have been two paths of propagation involved: one for the click that preceded the whistler and a second much longer path, over which the whistler and its echoes propagated.

Eckersley was able to show in 1931 that Earth's magnetic field permits a suitably polarized wave to pass completely through the ionosphere. This was in accordance with the *magnetoionic theory*. Tremellen had on one occasion observed that during a summer thunderstorm at night, every visible flash was followed by a whistler. This served as the first definite evidence of the relationship between lightning discharges and whistlers, the click being the lightning discharged that caused the whistler. As the Marconi workers continued their research, they also discovered a new type of atmospheric activity that sounded somewhat like the warbling of birds. Because the sound tended to occur most frequently at dawn, they gave it the name *dawn chorus*.

L.R.O. Storey of Cambridge University began an intensive study of whistlers in 1951. He confirmed Eckersley's law, which showed that most whistlers originate in ordinary lightning discharges, and found the path of propagation of the whistler to be along the lines of Earth's magnetic field.

Other interesting sounds include tweeks, which have been described by one listener as a cross between chirping birds and a hundred little men hitting iron bars with hammers. The ephemeral dawn chorus is a cacophony of sound that resembles nothing else on Earth. Tweeks are abrupt, descending notes that resemble pings. They are usually heard at night during the winter and early spring.

The dawn chorus has been variously described as sounding like birds at sunrise, a swamp full of frogs, or seals barking. In fact, the chorus varies constantly. Hooks, risers, and hisses are all part of the dawn chorus, but sometimes these sounds are also heard as solitary events. A hook starts out like a whistler but then abruptly turns into a rising tone. Risers increase in frequency from beginning to end. And hiss sounds just like the name. Appropriately enough, the dawn chorus is best heard around dawn, but it may occur at any time of the day or night. The dawn chorus is most likely to be heard during and shortly after geomagnetic storms.

Time of Listening

Most of the VLF signals, including clicks, pops, and tweeks, can be heard at almost any time of the day or night. The occurrence of whistlers is a function of both thunderstorm occurrence and propagation conditions through the ionosphere. The rate of whistlers is higher at night than during the day because of the diurnal variation in the D region absorption, which is highest when the sun is above the horizon. At night, the D layer does not exist because it is ionized by solar ultraviolet rays.

The whistler rate, or the number of whistlers heard per minute, has a marked dependence on sunspot activity; the rate increases with the sunspot number. Whistlers are heard only when sufficient ionization exists along the path to guide the waves toward Earth's magnetic field. This ionization is assumed to be supplied from the sun during a solar event, such as sunspot activity. To hear a whistler or other signals such as the dawn chorus, you will no doubt have to get away from power lines and industrial noise. The listening time for these signals will be from near local midnight to early morning hours.

Observing and Recording Hints

The best location to observe VLF signals is out in the countryside, in an area where few cars or trucks are likely to pass. You will want to get as far away from power lines as possible for optimum results. The lower the power line noise is, the stronger the VLF emission signals will be. Take a portable, battery-operated, cassette tape recorder with you and use a long audio cable to run to the tape recorder from the audio output of the VLF receiver.

Building a Whistler Receiver

Even though whistlers and related emissions occur at acoustical frequencies, they are radio signals. To hear whistlers, you must intercept their electromagnetic energy with an antenna and transform it to the mechanical vibrations to which our ears respond.

It is not easy to go out and buy a radio capable of tuning 1 to 10 KHz, but it is possible to build a sensitive receiver to permit you hear the sounds of natural radio waves. We will explore what causes these sounds and see how you can study them yourself.

The classic whistler receiving system consists of an antenna for signal collection, an amplifier to boost the signal level, and headphones or a speaker to transform the signal to sound waves. (A magnetic tape recorder can be substituted for the headphones or speaker.) A whistler receiver can simply be an audio amplifier connected to an antenna system. However, powerful manmade interference immediately above and below the frequencies of interest tend to seriously overload receivers of this sort and make reception of whistlers and related phenomena difficult.

To overcome these problems, a good whistler receiver circuit includes a circuit called a low-pass filter that attenuates all signals above 7 KHz. This greatly reduces interference from such sources. To escape interference from AC power lines and other forms of nonnatural radio noise, the receiver is designed for portable operation in the field, that is, away from such interference sources.

A sensitive dual FET whistler receiver is depicted in Figure 9-5. The antenna is first fed to an RF filter, composed of resistors R1 through R4 and capacitors C1 through C3. This filter allows only the desired frequencies to be sent to the next section of the receiver. The RF filter is next fed to the 60 Hz filter, or trap, which filters out 60 Hz power-line frequencies before the RF signals are amplified. The 60 Hz filter can be switched in and out of the circuit with switch S1:a. The dual-matched FET transistor Q1 along with transistor Q2 are used to amplify the RF signals of the desired band of frequencies. The amplified signals are then fed to the input of a set of selectable filters, which can be switched in and out of the circuit. Toggle switch S2 is used to switch in the high-pass filter circuit, whereas switch S3 is used to switch in the low-pass filter. You can select either or both filters if desired.

The output of the low-pass filter circuit is coupled to the final audio amplifier stage via capacitor C17. Potentiometer R12 controls the audio level entering the LM386 audio amplifier at U1. Note that R12 also contains a switch S5, which is used to supply power to the LM386 audio amplifier. The output of the LM386 is coupled to an audio output jack via capacitor C20. Capacitor C18 is coupled to a recorder output jack to allow a chart recorder or analog-to-digital converter to be used to monitor and record the output from the whistler receiver. The whistler receiver can be powered from a 9-volt battery for field receiving applications. Power to the circuit is supplied via power switch S4. For stationary reception and recording, you can power the circuit from a 9-volt wall-wart power supply.

The whistler receiver is constructed using good RF techniques on a glass epoxy circuit board with large ground plane structures, that is, using scrap pieces of circuit boards placed vertically acting as RF shielding (Figure 9-6). You could modularize the circuit building by using three small circuit boards: one for the RF filter and trap circuit, one for the high-pass and low-pass filter, and a third board for the audio amplifier and power circuit. Small metal shield covers could be built using scrap circuit board material. Keeping component leads short is a good idea with this circuit; it is an RF circuit but not critical because it is a low frequency RF.

When installing the electrolytic capacitors, be sure to observe the proper polarity to avoid damaging the circuit upon power-up. When handling the dual FET at Q1 and transistor at Q2, be very careful; the FET should have been shipped in a metalized foam that shorts all the pins together. The metalized foam prevents electrostatic damage to the front end of the FET. Use antistatic practices when handling Q1 and Q2. For example, make sure you are seated and use a wrist strap to prevent static damage on the FET. Also make sure you can identify the pinouts of both Q1

Figure 9-6   *Vertical circuit board shields*

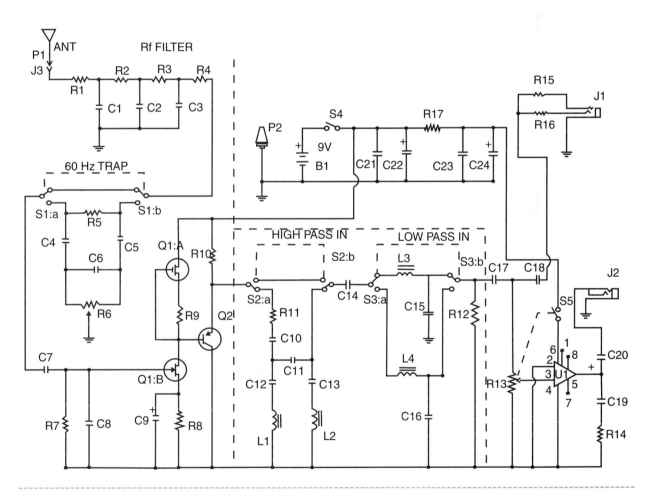

Figure 9-5   *Dual FET whistler receiver circuit*

and Q2 before installing them. The FET actually has two identical matched transistors in the same package. Each transistor in the package has three pins. Make sure you know each of the manufacturer's pin designations before installing the transistors.

A single IC in the receiver circuit and IC socket are recommended in the event of a circuit failure down the road. ICs generally have markings to indicate the proper orientation of the package. The LM386 should have either a small indented circle on the left side of the package or a cutout on the top center of the IC package. Pin 1 of the IC will be to the left of either the small cutout or the indented circle.

Once the circuit board has been completed, carefully check it over to make sure you find no shorting paths and no stray component leads that make bridges across the PC board circuit lands.

The whistler receiver prototype circuit board is housed in a cast $6 \times 5\frac{1}{2} \times 2$ inch aluminum box. The filter switches, gain control, and headphone jacks are mounted on the front of the box, as seen in the photo in Figure 9-6. The on-off switch and SO-238 UHF connector are mounted on the side panel of the enclosure. The ground binding post is mounted on the rear panel, opposite the front panel. The main circuit board can be mounted to the bottom of the enclosure with $\frac{1}{8}$-inch plastic standoffs.

The whistler receiver is designed to be used in open areas, with a vertical antenna of roughly 1 to 3 meters (3 to 12 feet). It can also be used in wooded or obstructed areas with wire antennas, roughly 50 to 200 feet in length and connected instead of vertical. Wire antennas should be insulated and supported off the ground as high as you can place them. The performance of a wire antenna improves the more you can get it in the clear and the more vertically it can be positioned. If you do use a wire antenna, you may have to reduce the value of the input resistor R1 for best results.

The whistler receiver also needs a ground connection, sometimes called a *counterpoise*. A simple but often effective ground can be provided by your body in contact with the ground terminal BP2. I have used an improvised finger ring attached to a wire connected to the ground terminal. Other places to make a ground connection include car bodies, wire fences, or other large metal objects not connected in any way to an AC power circuit. Another approach is to use short sections of copper tubing as ground stakes that are driven into the ground and connected to your receiver. Try several different grounding schemes to see which one provides the quietest reception. Don't be too surprised, however, to discover that you still have to make body contact with the ground terminal for quietest reception!

Using Your VLF Receiver

The prototype receiver, with a whip antenna, is mounted on a piece of wood approximately 2×60 inches. I installed the antenna at one end (it can be mounted on a bracket or just taped on) and attached the receiver about 1 foot down from the antenna using duct tape (two-sided foam tape or a mounting bracket could also be used). The monitor amplifier is mounted in a similar fashion, a short distance below the receiver. A length of wire runs from the receiver's ground terminal down the wood and is taped in place at several locations. Now you have a handy walking stick that was easy to carry and use! The base can be stuck into any convenient hole, jammed into a rock pile, clamped onto a fence or vehicle, or even held by hand.

For the best chance of hearing whistlers and related phenomena, you will need to put some distance between the whistler receiver and the AC power grid. The bare minimum is about one quarter mile from any AC power line; the more distance between the receiver and any AC power lines, the better your reception will be. You'll need to do some exploration to discover quiet sites near you. Once you've arrived at a potential listening site, set up your receiver, put in a ground connection, and see what happens. What you hear depends to some extent on the time of day, but you can always expect some sharp, crackling static. The intensity and volume of this static will depend on propagation conditions and where thunderstorms are in relation to your location. Chances are that you'll also hear some power-line hum, but hopefully it won't

Chapter Nine — Radio Projects

232   *Electronic Sensors for the Evil Genius*

be very loud. If your listening site is too noisy, you try another place.

If your ears have good high-frequency response, you will probably hear a continuing sequence of 1-second tones. These are from the OMEGA radio-navigation system. OMEGA is transmitted from 10 to 14 KHz, and OMEGA transmitters are very powerful. You may also hear anything that can produce an electrostatic discharge, particularly if the humidity is low. These miscellaneous noise sources can include wind, the buzz of insects flying near the antenna, dry leaves or grass moving in the wind, and even the electrostatic charges that build up on your clothing. Passing vehicles often emit noise from their electrical and ignition systems. And if you have a digital watch, keep it away from the antenna or you'll be listening to it instead of whistlers.

Anything you hear other than whistlers and related phenomena or OMEGA signals are likely to be the result of extraneous signals overloading your receiver. Burbling sounds (possibly mixed with OMEGA tones) are caused by military signals in the 15 to 30 KHz range, perhaps overloading the receiver. A ticking sound at a 10 Hz rate is from Loran-C radio-navigation signals at 100 KHz. The Loran-C navigation system was used extensively by ship captains, sailors, and boaters before the advent of global positioning system (GPS). In general, don't be surprised if you experience overloading from transmitters operating on any frequency, if they are within sight of the location where you're using your whistler receiver.

Some whistlers and related signals are so short lived and impossible to predict that you might want to consider using an unattended tape recorder to record the signals in the field. Any tape recorder with an external microphone input can be used. Best results are obtained with a recorder that has no *automatic level control* (ALC). The VLF receiver shown has a output jack that can be coupled to a recorder input for field recording. You may have to place the tape recorder away from the receiver to avoid picking up motor noise. Good luck and happy exploring!

VLF Receiver Parts List

Required Parts

R1, R4 10K ohm, 1/4-watt, 5% resistor

R2, R3 22K ohm, 1/4-watt, 5% resistor

R5 6.2-megohm, 1/4-watt, 5% resistor

R6 1-megohm potentiometer (trimpot)

R7 10-megohm, 1/4-watt, 5% resistor

R8 1K ohm, 1/4-watt, 5% resistor

R9 1k ohm, 1/4-watt, 5% resistor

R10 3.3K ohm, 1/4-watt, 5% resistor

R11, R12 820-ohm, 1/4-watt, 5% resistor

R13 10K ohm potentiometer (panel)

R14, R17 10-ohm, 1/4-watt, 5% resistor

R15, R16 2.2K ohm, 1/4-watt, 5% resistor

C1, C3 47 pF, 35-volt mica capacitor

C2 100 pF, 35-volt mica capacitor

C4, C5, C6 3.3 nF, 35-volt Mylar capacitor

C7, C15 0.01 uF, 35-volt ceramic disk capacitor

C8 27 pF, 35-volt mica capacitor

C9 1 uf, 35-volt electrolytic capacitor

C10 0.18 uF, 35-volt tantalum capacitor

C11 0.12 uF, 35-volt tantalum capacitor

C12 1.8 uF, 35-volt tantalum capacitor

C13 0.68 uF, 35-volt tantalum capacitor

C14 0.22 uF, 35-volt tantalum capacitor

C16 0.068 uF, 35-volt tantalum capacitor

C17, C18   0.22 uF,
   35-volt tantalum
   capacitor

C19   0.05 uF, 35-volt
   ceramic disc capacitor

C20, C22, C24   100 uF,
   35-volt electrolytic
   capacitor

C21, C23   0.1 uF, 35-
   volt ceramic disc
   capacitor

L1   120 mH coil (Mouser
   electronics)

L2   150 mH coil (Mouser
   electronics)

L3   18 mH coil (Mouser
   electronics)

L4   56 mH coil (Mouser
   electronics)

Q1   U401 Siliconix dual
   matched N-Channel FETs

Q2   MPSA56 transistor

U1   LM386 op-amp

S1   DPDT toggle switch
   (trap)

S2   DPDT toggle switch
   (high-pass)

S3   DPDT toggle switch
   (low-pass)

B1   9-volt transistor
   radio battery

J1, J2 1/8-inch mini
   phone jack with switch

J3   SO-238 UHF chassis
   jack

P1   PL259 UHF plug

P2   binding post
   (ground)

Miscellaneous   PC board,
   IC socket, solder,
   wire, antenna, battery
   holder, battery clip,
   hardware, chassis

Shortwave Radio

This shortwave radio project can open up a whole new world of possibilities to both young and old alike. Listening to radio broadcasts from far off lands is very exciting as well as interesting and can lead to new hobbies such as shortwave listening or perhaps amateur radio. The shortwave radio can also present many different avenues for research that perhaps you never thought about. You can use this shortwave receiver, for example, to study wave propagation by listening to the broadcasts from WWV, the time signal radio site in Fort Collins,, Colorado.

The simple, three-IC superheterodyne radio in Figure 9-7 can receive stations in the 4.5- to 10-MHz range from around the world with only a 10-foot antenna. A superheterodyne radio works by mixing the incoming RF signals with a *local-oscillator* (LO) signal to produce an IF, or *intermediate frequency*. The circuit then filters, amplifies, and diode-detects the IF signal to reproduce the audio signal contained in the RF input.

The transformer-capacitor circuit at the input provides impedance matching to the antenna; the T-C tuned circuit provides rough preselection for the 4.4- to 10-MHz RF signal. IC U1, a Philips NE602, contains the required RF stages. U1 also contains an active Gilbert-cell mixer and a transistor configured to provide the LO function (pins 6 and 7). The LO uses a simple Colpitts configuration. The L-C tank circuit determines the Colpitts configuration's frequency. The LO operates at 455 KHz above the incoming RF, thus producing a constant 455 KHz IF output at pin 5 of U1.

The TOKO ceramic filter removes any out-of-band responses. The 4 KHz passband of the filter provides surprisingly good audio quality and adjacent band rejection. The workhorse of the receiver is IC2, a Plessey ZN414, originally designed as a simple, one-chip AM radio. This IC provides more than 70 dB of IF amplification, an *automatic gain-control* (AGC) circuit, and a detector circuit in a TO-92 package. You can set the gain of the ZN414 by changing the bias on the device using the IF-gain trimpot. When U2 amplifies, provides AGC, and detects the IF, it produces baseband audio. U2 can directly drive high-impedance headphones, but this design uses an LM386 audio amplifier to drive a 3-inch loudspeaker.

This shortwave radio is constructed on a small printed circuit board. If a perf-board is used compo-

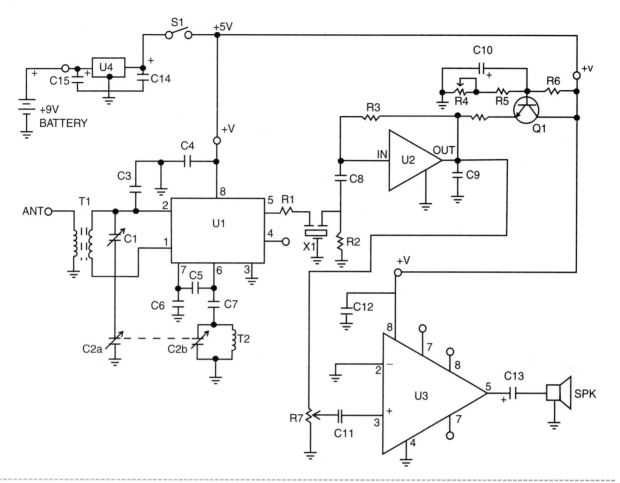

Figure 9-7   *Three IC shortwave receiver circuit*

nent leads should be kept as short as possible because this is an RF circuit. Be sure to use IC sockets for the ICs to avoid problems switching out components at a later date if there are defective parts. ICs need to be installed correctly to avoid destroying them. Most ICs have a notch cut out of the plastic at the top of IC package or they have an indented circle near pin 1 of the IC. Be careful when handling the ceramic filter and be sure to observe the center or ground wire when installing this device. When building the shortwave receiver be sure to observe the correct polarity of the electrolytic capacitors as well as that of the transistor.

Because this project is a radio project, you should choose a metal enclosure for the project. An aluminum chassis box measuring 6 × 6 × 2½ inches is best for this project. Tuning capacitors C1 and C2 are both mounted on the front panel of an aluminum chassis box, which was used to house the shortwave

receiver. Both potentiometers R4 and R7 are mounted on the front panel of the chassis box. A 2- or 3-inch speaker is also mounted on the front panel of the chassis along with the on-off switch at S1. The printed circuit board is mounted to the bottom of the chassis box with ¼-inch plastic stand-offs to lift the board of the metal floor of the chassis.

The front-end coil assembly at T1 is a hand wound coil consisting of a T37-2 ferrite core from Micrometals, in Anaheim, California. The input transformer coil T1 consists of two coils wound on the ferrite core. The primary coil consists of 24 turns of #26 gauge enamel wire closely wound on the ferrite core. The secondary coil consists of 5 turns of #26 gauge enamel wire wound on top of the primary coil. The coil at T2 consists of 22 turns of #26 gauge enamel wire closely wound on a ferrite core.

The IC at U2 is a Plessey Semiconductor RF, a complete amplifier package, which is no longer produced but is still available. A replacement device, an MK484, is readily available from Ocean State Electronics or Radio Laboratories as well as other suppliers. The balanced RF amplifier and oscillator, the NE602, at U1 is readily available from many sources such as Digi-Key Electronics or Ocean State Electronics. The ceramic filter module is available from TOKO Coils.

Powering-up and testing your new shortwave receiver is your next step. Check and recheck the circuit board to make sure that the components have been installed correctly and that you find no shorts or small wires bridging the PC lands on the PC board. Once you have completed checking your circuit, connect a 9-volt transistor radio battery. The current drain on the battery is only about 10 mA, so the battery should last for quite awhile.

Alignment of the receiver is simple. You should verify that the LO is oscillating from approximately 5 to 10 MHz when you tune C2. You can perform this verification by placing a direct or ×1 probe or antenna from a frequency counter on or near the NE602 chip. Don't connect the counter directly to the NE602 pins, because the added capacitance changes the oscillation frequency.

Next, set the IF-gain trimpot near the center of its range and add an antenna to the circuit. You could use 10 or 20 feet of wire strung randomly about the house, but a longer outdoor antenna substantially improves performance. Tune to the center of the receiver's range and carefully listen for a station. You can usually hear a Department of Defense fax transmission on 8080 KHz at night from anywhere in the United States. (The transmission sounds like a scratchy record.) Once you hear a station, carefully tune C2 to peak the signal. During the peaking adjustment, you may need to adjust the IF gain or volume for optimum reception. Now, tune to the upper portion of the band. You can hear WWV, the National Institutes of Standards and Technology standard time-and-frequency station, at 10 MHz. The station transmits time information 24 hours a day. Repeat the adjustment of C1 for optimum reception.

You may need to adjust C2 at the low, center, and high portions of the band for a best compromise. Now, tune to any moderately strong station and set the IF-gain trimpot for maximum signal with minimum audio distortion. A large increase in the output noise level accompanies the onset of distortion.

You will have fun learning about other interesting things on WWV at certain times besides just the time function feature. Features such as accurate tone frequencies, geophysical alerts (solar activity reports), marine storm warnings, *global positioning system* (GPS) reports, and OMEGA navigation system status reports are broadcast also.

Figure 9-8 shows the broadcast schedule for WWV in Fort Collins, Colorado. A similar schedule is available for WWVH in Hawaii. This diagram is extracted from a National Institute of Standards and Technology (NIST) publication. (NIST publications are available for a modest charge from the Government Printing Office, Washington DC 20402. Write and ask for a current list of available publications.) The two features of most interest are precise frequency of the carrier signal to calibrate radio and test equipment, and the precise time information given each minute as a voice transmission. But you will find many other interesting parts of their transmissions.

Various tones can be and are used to check audio equipment or musical instruments. These tones can be captured to detect the start of each hour or each minute. At specific intervals, voice announcements are made on WWV for the benefit of other government agencies. The predominant ones are as follows:

> **Marine storm warnings** that are of interest to the U.S. government are prepared by the National Weather Service and broadcast for areas of the Atlantic and Pacific Oceans.
>
> **Global positioning system** status announcements are prepared by the U.S. Coast Guard to give current status information about the GPS satellites.
>
> **OMEGA navigation system** reports are prepared by the U.S. Coast Guard to give the status of the eight OMEGA transmitting stations in the 10 to 14 KHz frequency range. These serve as navigation aids.

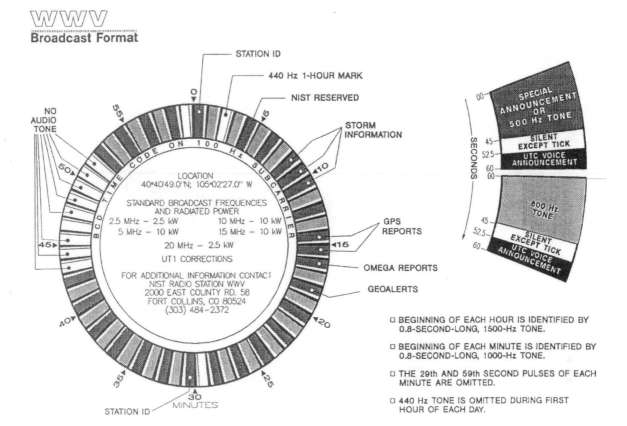

WWV Broadcast Format

Figure 9-8   *Broadcast schedule for WWV, Fort Collins, Colorado*

Geophysical alerts are prepared by the Space Environment Service Center of the *National Oceanic and Atmospheric Administration* (NOAA). They are broadcast on the 18th minute of the hour and give information of interest to amateur radio operators and various scientific organizations regarding solar activity, geomagnetic fields, solar flares, and other geophysical statistics. This propagation information can help you decide when the DX signals will be good.

Specific timing transmissions are made to allow listeners to synchronize equipment:

- The beginning of each hour is identified by 0.8-second long, 1,500 Hz tone.
- The beginning of each minute is identified by 0.8-second long, 1,000 Hz tone.
- The 29th and 59th second pulses of each minute are omitted.
- A 440 Hz tone is omitted during the first hour of each day.

Frequency Calibration

How are frequency calibrations made with a WWV receiver? It is easy to calibrate a signal generator using the shortwave receiver. Simply couple the signal generator with a *continuous wave* (cw) signal loosely to the antenna input with an antenna connected, and adjust the signal generator level to get a beat note on the receiver. Then, adjust the frequency of the signal generator for zero beats. The signal generator can then be used to calibrate other equipment, such as a frequency counter.

Shortwave Receiver Parts List

Required Parts

R1 100-ohm, 1/4-watt resistor

R2 1.5K ohm, 1/4-watt resistor

R3 100K ohm, 1/4-watt resistor

R4 10K ohm potentiometer (chassis mount, IF gain)

R5 10K ohm, 1/4-watt resistor

R6 15K ohm, 1/4-watt resistor

R7 10K ohm potentiometer (chassis mount, volume)

C1 10 to 180 pF tuning capacitor

C2 10 to 35 pF dual gang tuning capacitor

C3, C8 0.01 uF, 35-volt disc capacitor

C4, C11 0.22 uF, 35-volt Mylar capacitor

C5, C6 380 pF, 35-volt Mylar

C7, C12 100 pF, 35-volt Mylar capacitor

C9, C15 0.1 uf, 35-volt disc capacitor

C10, C14 1 uF, 35-volt electrolytic capacitor

C13 220 uF, 35-volt electrolytic capacitor

Q1 2N3904 transistor

X1 TOKO HDFM2-456 ceramic filter

T1 T37-2 ferrite core with two coils (see text)

T2 T37-2 ferrite core with single coil (see text)

U1 NE602 Phillips balanced RF amplifier IC

U2 ZN414 IF module or newer MK484

U3 LM 386 audio amplifier module

U4 LM 7805 regulator IC

S1 SPST toggle switch

SPK 2-inch or larger, 8-ohm speaker

Miscellaneous PC board, IC sockets, wire, connectors, etc.

Jupiter Radio Telescope

The Jupiter radio telescope is a special shortwave receiver that will pick up radio signals from the planet Jupiter and also from the sun. It is a very exciting project. This project will allow you to do amateur research into radio astronomy in your own back yard.

Jupiter is the fifth planet from the sun and is the largest one in the solar system. If Jupiter were hollow, more than one thousand Earths could fit inside. It also contains more matter than all of the other planets combined. It has a mass of 1.9×1027 kilograms and is 142,800 kilometers (88,736 miles) across at its equator. Jupiter possesses 28 known satellites, four of which were observed by Galileo as long ago as 1610. They are Callisto, Europa, Ganymede, and Io. Another 12 satellites around Jupiter have been recently discovered and given provisional designators until they are officially confirmed and named. Jupiter has a ring system, but it is very faint and totally invisible from Earth. (The rings were discovered in 1979 by Voyager 1.) Its atmosphere is very deep, perhaps comprising the whole planet, and is somewhat like the sun. It is composed mainly of hydrogen and helium, with small amounts of methane, ammonia, water vapor, and other compounds.

Colorful latitudinal bands, atmospheric clouds, and storms illustrate Jupiter's dynamic weather systems (see Figure 9-9). The cloud patterns change within hours or days. Jupiter's *Great Red Spot* is a complex storm moving in a counterclockwise direction. At the outer edge of the spot, material appears to rotate in 4 to 6 days; near the center, motions are small and nearly random in direction. An array of other smaller storms and eddies can be found throughout Jupiter's banded clouds.

Figure 9-9 *Jupiter's colorful latitude bands, atmospheric clouds, and storms*

Auroral emissions, similar to Earth's northern lights, were observed in the polar regions of Jupiter. The auroral emissions appear to be related to material from Io that spirals along magnetic field lines and fall into Jupiter's atmosphere. Cloud-top lightning bolts, similar to superbolts in Earth's high atmosphere, have also been observed.

Radio signals from Jupiter are very weak; they produce less than a millionth of a volt at the antenna terminals of the receiver. These weak RF signals must be amplified by the receiver and converted to audio signals of sufficient strength to drive headphones or a loudspeaker. The receiver also serves as a narrow filter, as it can be tuned to a specific frequency to hear Jupiter while at the same time blocking out strong Earth-based radio stations on other frequencies. The receiver and its accompanying antenna are designed to operate over a narrow range of shortwave frequencies centered on 20.1 MHz. This frequency range is optimum for hearing Jupiter signals.

The block diagram shown in Figure 9-10 illustrates the various parts of the Jupiter radio telescope. First the antenna intercepts weak electromagnetic waves that have traveled some 500 million miles from Jupiter to the Earth. When these electromagnetic waves strike the wire antenna, a tiny RF voltage is developed at the antenna terminals. Signals from the antenna are delivered to the antenna terminals of the receiver by a coaxial transmission line. Next the signal is delivered to the RF bandpass filter and preamplifier, where

they are filtered to reject strong out-of-band interference and are then amplified using a *junction field effect transistor* (JFET). This transistor and its associated circuitry provide additional filtering and amplify incoming signals by a factor of 10. The receiver input circuit is designed to efficiently transfer power from the antenna to the receiver while developing a minimum of noise within the receiver itself.

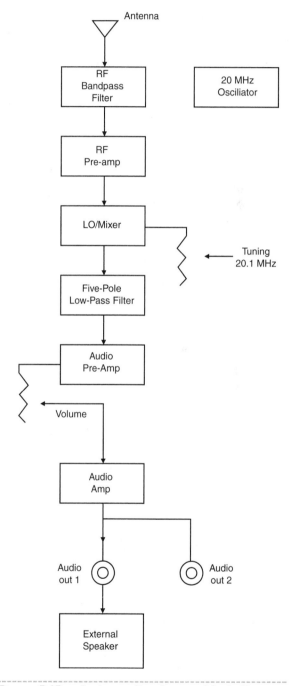

Figure 9-10 *Jupiter radio telescope block diagram*

Next the signals move to the *local oscillator* (LO) and mixer, which perform the important task of converting the desired RF signals down to the range of audible frequencies. The LO generates a sinusoidal voltage wave form at a frequency in the vicinity of 20.1 MHz. The exact frequency is set by the front-panel tuning control. Both the amplified RF signal from the antenna and the LO frequency are fed into the mixer. The mixer develops a new signal that is the arithmetic difference between the LO and the incoming signal frequency. Suppose the desired signal is at 20.101 MHz and the LO is tuned to 20.100 MHz. The difference in frequencies is therefore $20.101 - 20.100 = 0.001$ MHz, which is the audio frequency of 1 kilohertz. Or if a signal were at 20.110 MHz, it would be converted to an audio frequency of 10 KHz. Because the RF signal is converted directly to audio, the radio is known as a direct-conversion receiver.

To eliminate interfering stations at nearby frequencies, a low-pass filter is used next; this operates like a window only a few kilohertz wide through which Jupiter signals can enter. When listening for Jupiter or the sun, the radio will be tuned to find a clear channel. Because frequencies more than a few kilohertz away from the center frequency may contain interfering signals, these higher frequencies must be eliminated. This is the purpose of using the low-pass filter following the mixer. It passes low (audio) frequencies up to about 3.5 KHz and attenuates higher frequencies.

Finally the low-pass filter feeds an audio amplifier stage, which takes the very weak audio signal from the mixer and amplifies it enough to drive headphones directly or enough to drive an external amplified speaker assembly.

Circuit Diagram

The block diagram of the Jupiter radio receiver in Figure 9-10 shows the radio as a group of functional blocks connected together. Although this type of diagram does not show individual components like resistors and capacitors, it is useful in understanding signal flow and the various functions performed within the radio. The next level of detail is the schematic diagram. A schematic is used to represent the wiring connections between all of the components that make up a circuit. A schematic diagram uses symbols for each of the different components rather than using pictures of what the components actually look like.

A schematic diagram of the complete receiver is seen in Figure 9-11. On this schematic, the part types are numbered sequentially. For example, inductors are denoted L1 through L7, and resistors are denoted R1 through R31. Signal flow as shown in the schematic is as follows. The signal from the antenna connector (J2) is coupled to a resonant circuit (bandpass filter L1, C2, C3) and then to the J-310 transistor (Q1) where it is amplified. The output of the J-310 goes through another resonant filter (L3, C6) before being applied to the resonant input circuit (L4, C9, C10) of the SA602 IC (U1), which serves as the local oscillator and mixer. The center frequency of the local oscillator is set by inductor L5 and adjusted by the tuning control R7. The audio output from U1 passes through the low-pass audio filter (L6, L7, C20, C21, and C22). The audio signal is next amplified by U2 (an NTE824) before going to the volume control R15. The final audio amplifier stages comprise U3 (another NTE824) and the output transistors Q2 (2N-3904) and Q3 (2N-3906). After the receiver has been assembled, the variable capacitors C2 and C6 and variable inductors L4 and L5 will be adjusted to tune the receiver for operation at 20.1 MHz.

PC Board Assembly

After identifying and locating all the components for the Jupiter radio and preparing your printed circuit board, you are ready to begin building the project. When installing the components, take some extra time to insert them carefully and correctly in their proper holes. IC sockets and transistor sockets are highly recommended, and can avoid major problems if the circuit ever fails and components need to be

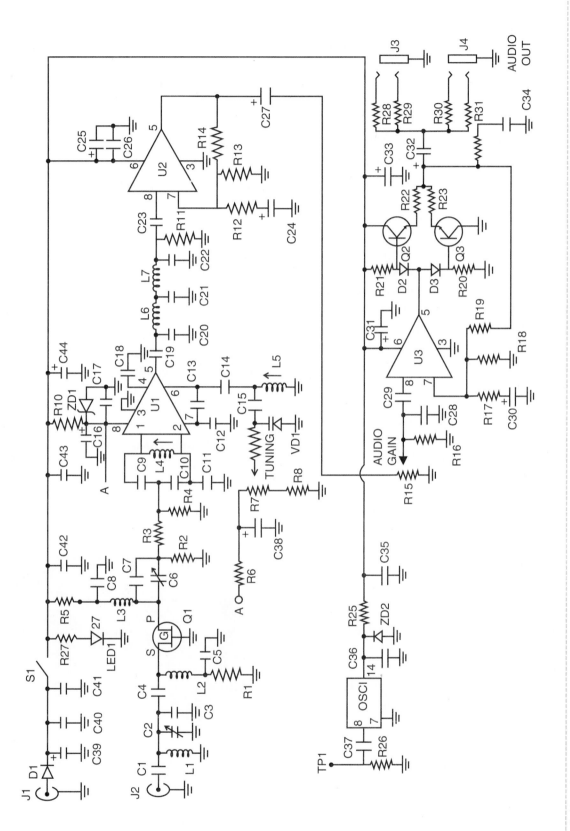

Figure 9-11   *Jupiter radio telescope circuit*

replaced. In order to avoid damaging the receiver it is important to install the capacitors, diodes, and semiconductors with respect to proper polarity marked on each of them. Electrolytic capacitors have either a plus or minus marking on them and must be installed correctly by observing the polarity of the capacitor with respect to PC board. All diodes including varactor diodes have polarity, so be aware of their polarity when installing the diodes. (A varactor diode, or tuning diode, is a type of diode used principally as a voltage-controlled capacitor.) The band usually denotes the cathode end of the diode. Transistors generally have three leads: a base, a collector, and an emitter; pay particular attention to these leads when installing the transistors. ICs always have some type of markings indicating their orientation. You will find either a plastic cutout on the top of the IC package or an indented circle next to pin 1 of the IC. Be sure to use these markings when installing the ICs. A number of coils are used in this project; although there is no polarity to the coils, you will need to pay attention to the values of the coils as well as which ones are adjustable and which are fixed before you install them on the PC board. Note on the schematic the two points marked A; one is at R6 and the other at C16. Both of these points are connected together. Note the test point marked TP1 at the OSC-1 module, which can be used to ensure that the oscillator module is working correctly. Power is brought to the circuit via the coaxial power input jack at J1. The antenna connection is shown at J2. This connector can be either a BNC or an F type chassis-mounted connector. The receiver circuit has two audio output jacks, both at J3 and J4. Two circuits of mini 1/8-inch stereo jacks are used in the prototype.

Once the printed circuit has been assembled, recheck the solder joints to make sure they are smooth and that no cold solder joints exist. Also observe the PC circuit lines to be sure that no bridges or shorts have formed by cut component leads. Once the circuit has been checked, you are ready to install the circuit board in the chassis.

A metal enclosure is chosen because this is a sensitive RF receiver and you will want to keep outside interference from effecting the circuit operation

(see Figure 9-12). Wire the four jacks J1 through J4 to the circuit board. The prototype has the power and antenna jacks mounted on the rear panel of the chassis box. One audio jack is mounted on the front panel, and the second audio jack is mounted on the rear chassis panel. The power switch S1 is wired to the PC and mounted on the front panel of the receiver along with the power-indicating LED. The tuning control R7 and volume control R15 are both mounted on the front panel of the chassis box.

Once you are satisfied that your circuit board is wired correctly, you are ready to apply power to the circuit and begin testing for proper operation.

Testing and Alignment

The receiver requires 12 volts DC (vdc), which may be obtained from a well-regulated power supply or from a battery. Current drain is approximately 60 milliamps (ma). The power cable supplied with the kit has a female power plug on one end and stripped leads on the other. Notice that the power cable has a black stripe, or tracer, along one of the wires. This is the wire that is connected to the center conductor of the plug and must be connected to the plus (+) side of the power source. The Radio Shack RS 23-007 (Eveready) 12-volt battery or equivalent is suitable.

Figure 9-12   *A metal enclosure for the circuit*

Next, turn the Jupiter receiver's power switch to off. Connect either headphones or an amplified speaker (Radio Shack 277-1008C or equivalent) to the receiver audio output (J3 or J4). These jacks accept 3.5 millimeter (¹⁄₈-inch) monaural or stereo plugs.

If you are using a Radio Shack amplified speaker, turn it on and adjust the volume control *on the speaker* up about ¹⁄₈ of a turn. If you are using headphones, hold them several inches from your ear as there may be a loud whistle due to the internal test oscillator. Turn the receiver on. The LED should light up. Set the volume control to the 12 o'clock position. Allow the receiver to warm-up for several minutes.

Set the tuning control to the 10 o'clock position. *Carefully* adjust inductor L5 (Figure 19-12) with the white tuning stick until a loud low-frequency tone is heard in the speaker (set volume control as desired). **Caution**: Do not screw down the inductor slugs too far, as the ferrite material could crack. By adjusting L5 to hear the tone, you are tuning the receiver to 20.00 MHz. The signal that you hear is generated in OSC1, a crystal-controlled test oscillator built into the receiver. Once L5 has been set, **do not** readjust it during the remainder of the alignment procedure. (When the receiver tunes 20.00 MHz with the knob set to the 10 o'clock position, it will tune 20.1 MHz with the knob centered on the 12 o'clock position.)

The following steps involve adjusting variable capacitors (C2 and C6) and a variable inductor (L4) to obtain the maximum signal strength (loudest tone) at the audio output. For some, it is difficult to discern slight changes in the strength of an audio tone simply by ear. For this reason, three different methods are described, each using a form of test instrument. In the event no test equipment is available, a fourth method—simply relying on the ear—is possible. In all cases, adjust the receiver tuning knob so that the audio tone is in the range of about 500 to 2,000 Hz. Use procedure A as a method to tune up your receiver.

If no test equipment is available, simply tune by ear for the loudest audio signal. Listen to the tone and carefully adjust the tuning knob to keep the pitch constant. If the pitch changes during the alignment, it indicates that the receiver has drifted off frequency.

As you make adjustments, the signal will get louder. Reduce the receiver volume control as necessary to keep the tone from sounding distorted or clipped.

To test the receiver on the air, simply connect the antenna. For best performance, use a 50-ohm antenna designed to operate in the frequency range of 19.9 to 20.2 MHz. At certain times of the day, you should be able to hear WWV or WWVH on 20.000 MHz. The Jupiter radio telescope direct-conversion receiver design does not allow clear reception of *amplitude modulated* (AM) stations like WWV. So the voice will probably be garbled unless you tune very precisely. The receiver does work well on *single sideband* (SSB) signals and CW signals.

The Jupiter Radio Telescope Antenna

The antenna intercepts weak electromagnetic waves that have traveled some 500 million miles from Jupiter to the Earth or 93 million miles from the sun. When these electromagnetic waves strike the wire antenna, a tiny RF voltage is developed at the antenna terminals. A basic dipole antenna is shown in Figure 9-13. The recommended antenna design for the Jupiter radio telescope is the dual dipole shown in Figure 9-14. Signals from each single dipole antenna are brought together with a power combiner via two pieces of coaxial cable. The output of the power combiner is delivered to the receiver by another section of coaxial transmission line. The antenna system requires a fair-size area for setup: minimum requirements are a 25 × 35 foot flat area that has soil suitable for putting stakes into the ground. Because the antenna system is sensitive to noise, it is best not to set it up near any high-tension power lines or close to buildings. Also for safety reasons, please keep the antenna away from power lines during construction and operation. The best locations are in rural settings where the interference is minor. Because many of the observations occur at night, it is wise to practice setting up the antenna during the day to make sure the site is safe and easily accessible.

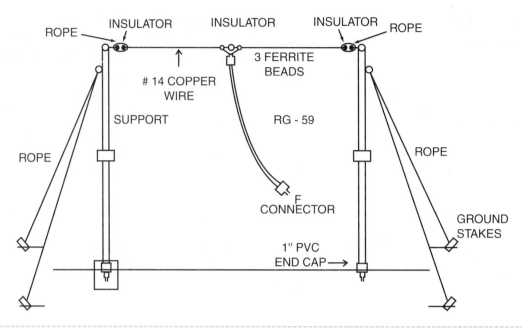

Figure 9-13   *Basic dipole*

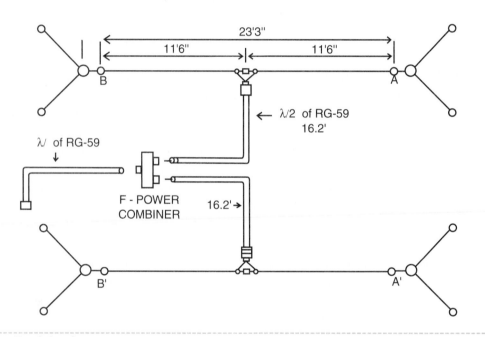

Figure 9-14   *Dual-dipole antenna*

If you want to locate your receiver indoors and your antenna is outside or on the rooftop, and if your antenna lead will not reach, you must use a longer coax cable. **Do not use just any length of coax.** The cable going from the power combiner to the receiver should be a multiple of half a wavelength long. The cable supplied with the kit is one wavelength: that is, 9.85 meters long (taking into account the 66 percent velocity factor of the RG-59/U cable). The maximum recommended cable length is five wavelengths.

There are many different manufacturers and qualities of coaxial cable. The 75-ohm cable supplied with the kit is manufactured by Belden and has a solid center conductor and velocity factor of 66 percent. Radio Shack does not carry RG-59/U cable, but they do have RG-6 and the higher grade RG-6QS (quad shield), which is also 75-ohm cable. Both of these cables have a velocity factor of 78 percent. One wavelength at 20.1 MHz in RG-6 cable is 11.64 meters. If you are going to put in a longer feedline, we recommend that you completely replace the existing one-wavelength piece, rather than splicing another length of cable onto the end.

Observing Tips: Jupiter's Radio Emissions

Most Jovian radio storms received via the Jupiter radio telescope will likely not be very strong. You will need to maximize your chances of receiving the Jovian storms by picking the best opportunities, selecting a good observing site, and using good observing techniques.

During the daytime, Earth's ionosphere blocks Jupiter's radio emissions. Therefore you will have to confine your investigations to nighttime listening. Radio noise from electrical sources, such as power lines, fluorescent lights, computers, and motors, can mask the signals from Jovian storms.

If possible try to select an observing site away from buildings and power lines. Also try to use portable power supplies such as a battery power supply. Note that Jovian radio storms may be short in duration, so persistence and luck are needed to capture Jupiter's radio signals.

Nighttime observations from a temporary field setup can be dangerous, so **plan carefully**! Get permission in advance to use the site, set up before dark, and use caution when walking around the site at night. The Jupiter radio telescope has two audio outputs: one channel can be used for ordinary listening and the second output jack can be used for recording. You can elect to use a portable audio tape recorder, however note that nearly all common portable tape recorders use an *automatic recording level control* (ALC) circuit. The ALC can mask the variations in the Jovian L bursts. (The instantaneous bandwidth of an L-burst is very large, greater than 5 MHz. They have a slow undulating quality that reminds many listeners of waves crashing on a distant beach.) Use a tape recorder that allows you to switch off the ALC. Keep the ALC turned off.

Another alternative for recording Jovian storms is the use of a computer or laptop to record the signals from your Jupiter radio telescope. Using a laptop computer to record the Jovian storms has the distinct advantage of allowing you to analyze the audio recording with special spectral software that can help you make sense of the complex signals received.

Note that you could also elect to use an audio tape recorder to record the Jovian signals in the field, and later when you return home play the tape recording into your laptop or desktop PC. Many audio analyzing software programs are available on the Internet. And many are free. Many programs allow *fast Fourier transform* (FFT) and frequency division analysis.

Jupiter Radio Telescope Parts List

Required Parts

R1	68-ohm resistor
R2	294-ohm resistor
R3	17.4-ohm resistor
R4	294-ohm resistor
R5	100-ohm resistor
R6	2.2K ohm resistor
R7	10K ohm linear potentiometer
R8	2.2K ohm
R9, R19	100K ohm
R10	220 ohm
R11	1.5K ohm
R12, R20, R21, R27	1K ohm
R13, R18	27K ohm
R14	100K ohm
R15	10K ohm potentiometer /switch
R16	10K ohm
R17	1.5K ohm

R22, R23 2 ohm

R24 1 ohm

R25 220 ohm

R26 47 ohm

R28, R29, R30, R31, R32
10 ohm

C1 39 pF, 35-volt disc
ceramic

C2 4-40 pF, variable
capacitor

C3 56 pF, 35-volt disc
ceramic

C4 22 pF, 35-volt disc
ceramic

C5, C8, C11, C14 0.01
mF, 35-volt dipped
ceramic

C6 4-40 pF, variable
capacitor

C7 not used

C9, C12, C13 47 pF, 35-
volt disc ceramic

C10 270 pF, 35-volt
disc ceramic

C15 10 pF disc ceramic

C16, C24, C25 10 mF,
35-volt DC,
electrolytic

C17, C18, C21, C23, C26,
C29 0.1 mF, 35-volt
dipped ceramic
capacitor

C19 1 mF, 35-volt metal
polyester capacitor

C20, C22 0.068 mF, 35-
volt, 5% metal film
capacitor

C27 10 mF, 35-volt DC,
tantalum, stripe,
capacitor

C28 220 pF, 35-volt
ceramic disc capacitor

C30, C31, C33 10 mF,
35-volt DC, elec-
trolytic capacitor

C32 330 mF, 35-volt DC,
electrolytic capacitor

C34, C35, C36 0.1 mF,
35-volt dipped ceramic
capacitor

C37 10 pF, 35-volt
ceramic disc capacitor

C38 10 mF, 35-volt DC,
electrolytic capacitor

C39 100 mF, 35-volt DC,
electrolytic capacitor

C40, C41, C42, C43 0.1
mF, 35-volt dipped
ceramic capacitor

C44 10 mF, 35-volt DC,
electrolytic capacitor

D1 1N4001 diode

D2, D3 1N914 diode

LED red LED

VD1 MV209, varactor
diode

ZD1 1N753 6.2-volt,
zener diode, 400
megawatts

ZD2 1N5231, 5.1-volt,
zener diode, 500
megawatts

L1 0.47 mH, (gold, yel-
low, violet, silver)

L2 1 mH, (brown, gold,
black, silver)

L3 3.9 mH, (orange,
gold, white, gold)

L4, L5 1.5 mH,
adjustable inductor

L6, L7 82 mH, fixed
inductor

Q1 J-310 junction field
effect, transistor
(JFET)

Q2 2N-3904 bipolar, NPN
transistor

Q3 2N-3906 bipolar, PNP
transistor

U1 SA602AN mixer /
oscillator IC

U2 LM387 audio pream-
plifier IC

U3 LM387 audio pream-
plifier IC

OSC1 20 MHz crystal
oscillator module

J1 power jack, 2.1
millimeter

J2 F female chassis
jack connector

J3, J4 3.5 millimeter
stereo jack, open ckt

Miscellaneous PC board,
enclosure 5×7×2,
knobs, solder lugs,
wire

Chapter Ten

Radiation Sensing

The radiation spectrum is usually broken down into electromagnetic radiation and ionizing radiation. Electromagnetic radiation includes visible light and longer wave bands like heat waves and radio waves. The bulk of the energy from the sun is in this range. We use this light to see, grow plants, and to power solar cells. The second type of radiation is called ionization radiation.

Ionizing radiation is usually thought of as high-energy and high-speed particles, though sometimes as very short wavelength waves. The particles can be photons, electrons, protons, and ionized elements, such as helium and iron. The ionized elements have been stripped of their electrons. When these high-speed particles pass through matter, they can do damage, possibly deep inside the matter. As they go through, they leave behind them a trail of ionized particles (that is, particles with missing electrons). The number of ionized particles per centimeter of path depends on the type of particle and its velocity. A bigger and higher-speed particle will do more damage. The ionizing radiation near Earth comes from our sun and also distant parts of the universe.

In this chapter, you will learn how to construct and utilize a cloud chamber for detecting low-ionizing alpha particles. You will discover how to detect ionizing radiation using the low-cost electronic ion chamber, which can be constructed using four commonly available transistors. Junior scientists can construct their own advanced electronic ion chamber that will permit more serious radiation study. Rock hounds can build a battery-powered Geiger counter that can be used for rock collection and field radiation studies.

Space Radiation

Three main sources of radiation can be found in space: the first is from *cosmic radiation*, the second is from *solar particle events*, and the final source is radiation from the *Van Allen belts*. Cosmic radiation comes from distant parts of the universe. The word *cosmos* means universe. Initially cosmic radiation was called *cosmic rays*, but now that we know it is really particles not rays. It is also sometimes called *galactic cosmic radiation*. We now know it is coming uniformly from all over space and not just from our galaxy, so it does not make sense to say *galactic*.

Shielding against cosmic radiation is difficult because its particles are so high energy. The energy of cosmic radiation can penetrate through meters of shielding. Fortunately, the level is low enough that it is not a problem for a short trip into outer space. However, cosmic radiation could be a significant problem for long-term space travelers, for example those who might staff a space station for more than a couple years.

The sun gives off a regular *solar wind* and a few times a year a *solar particle event* (SPE). These SPEs are sometimes called *solar flares*, but they are really more linked to *coronal mass ejections*. SPEs send out electrons and larger particles. The electrons arrive well in advance of the larger particles (on the order of half a day). The electrons are not so dangerous and can be used to predict when the larger particles will be coming. Also observations of the sun can indicate when things are happening. NASA has a Web page called Space Weather Now where you can see predictions for the sun's activity.

Some of the particles from the sun get trapped and collected by the earth's magnetic field. Inside this area the particles build up to higher concentrations than they do in open space. The atmosphere has an inner concentration of protons and an outer concentration of electrons. Each group circles the earth like a doughnut or belt. This area is known as the *Van Allen belts* after the man that discovered them in 1958. Probably the *Van Allen doughnuts* does not sound as good. Below about 1,000 kilometers altitude, the radiation is reduced to trace levels.

Radiation Sources on Earth

There are many natural sources of radiation on Earth, such as radon gas, which is often found leaking from the ground in certain areas of the country. Radium and uranium can be found in many locations in the Southwest. Radioactivity can also be found in Coleman-style camping lamp mantels, and many smoke detectors emit alpha particles that can be readily detected.

Radiation Caveats

Radiation can be confusing because so many different terms are used to describe it. You can talk about how many *million electron volts* (MeVs) of energy particles have and what their flux is (numbers like X particles per square centimeter per second). Because a range of energies usually exist, you need to know the shape of the distribution curve showing the number of particles at each energy level. There are also many different types of particles, so you need to know that distribution as well. This gets complicated fast.

People generally worry about figures relating to *absorbed radiation dosage*. This term indicates how much radiation energy is absorbed per kilogram of mass. Absorbed radiation dosage starts out as a single simple number. And although it soon gets a bit more complicated, it is still much simpler than dealing with flux information.

Working with radiation and radioactive materials requires great care and can be hazardous to your health if proper handling techniques are not used. Make sure that you are well informed as to radiation safety methods before handling radioactive substances.

Fun with a Cloud Chamber

A cloud chamber is a device used to detect elementary particles and other ionizing radiation. A cloud chamber consists essentially of a closed container filled with a supersaturated vapor, for example, water in air. When ionizing radiation passes through the vapor, it leaves a trail of charged particles (ions) that serve as condensation centers for the vapor, which then condenses around these particles. The path of the radiation is thus indicated by tracks of tiny liquid droplets in the supersaturated vapor.

The cloud chamber was invented in 1900 by C.T.R. Wilson. The type of cloud chamber he devised, often called the *Wilson cloud chamber*, is filled with air or another gas that is saturated with water vapor and enclosed in a cylinder fitted with a transparent window at the top and a piston or other pressure-regulating device at the bottom. When the pressure in the chamber is suddenly reduced (e.g., by lowering the piston), the gas-vapor mixture is cooled, producing a supersaturated gas. Cloud chambers of this design are also called the *pulsed cloud chamber*, because they do not maintain a continuous state of supersaturation of the vapor.

A more-recent design is the *diffusion cloud chamber* (see Figure 10-1). In this device a large temperature difference is maintained between the top and bottom of the chamber, usually by cooling the bottom of the chamber with dry ice. The gas in the chamber, usually air, is saturated with a vapor, usually alcohol. The air-vapor mixture cools as it diffuses toward the cool bottom, becoming supersaturated. If

Figure 10-1   *Diffusion cloud chamber*

Because the vapor is at a temperature where it normally can't exist, it will very easily condense into liquid form. When an electrically charged cosmic ray comes along, it ionizes the vapor. That is, the cosmic ray tears away the electrons in some of the gas atoms along its path. This leaves these atoms positively charged (because it removed electrons that have a negative charge). Other, nearby atoms are attracted to this newly ionized atom. This is enough to start the condensation process. So, you see little droplets forming along the path the particle took through the chamber.

the gas is kept saturated with a fresh supply of vapor (e.g., by an alcohol-soaked pad inside the top of the chamber), the operation of the chamber can be essentially continuous. One disadvantage of the cloud chamber is the relatively low density of the gas, which limits the number of possible interactions between ionizing radiation and molecules of the gas. For this reason physicists have developed other particle detectors, notably the *bubble chamber* and the *spark chamber*.

How Does the Cloud Chamber Work?

Because so much alcohol exists in the alcohol-soaked liner or pad, the chamber is saturated with alcohol vapor (the gaseous form of alcohol). The dry ice keeps the bottom very cold, while the top is still at room temperature. The higher temperature at the top means that the alcohol in the felt produces much vapor, which tends to fall downward. The vapor first begins to rise, then once confined by the chamber walls tumbles down along the chamber walls and mixes with the heavier cold air near the bottom of the chamber; it then tends to stay there.

The low temperature at the bottom means that once the vapor has fallen, it is supercooled. That is, it is vapor form, but at a temperature at which vapor normally can't exist. It's as if you had made steam at 95° C.

Cloud Chamber Parts List

Required Parts

A clear, see-through container with an open top, about 6 × 12 inches, and about 6 inches high. Make sure it is boxlike with flat sides, rather than being round.

A slide projector or other very strong light

A sheet of metal to cover the top of the container

A piece of thin cardboard (from a notebook or cereal box) the same size as the metal sheet

Black electrical tape

Felt or tissue for lining the container

A box a little bit bigger than the metal sheet

4 binder clips

Constructing the Cloud Chamber

First, line the sides of the container near the bottom with felt or tissue. This lining will be soaked with alcohol when you run the chamber, so do not use alcohol-soluble tape or glue to attach it. Next, cover one side of the cardboard with the black electrical tape; this will make the particle tracks easier to see. Place the cardboard, tape-side up, on the sheet of metal, then cover the container with the metal and the cardboard, so that the tape is facing the inside (see Figure 10-2).

Use the binder clips to secure the container to the metal/cardboard top. This is to prevent air leaks, so be sure it is tight. Turn the container over so that the metal is on the bottom and the felt is at the top. Place the container into the box. Place the slide projector against one side of the chamber so that it shines in. This is the dry configuration of the chamber. You won't see anything yet, but now you are ready to go.

In order to operate the cloud chamber, you'll need to obtain some pure (not 70%) isopropyl alcohol and 1 pound of dry ice. (You can usually get dry ice at ice cream stores.) Cut the dry ice into thin slices. Place the dry ice in the box underneath the chamber, between the box and the metal plate. Make sure that the slice of dry ice is shorter than the sides of the box. Note: Please use gloves and proper precautions when handling dry ice to avoid serious burns.

Remove the container from the box, open it, and soak the felt with the alcohol. Also place enough alcohol on the tape so that it is covered with a thin layer of liquid. Clip the metal and cardboard back into place, then replace the chamber on top of the dry ice. Be sure that the metal plate is resting directly on the dry ice. Turn on the slide projector lamp.

At first, you will see only a rain-like mist of alcohol. But after about 15 minutes, you should begin to see the tracks of particles passing through. The tracks look a little like spider's threads going along the chamber floor. Turn off the room lights; it may help you to see the tracks in a darkened room. For a 6 × 12 inch chamber, you should see about one track per second. Once you begin to see tracks, look for a track that goes straight, then kinks off to the left or right sharply. This is an indicator of *muon decay*. You might also see three tracks that meet at a single point. In these events, one track is an incoming cosmic ray. This particle hits an atomic electron. The electron and the outgoing cosmic track are the two other tracks. Look for a very windy, jagged track. This is an indicator of *multiple scattering*, as a low-energy cosmic ray bounces off of one atom in the air and runs into the next. While looking at tracks, you might notice that some tracks are very distinct and thick and others are very faint.

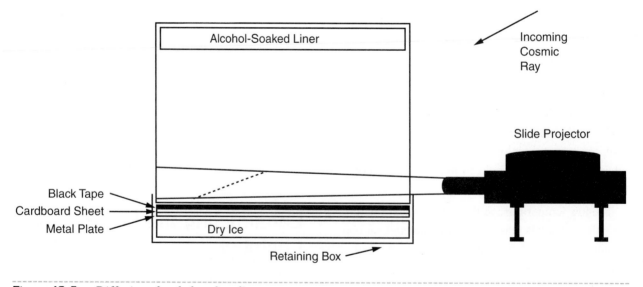

Figure 10-2 *Diffusion cloud chamber diagram*

Once you have some experience seeing these tracks, try some other experiments such as using a Polaroid camera to record the events. Or you could place a very strong magnet underneath the chamber. You will then see the particles bend when they are near the magnet. You might also try to place several plates of metal upright in the chamber, one behind the other. With this experiment you can see how many plates the tracks can go through.

Troubleshooting Your Cloud Chamber

Like any experiments, you may find yourself having difficulties in setting up or maintaining your cloud chamber. Here are a few common problems and their possible solutions.

- "I don't see anything at all!" Solution: Be sure the light is well placed. Make sure the dry ice is neatly packed and in good contact with the metal plate. Try adding more alcohol.

- "I only see mist, and no tracks." Solution: Wait. It takes about 15 minutes for the chamber to get to the right temperature.

- "I waited 15 minutes, and still nothing!" Solution: Be sure the light is well placed and shining into the chamber. Check that the chamber is airtight.

- "It's airtight, and there's good light, but I still see nothing." Solution: if you see only a very thick mist, try opening the chamber, letting some vapor escape, then starting over. If nothing works, try a new container that is a little shorter or taller.

- "I see big clouds at the edges of the chamber." Solution: This probably means you have an air leak. Be sure that the chamber is tightly sealed.

Low-Cost Ion Chamber

When ionizing radiation (ultraviolet light, X-rays, etc.) pass through a gas, collisions with the gas molecules produce ion pairs, typically charged molecules and free electrons. If an electric field is present, the ions will move apart in opposite directions along the electric field lines until they encounter the conductors that are producing the electric field. An ion chamber is an extremely simple device that uses this principle to detect ionizing radiation.

The basic chamber is simply a conducting can, usually metal, with a wire electrode at the center, well insulated from the chamber walls. The chamber is most commonly filled with ordinary dry air, but other gasses like carbon dioxide or pressurized air can give greater sensitivity. A DC voltage is applied between the outer can and the center electrode to create an electric field that sweeps the ions to the oppositely charged electrodes. Typically, the outer can has most of the potential with respect to ground so that the circuitry is near ground potential. The center wire is held near zero volts, and the resulting current in the center wire is measured.

This voltage required to sweep the ions apart and to the center wire or outer can before a significant number of them recombine or stick to a neutral molecule is usually under 100 volts and is often only a few volts. In fact, if the voltage is above a couple hundred volts, the speeding electrons will produce additional ion pairs called *secondary emissions* giving an enhanced response. Geiger tubes operate at even higher voltages with a special mixture of gasses and exhibit a sudden and very large discharge for each ionizing particle. But below 100 volts, only the ions produced by the radiation produce any current. The resulting current is extremely low in most situations, and detecting individual X-rays is difficult, especially with ordinary air at atmospheric pressure. Usually the capacitance of the electronics connected to the center wire smoothes the individual pulses too much for detection even when feedback is used to greatly reduce the time constant. These room-pressure chambers therefore respond to the average level of ionizing radiation and do not provide clicks like those produced in a Geiger counter tube.

Sensitive homemade ion chambers used for detecting nuclear radiation are fairly easy to build but the circuitry is tricky and should be attempted only by seasoned experimenters

Low-Cost Ion Chamber Radiation Detector

Is it possible to build a sensitive yet inexpensive radiation detector with a couple of transistors? The answer is a definite yes! A simple radiation detector was fabricated around a small ion chamber with four darlington transistors acting as current amplifiers. A few extra components were required also (see the photo in Figure 10-3).

The basis for this experiment is a single darlington transistor, with its base connected directly to the sense wire. In this setup, there was almost no collector current. Some leakage was expected with a floating base and a gain in the tens of thousands that can be expected from a darlington. The MPSW45 NPN darlington appears to be a good choice, as the leakage current is amazingly low and the gain seems very high, perhaps 30,000 with only a few tens of

picoamperes of base current. The gain was tested with a 100,000 megohm test resistor connected to a variable supply. A third and fourth darlington transistor is added to the final sensing circuit, as shown in Figure 10-4. (The original proof-of-concept radiation detector had only two darlingtons to test the principal.)

The ion chamber at RC-1 is constructed by using a 4.5 × 4 inch diameter peanut can, as seen in Figure 10-5. One end of the peanut can is removed, and the opposite end is drilled with a ³/₈-inch hole. The sensing transistors and other components are placed in their own compartment, as shown in Figure 10-6. For

Figure 10-4 *Peanut can used to make an ion chamber (Courtesy Charles Wenzel)*

Figure 10-5 *Sensing transmitters and other components of the ion chamber placed in their own compartment (Courtesy Charles Wenzel)*

Figure 10-6 *End cap and connector pins (Courtesy Charles Wenzel)*

Figure 10-3 *Simple radiation detector (Courtesy Charles Wenzel)*

this compartment, a steel wheel from the center of an electronic component reel is used, but any small can would work here. The small can chosen should fit over the peanut can, so it can be soldered together at a later time (see Figure 10-7). A large hole is then drilled into the component reel can, so you could insert an eight-pin glass-to-metal header. Note that you could also use an eight-pin microphone connector in place of the glass-to-metal header (see Figure 10-8). The header is epoxied into the hole in the small can; if you use a microphone connector, you will need to drill a suitable hole in the component reel can.

The ion chamber radiation detector circuit is depicted in Figure 10-4. Note that all the components are soldered around the eight-pin glass header or microphone connector. Notice also that the battery and the switch are wired in series across the rails of the circuit, so only two wires need to be brought out for power. The points marked X and Y on the circuit

diagram are actually pins of the glass-to-metal header or microphone connector, which allow a remote 100 uA meter to be wired to the circuit. A 4-inch long sensor wire is soldered to the base of transistor Q1. After all the components are soldered inside the small can, it is picked up and rotated 180 degrees and placed over the 3/8 inch hole in the peanut can. The sensor wire then goes through the 3/8-inch hole, almost to the opposite end of the peanut can. No insulator is used; just the air gap is used. But make sure the sensor wire does not touch the sides of the can when the reel can is mounted atop the peanut can. The small canister or reel can will eventually be soldered to the peanut can as shown. The opposite or open end of the peanut can is covered with aluminum foil to keep out air currents and electric fields but to allow less energetic or larger particles in. Actually, foil is a bit too thick, and other choices such as Mylar foil or plastic wrap might be better. Whichever material you choose, it can be held on with a large rubber band. The chamber is at room pressure so the membrane does not need strength.

Once you have built the ion chamber circuit in the component or reel can, recheck your wiring to make sure you haven't made any mistakes. Now you can tack-solder the component can housing to the peanut can as shown. Solder at only two points so that you can disassemble the can if a mistake was made or if the circuit doesn't work initially.

Figure 10-7 *End cap inside view (Courtesy Charles Wenzel)*

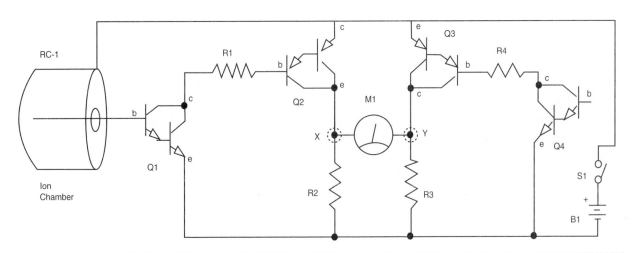

Figure 10-8 *Ion chamber I (Courtesy Charles Wenzel)*

Connect your meter to the header and then connect your battery power supply. A 12-volt DC battery source consisting of either eight AA cells or a 12-volt lantern battery can be used to get sufficient potential in the ion chamber for it to work efficiently. The peanut can is secured to a wooden baseplate as shown. A toggle switch can be mounted to the wooden base via a small metal mounting plate and wired between the battery and the header on the ion chamber. You can mount two four-AA battery holders to the wooden base plate or simply attach a 12-volt lantern battery to the ion chamber header connector.

Connect the battery and apply power to the circuit. If the meter goes negative, switch the pickup wire to the other transistor base and/or reverse the meter leads. If the voltage on the 2.2K resistors isn't low, maybe under a volt, try cleaning everything with a solvent and thoroughly drying it. If the voltage is still high, your darlingtons may not be good enough; try another type. Once the reading is low and steady, hold a radioactive source, such as a lantern mantle, near the foil window and watch the meter quickly climb. Note that you could use a digital voltmeter with a 1-volt scale instead of the 100 uA current meter shown.

The ion chamber circuit actually works very well and, after giving it 5 or 10 minutes to settle down, it could detect the lantern mantle from several inches away. The circuit is somewhat temperature sensitive, and the meter will move up the scale slightly with any increase in room temperature. The performance of this circuit is quite amazing. You can easily detect a Coleman lantern mantle and watch the meter movement as the mantle is brought closer to the detection chamber. This low-cost ion chamber would make an ideal demonstration project or science fair project for the junior scientist in the family.

Low-Cost Ion Chamber Radiation Detector Parts List

Required Parts

R1 10K ohm, 1/4-watt resistor

R2, R3 2.2K ohm, 1/4-watt resistor

R4 10K ohm, 1/4-watt resistor

Q1, Q4 MPSW45A dual NPN transistor

Q2, Q3 MPSA64 dual PNP transistor

M1 100 uA meter

S1 SPST toggle switch (power)

B1 battery

RC-1 radiation chamber (see text)

Miscellaneous perf-board circuit, wire, peanut can, mounting hardware, battery clips, wood baseplate, etc.

Advanced Ion Chamber Radiation Detector

Sensitive homemade ion chambers for detecting nuclear radiation are fairly easy to build, but the circuitry is tricky and should be attempted only by seasoned experimenters. The currents are likely to be well below 1 pA unless a serious nuclear war is in progress! Special electronics are needed at the front end. Typically an ion chamber detector is a type of special *electrometer* circuit, which produces an output voltage in proportion to the input current. The electrometer must have a very low bias or leakage current to avoid masking the desired signal and the intrinsic impedance of the amplifier must be extremely high. The input impedance of the electrometer may be fairly low, however, using feedback to convert the tiny current into a usable voltage.

Older ion chamber designs used special electrometer tubes like the 5886, which requires only 10 mA at 1.25 volts for the filament and about 10 volts for the plate. These tubes are great for the experimenter because they are relatively immune to static discharge, and they consume about the same amount of power as a typical transistor stage. Some electrometers use vibrating capacitors or mechanical choppers to convert the tiny DC currents into AC before amplification to avoid DC bias and leakage problems. Newer circuits typically use MOSFETs or electrometer grade JFETs in the front-end. MOSFET op-amps usually contain protection diodes that can be responsible for several picoamperes of leakage at room temperature and a fairly steep increase in leakage as the temperature increases, but in some ion chamber applications this extra leakage is tolerable. Nonprotected MOSFET front ends are easily damaged by static electricity, and special low-leakage protection diodes are usually added. Low-current JFETs like the 2N4220 give respectable performance, and the types intended for electrometer applications like the 2N4117A are quite impressive, exhibiting leakage well below 1 pA. These low-current JFETs have the added benefit of being significantly less sensitive to static electricity than do unprotected MOSFETs. Full *electrostatic discharge* (ESD) precautions must be observed with any of these approaches!

As mentioned previously, most electrometer circuits use feedback to reduce the effective input impedance and to direct the tiny input current through a very large feedback resistor such that a reasonable voltage is produced at the output. The feedback resistor must be quite large, however. If the input current is 1 pA and the feedback resistor is 100 megohms, the output voltage will only be 100 uV. Special resistors measuring in the millions of megohms are available but are usually difficult for the experimenter to obtain (see www.ohmite.com/catalog/v_rx1m.html, for example, or Victoreen). Lower value resistors may be bootstrapped to increase their effective value by a large factor, perhaps as much at 1,000, but that factor would only bring 10 megohm up to a mere 10,000

megohms. Actually, that is a workable value for most situations, but the circuit can require careful adjustment because that amount of bootstrapping is pushing the limits of practicality.

A very sensitive advanced experimental ion chamber is shown in Figure 10-9. The sensitive ion chamber can easily see background radiation. A shorter 4 × 3 inch diameter peanut can is used to house the sensitive ion chamber. A steel wheel from the center of an electronic component reel is used to house the electronics. This small can fits nicely on the bottom of the peanut can as shown. A ring of wire is used for the center electrode (see the schematic in Figure 10-11). The 4.7 uF capacitor connected to the ion chamber should be a nonpolar film type with a voltage rating above the voltage used (45 volts in the schematic). A nonpolar type allows the voltage polarity to be reversed for experimental purposes. The insides should be washed well with a solvent and then dried with a hot air gun before the base is added and tack-soldered. A hole is then drilled in the bottom of the peanut can for the electrode. No insulator is used —just the air gap.

The input detector, a sensitive FETs transistor, is mounted inside the chamber under the theory that this will eliminate the problem of connecting the extremely high-impedance probe to the outside world without creating leakage paths to ground. The

Figure 10-9 *Sensitive advanced experimental ion chamber (Courtesy Charles Wenzel)*

problem with this concept is that the transistor bodies and leads compete with the wire for the free ions! Carefully painting the transistor bodies and legs with conformal coating helps, but the circuit will not tolerate the coating around the base of the transistor—it is too conductive! The pickup wire should be thin and near the center of the can to keep the capacitance low so the response time is as short as possible.

The circuit for the second more sensitive ion chamber, shown in Figure 10-11, includes several improvements over the original circuit. A sensitive FET is now used at the front end of the detector circuit. A Victoreen 100,000 megohm resistor, which is the long glass tube in Figure 10-10, is used. You may have some difficulty in obtaining this resistor; contact IRC Technologies Division (see contact information in appendix). See text elsewhere. A zener diode is added to the emitter of the 2N4401 to increase the loop gain, and a 0.01 uF Miller capacitor is added to reduce the amplifier frequency response (for stability and to reduce 60 Hz gain). An op-amp (OP-07) is added to boost the output by a factor of 100. The zero pot is used to set the output to a few volts because the OP-07 cannot swing below 1 or 2 volts without a negative supply. This pot must be able to be adjusted to the gate voltage, and with some FETs the voltage may not go low enough. The symptom will be a high op-amp output voltage. If you have this situation, lower the 10K resistor or add a 1K resistor above the pot. An additional zero pot for the meter could be added as in the first schematic to get a near-zero reading for the background radiation, if desired.

Figure 10-10 *Sensitive ion chamber (Courtesy Charles Wenzel)*

The drain resistor on the FET is a 125K resistor. This value was experimentally determined by finding the drain current that gives the 2N4117A a near-zero temperature coefficient. The test circuit is simple: Connect a sensitive current meter from +10 volts to the drain, ground the gate, and connect the source to ground through a 500K pot. The current is observed at room temperature, then the FET is warmed and the current change is noted. The pot is adjusted until little or no change occurs. I heated the FET by touching a warm *positive temperature coefficient* (PTC) to the can, forcing it up to about 65 degrees Celsius, and the final current change was below the current change caused by a 100 uV gate voltage change. (This corresponds to less than 1 fA ion chamber current for 40 degrees.) Room temperature may vary by $+/-4$ degrees, corresponding to a variation of 0.1 fA, which is well below the 40 fA background current from the chamber. The bias current that gives this wonderful temperature compensation is 40 uA, and because the drain resistor will have 5 volts across it, the desired resistor value is $^5/_{40}$ uA $= 125k$. Your results may vary.

The output of the sensitive ion chamber is a voltmeter because this circuit has more gain and a higher output than the previous circuit with the use of the op-amp at U1. You can elect to use a digital voltmeter, a 10-volt panel meter. The potentiometer at R9 acts as a zero set control. The output of the op-amp at pin 6 is fed directly to a voltmeter.

Note that this circuit has two batteries. The 15-volt battery is used to supply power to the transistors and the op-amp, whereas the 45-volt battery is used to capture the electrons and is applied to the metal chamber. A 22.5 volt battery was insufficient to capture all of the ions but two batteries (totaling 45 volts) seemed to work fine. Higher voltages may be desired for observing individual events, however, because the ions will be swept to the electrode faster.

The ion chamber II circuit was housed in the metal can as shown in Figure 10-12. The opposite end of the peanut can was sealed with aluminum foil to keep out air currents and electric fields but to allow less energetic or larger particles in. Actually, foil is really too thick and other choices might be better. For

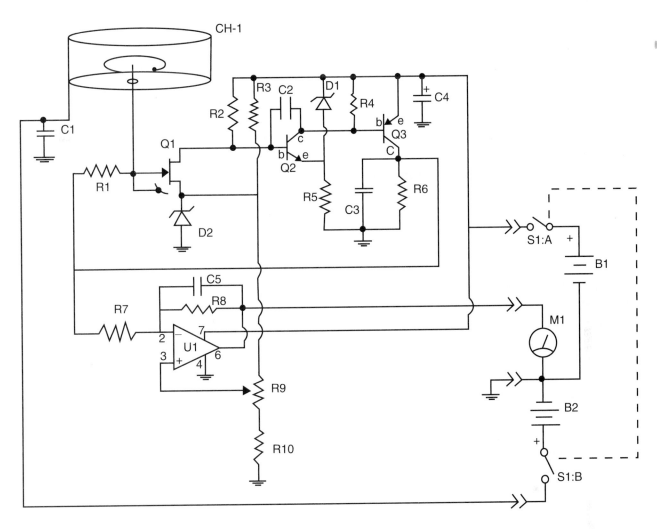

Figure 10-11   *Ion chamber II (Courtesy Charles Wenzel)*

example, you could use Mylar attached over the end of the can with a rubber band. The chamber is at room pressure so the membrane does not need added strength. Little feet are added to the bottom of the can so that you could easily slip the radiation disk underneath without disturbing the chamber. Ion chamber II produces some gratifying results: A very minute radiation source gives an output voltage change of about 70 mV, which is very large compared to the meter wander of about 2 mV.

Figure 10-12   *Ion chamber housed in a metal can*

When power is first applied, it can take a very long time, maybe 20 minutes, before the circuit settles down to a steady reading. The 15-volt power supply voltage should be fine for most ion chamber sizes. If the voltage is too low, the readings will be low, as the ions have time to recombine before being swept to the electrodes.

The radioactive element from a smoke detector can be held up to the Geiger counter so you can watch the count soar to about 22,000 *counts per minutes* (cpm). Next try placing a piece of aluminum foil in between to watch the count drop to 200 cpm. The ion chamber should give a reading of 200 mV, which is in perfect agreement. I didn't expect agreement in my test because the source is small relative to the Geiger tube. The Mylar window on the Geiger tube blocks the alpha particles some, and this may account for the agreement. These calibrations are really coarse! By using the Geiger counter to measure the background radiation, I was able to determine that the ion chamber should be indicating 13 mV, but because the zero setting is arbitrary, it was hard to confirm this level. Reversing the polarity on the outer can caused a shift of about 30 mV (after several minutes of settling), which is about what is expected if the background is near 15 cpm (the range from plus 15 to minus 15 is a total of 30).

Advanced Ion Chamber Parts List

Required Parts

R1 100,000-megohm glass resistor (Victoreen)

R2 125K ohm, 1/4-watt resistor

R3 8.2K ohm, 1/4-watt resistor

R4, R7 47K ohm, 1/4-watt resistor

R5, R6, R10 10K ohm, 1/4-watt resistor

R8 4.7-megohm, 1/4-watt resistor

R9 1K ohm, 1/4-watt resistor

C1 4.7 uF, 50-volt electrolytic capacitor

C2, C3 0.01 uF, 35-volt disc capacitor

C4 10 uF, 35-volt electrolytic capacitor

C5 0.1uF, 35-volt disc capacitor

D1 1N751 Zener diode

D2 1N753 Zener diode

Q1 2N4117A FET transistor

Q2 2N4401 NPN transistor

U1 OP-07 low-noise op-amp

M1 voltmeter

S1 DPST toggle switch

B1 15-volt power supply

B2 45-volt battery or power supply

CH-1 ion chamber (see text)

Miscellaneous perfboard, wire, IC socket, hardware, battery clips, connectors, etc.

Experimenting with a Geiger Counter

Everything on Earth is constantly bombarded by various kinds of nuclear radiation. These invisible particles and rays are collectively known as *background radiation*. Electromagnetic radiation includes everything from kilometers-long radio waves to tiny X-rays and gamma rays. Therefore, the term *radiation* can refer to the electromagnetic radiation emitted by a candle flame or the subatomic particles emitted by uranium ore.

Background radiation can strip electrons away from atoms or molecules, thereby transforming them into ions. Therefore, background radiation is known as *ionizing radiation*. X-rays, gamma rays, and both alpha and beta particles are all forms of ionizing radiation. Three forms of energy are associated with radioactivity: alpha, beta, and gamma radiation. The classifications were originally made according to the penetrating power of the radiation.

Gamma rays and X-rays are made of high-energy photons and have wavelengths much shorter than a wavelength of light. They travel at the speed of light and have enormously high penetrating power. They can pass entirely through a human body and are stopped only by thick layers of soil and concrete. Gamma rays are able to pass through several centimeters of lead and other dense materials and can still be detected on the other side.

Beta rays were found to be made of electrons, identical to the electrons found in atoms. Beta rays have a net negative charge. Beta rays have a greater penetrating power than do apha rays and can penetrate 3 millimeters of aluminum. A beta particle can pass through a few centimeters of human tissue.

Alpha rays were found in the nuclei of helium atoms, where two protons and two neutrons bound together. Alpha rays have a net positive charge. Alpha particles have weak penetrating ability, a couple of inches of air or a few sheets of paper can effectively block them. Your skin protects you from alpha radiation, but a source of alpha particles can prove dangerous if it is inhaled or swallowed.

Sources of Background Radiation

Various kinds of background radiation that arrive from space are known as cosmic rays. Only the most energetic of these powerful, poorly understood rays actually reach the earth's surface. Nearly all types strike molecules of air in the upper atmosphere, where they release showers of subatomic particles.

A major source of terrestrial background radiation is radon, an odorless, invisible gas that is heavier than air. Radon is emitted by uranium and thorium in soil and rock around the earth, with some locations being especially hot, or high in concentration of radon. For example, the radiation dose on a hill near Pocos de Caldas in Brazil is 800 times higher than the average. And the thorium in the beach at a nearby coastal town causes a radiation dose some 560 times higher than average.

In some areas radon seeps into basements and lower stories of houses, office buildings, factories, and schools. A study of some houses in Helsinki, Finland, revealed that radon levels inside were more than 5,000 times higher than outside.

Well water sometimes contains radon, and a Canadian study found that radon increased dramatically in a closed bathroom during a shower with contaminated water. It took more than an hour for the bathroom to return to the level that it was before the shower. Houses have been built on or near the radon-emitting tailings from uranium mines. And some building materials emit radon. For example, bricks are sometimes made from radioactive slag and fly ash from coal-fired power plants. And radioactive substances such as phosphogypsum and alum shales have been used to make plasterboard, blocks, cement, and otherbuilding materials. Ordinary sand and gravel canbe slightly radioactive. Nuclear power plants,atmospheric nuclear explosions, radioactive waste, and medical and dental X-ray machines are also sources of radiation.

Depending on where you live, perhaps half the background radiation passing through this page, and through your body, comes from outer space. The rest comes from soil, rocks, building materials, air, and even your own bones.

Detecting Radiation

Many different methods for detecting radiation have been developed. Some provide an instantaneous indication of the presence of radioactivity. Others indicate dose received over a period of time.

Photographic film is a good example of the latter. It is the oldest method for detecting radiation and is still widely used. Radiation can also be detected with an electroscope, various kinds of semiconductors, and by sensitive light detectors that respond to light generated when radiation passes through certain materials.

Geiger counters are instruments that can detect and measure radioactivity. H. Geiger and E.W. Muller invented the Geiger counter in 1928. With a Geiger counter you can check materials and environments for radioactivity. You could go prospecting for uranium, or you could check for radon in your basement. Radon gas would show itself by increasing the radiation counts over the background radiation.

Geiger counters detect radiation by means of a *Geiger Mueller* (GM) tube. A typical GM tube is a sealed, electrically conductive tube that contains argon, air, or other gases and a single-wire electrode. If alpha particles are to be detected, the tube is equipped with a thin window of mica or Mylar.

The GM tube hasn't changed much since it was invented in 1928. The operating principle is the same. A cutaway drawing of the tube is shown in Figure 10-13. The wall of the GM tube is a thin metal cylinder (cathode) surrounding a center electrode (anode). It is constructed with a thin mica window on the front end. The thin mica window allows the passage and detection of alpha particles. The tube is evacuated and filled with neon plus argon plus halogen gas. It is interesting to see how the GM tube

detects radioactivity. A 500-volt potential is applied to the anode (center electrode) through a 10 megohm current-limiting resistor. To the cathode of the tube, a 100K ohm resistor is connected.

In the initial state, the GM tube has a very high resistance. When a particle passes through the GM tube, it ionizes the gas molecules in its path. This is analogous to the vapor trail left in a cloud chamber by a particle. In the GM tube, the electron liberated from the atom by the particle and the positive ionized atom both move rapidly toward the high-potential electrodes of the GM tube. In doing so they collide with and ionize other gas atoms. This creates a small conduction path allowing a momentary surge of electric current to pass through the tube.

This momentary pulse of current appears as a small voltage pulse across R2. The halogen gas quenches the ionization and returns the GM tube to its high-resistance state making it ready to detect more radioactivity. The output from the Geiger counter may be an analog meter that indicates radiation in terms of *milliroentgens per hour* (mR/hr) or a digital readout that gives the number of *counts per minute* (cpm). In the latter case, the instrument might be supplied with a calibration number to convert the cpm rate to mR/hr.

Count Rate versus Dose Rate

Each output pulse from the GM tube is a count, and the counts per second give an approximation of the strength of the radiation field. In the count rate versus dose rate chart shown in Figure 10-14, the GM tube has been calibrated using a cesium-137.

The Geiger counter project shown in Figure 10-15 is sensitive to gamma, beta, and alpha radiation sources. The Geiger counter circuit depicted in Figure 10-16 literally starts with the 4049 hex inverting buffer at the left of the diagram. The hex inverters are set up as a square wave generator. The 4.7K ohm potentiometer adjusts the width of the square wave. The power MOSFET IRF830 switches the current on

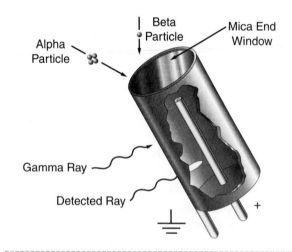

Figure 10-13 *Cutaway drawing of a Geiger Mueller (GM) tube*

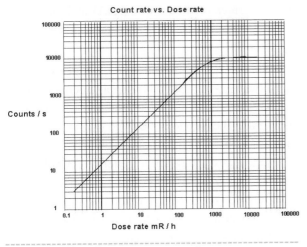

Figure 10-14   *Count rate versus dose rate*

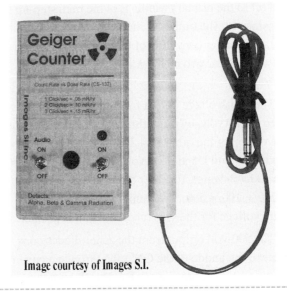

Figure 10-15   *Geiger counter*

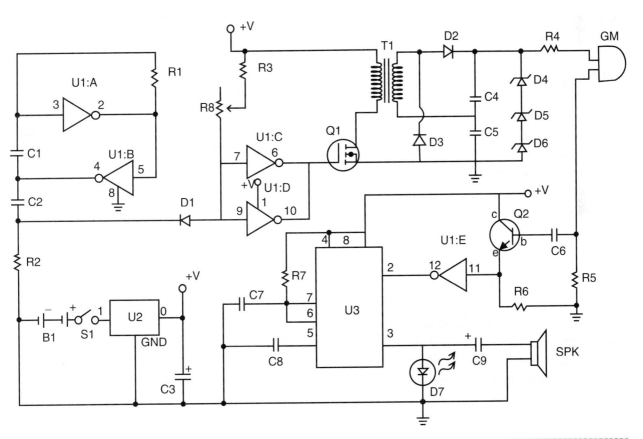

Figure 10-16   *Geiger counter circuit*

and off to the primary windings of the mini step-up transformer. The output of the transformer is fed to a voltage doubler consisting of two high-voltage diodes D2 and D3 and two high-voltage capacitors C4 and C5.

The high-voltage output from this stage is regulated to 500 volts by using three zener diodes stacked one on top of the other (D4, D5, and D6). Zener diodes D4 and D5 are 200-volt zeners, and diode D6 is a 100-volt zener. Together they equal 500 volts (200 + 200 + 100 = 500). Five hundred volts is the optimum voltage for the GM tube used in this project.

The 500-volt output from the zener diodes connects to the anode of the GM tube through a current-limiting 10-megohm resistor R4. The 10-megohm resistor limits the current through the GM tube and helps the avalanched ionization to be quenched.

The cathode of the tube is connected to a 100-kilohm resistor (R5). The voltage pulse across R5 generated by the detection of radiation feeds to the base of a 2N2222 NPN transistor, through a 1 uF capacitor (C6).

The NPN transistor clamps the output pulse from the GM tube to Vcc and feeds it to an inverting gate on the 4049. The inverted pulse signal from the gate is a trigger to the 555 timer. The timer is set up in monostable mode that stretches out the pulse received on its trigger. The output pulse from the timer flashes the LED and outputs an audible click to the speaker via pin 3.

Circuit Construction

None of the specifications to the elements in this circuit are critical; it may be hand wired on a prototyping PC board or built on a printed circuit board. The printed circuit board makes the construction much easier and is available for only $15 (see parts list).

Begin construction by wiring the square-wave generator and pulse-shaping circuit using the 4049. Place and wire R1, R2, R3, V1, C1, C2, and D1. Now construct the high-voltage section consisting of the step-up transformer T1, diodes D2 and D3, and capacitors C3 and C4. To this add the 5-volt 7805 regulator and capacitor C9.

At this point in the construction, you need to adjust the high-voltage power supply. Set up a VOM to read 500 to 1,000 volts. Place the positive lead of the VOM at the junction of C3 and D2. Apply power and use V1 to adjust peak voltage output. You should be able to adjust the circuit to generate 700 volts or more. Once the circuit is adjusted, turn off the power. Add the three zener diodes used for voltage regulation. Apply power again, with the positive lead of the VOM still attached to the junction of C3 and D2 you should read a voltage of 500 volts. If you're not getting a proper reading, check the zener diodes to make sure you have them orientated in the right direction. Finish the construction by installing the 555 timer for pulse stretching, current-limiting resistors, transistor, LED, and speaker. The input leads for the GM tube is connected to a $\frac{1}{8}$-inch phone jack.

The circuit board and components may be mounted in any plastic enclosure large enough to hold everything. On the top of the enclosure, drill two holes: one for the on-off switch and the other for the LED. Drill a larger hole for the speaker. A flat piezo-electric speaker with a plastic frame is used as the clicks or counts sounder. The speaker is glued to the top of the plastic enclosure. Finally, drill a hole in the side of the enclosure for the $\frac{1}{8}$-inch phone jack.

Geiger Tube Assembly

The Geiger Muller tube is delicate and needs to be protected in an enclosure. However the tube enclosure should allow the front end of the tube to remain open so that alpha particles can make their way through the thin mica window and be detected. A suitable enclosure may be made from plastic PVC pipe. PVC pipe about 6 inches long with a $\frac{1}{2}$-inch *inside diameter* (ID) will accommodate the GM tube. Louvers are cut in one side of the PVC pipe to allow beta radiation access to the GM tube unimpeded.

Obtain a length of shielded cable. Solder the center wire of the cable to the solder clip attached to the center electrode of the tube. Solder the shielded wire portion of the cable to the wire connected to the metal wall of the GM tube. To prevent a short between the two GM tube wires, cover one or both with shrink tubing, silicone sealant cement, or corona dope.

In the prototype, a ¼-inch phone plug is attached to the opposite end of the cable. Before attaching the phone plug place a rubber grommet with an *outside diameter* (OD) of ¼ inch on the shielded cable. This will be used to close the PVC pipe holder.

Insert the GM tube inside the PVC pipe so that the tube lies ¼ inch away from the end with the louvers. Glue the GM tube in position with silicone sealant cement, the type that is used to construct and repair aquariums. This type of cement fills spaces between surfaces, and when it cures still retains a degree of flexibility. Apply the glue only to the outer perimeter of the top of the GM tube and the matching surface of the PVC pipe. Do the same for the bottom of the tube. To reach inside the PVC pipe to the bottom of the GM tube, you will need to place a small amount of silicone glue on an extension. When this glue has cured, you are ready for the final assembly. Slide the grommet up to and inside the PVC pipe. Secure the grommet in position with a generous amount of silicone sealant. This will form a good strain relief.

Attach a piece of velcro to the side of the PVC pipe and its mating surface to the side of the case. This allows you to attach the GM tube wand to the case.

Radiation Sources

Ionization types of smoke detectors use a radioactive alpha source, typically americium-241. To use the americium, you need to remove it from the smoke detector. Alpha particles travel only a few inches in open air, so you need to get pretty close to the source

with the GM tube to detect anything. The mantle in some Coleman lanterns is radioactive. Bring your Geiger counter to a local hardware store and check them out. Uranium ore from a mineral or a rock store should also emit sufficient radiation to trigger the counter.

For a more reliable source, you may want to purchase radioactive material. Small amounts of radioactive materials are available for sale encased in 1-inch diameter by ⅛-inch thick plastic disks. The disks are available to the general public even without a license. This material outputs radiation in the microcurie range and has been deemed by the federal government as safe. The cesium-137 is a good gamma ray source and has a half-life of 30 years.

Turn on the Geiger counter. If you have a radiation source, bring the GM tube close to it. The radiation will cause the Geiger counter to start clicking. The LED will pulse with each click. Each click represents the detection of one of the radioactive rays: alpha, beta, or gamma. Background radiation, from natural sources on Earth and cosmic rays will cause the Geiger counter to click. In my corner of the world, I have a background radiation that triggers the counter 12 to 14 times a minute.

Surveying Your Home

A Geiger counter can be used to check for radioactive materials around your home. First, measure the average background count for at least 5 minutes. Then place a Geiger counter near suspect items for 5 minutes and compare the difference. When I tried, a background count averaged 11 cpm, a glazed brick gave a reading of 40 cpm, and a ceramic tile entry gave a reading of 16 cpm. An operating color television gave a reading of 28 cpm on a day when the background count averaged 16 cpm.

Other household materials and items that might be slightly radioactive include bricks, stone (especially granite), and watches and clocks whose hands are painted with radium impregnated luminescent paint. Earthenware may be glazed with orange or red pigment that contains uranium oxide. Ionization-type smoke detectors have a tiny bit of radioactive material, but I've not been able to detect its presence with a Geiger counter.

Camping lantern mantles that contain thorium are radioactive. A Coleman mantle will give a reading of from 0.1 to 0.2 milliroentgens per hour (mR/hr) when the end window of a Geiger counter is placed directly adjacent to the mantle. When the mantle is rolled into a tight bundle, the reading will increase to around 0.5 mR/hr. Because the count rate drops only slightly when a piece of paper is placed between a lamp mantle and a G-M tube, it appears that most of the radiation is in the form of beta particles emitted by the radium-228 byproduct of the thorium in the mantle.

Using a Geiger Counter to Monitor Solar Flares

Solar flares are explosive storms in active areas on the surface of the sun. Most of the time they occur in the locality of sunspot clusters. The duration of flares can range from as little as 3 minutes to as long as 8 hours.

Flares are classified by intensity. An *M-class flare* is ten times as intense as a *C-class flare*, and an *X-class flare* is ten times as intense as an M-class flare. There are five different classes of X-ray flares. The purpose of this project is to detect M- and X-class solar X-ray flares from the surface of the earth using a Geiger counter to detect changes in the background count. The background count of a Geiger counter is caused by cosmic rays and radioactive materials in the earth's crust. Cosmic rays come from deep space and from solar flares on the sun. Radium is found in the ground and can be located by using a Geiger counter to detect radon seeping from soil and rocks. For this project, I used a Geiger counter to detect the background count on an hourly basis. I then made graphs of the background count and of both M- and X-class solar X-ray flares. The detection

of M- and X-class solar X-ray flares in correlation with background count is a fascinating topic to study and would make a great science fair project.

Commercial Geiger Counters

The Radalert 50 is a small self-contained digital Geiger counter made by International Medcom, which sells for about $185. This rugged instrument indicates either the accumulated count or the counts per minute. One of its most important features is an output jack that permits the instrument to be coupled to an external circuit, recorder, or computer. This will allow you to collect data over a period of time and analyze the results later. Many researchers have used this Geiger counter along with an HP-95LX minicomputer. The RM-60 from Aware Electronics is another Geiger counter and contains an output that can be connected to a computer. This counter with software sells for about $149.00.

Geiger Counter Parts List

Required Parts

R1 5.6K ohm, 1/4-watt resistor

R2 15K ohm, 1/4-watt resistor

R3 1K ohm, 1/4-watt resistor

R4 10-megohm, 1/4-watt resistor

R5, R7 100K ohm, 1/4-watt resistor

R6 10K ohm, 1/4-watt resistor

R8 5K potentiometer

C1 0.0047 uF, 35-volt Mylar capacitor

C2, C8 0.01 uF, 35-volt disc capacitor

C3 100 uF, 35-volt electrolytic capacitor

C4, C5 0.001 uF, 35-volt disc capacitor

C6 1 uF, 35-volt tanta-
    lum capacitor

C7 0.047 uF, 35-volt
    Mylar capacitor

C9 220 uF, 35-volt
    electrolytic capacitor

D1 1N914 silicon diode

D2, D3 1N4007 silicon
    diode

D4, D5 1N5281B zener
    diode

D6 1N5271B zener diode

D7 LED

Q1 IRF830 power FET
    transistor

Q2 2N2222 transistor

U1 CD4049 CMOS inverter

U2 LM7805 5-volt
    regulator

U3 LM555 timer IC

T1 8- to 1,200-ohm,
    audio interstage
    transformer

SPK-1 8-ohm speaker

S1 SPST power switch

B1 9-volt transistor
    radio battery

GM-1 500-volt Geiger
    Mueller tube

Miscellaneous chassis,
    PC board, box, wire,
    IC socket, battery
    clip, hardware, etc.

Specialized Components—All Available from Images SI, Inc.

Required Parts

GM tube $54.95

Mini step-up transformer $12.00

Geiger counter PCB $10.00

Radioactive sources:
    uranium ore $14.95;
    cesium-137 source
    $94.00

Complete Geiger counter
    kit (all components)
    GCK-01 $176.95

Note: Other models
    available including
    remote data logging

(see appendix for
    contact information)

Helpful Contact Information

Acculex, Inc.
PO Box 10300
Bedford, NH 03110
440 Myles Standish Boulevard
Taunton, MA 02780
Maker of Acculex DP-650 digital panel meter

Allegro MicroSystems, Inc.
115 Northeast Cutoff
Worcester, MA 01615
(508) 853-5000 phone
UGN3503U Hall effect Sensor

AmaSeis seismic software
available from Dr. Alan Jones
www.binghamton.edu/faculty/jones/AmaSeis.html

The Amateur Seismologist
2155 Verdugo Boulevard, PMB 528
Montrose, CA 91020
(818) 249-1759
info@amateurseismologist.com
Complete AS-1 system and parts

Andover Corporation
4 Commercial Drive
Salem, NH 03079

Barr Associates
PO Box 557
Westford, MA 01886
Optical filters

CityTech
www.citytech.com/
Pellistor combustible gas sensors
Larry Cochran
lcochrane@webtronics.com
WWW.seismicnet.com/serialtod.html
(650) 365-7162

DATAQ Instruments, Inc.
241 Springside Drive
Akron, OH 44333
(330) 668-1444
A/D dataloggers

Digi-Key Corporation
PO Box 677
Thief River Falls, MN 56701
Clairex's CLD71 silicon photo diode

Edmund Scientific Company
101 East Gloucester Pike
Barrington, NJ 08007
Optical filters

EG&G Judson
221 Commerce Drive
Montgomeryville, PA 18936
Interference filters
DF-5000 UV detectors

EyeThink, Inc.
115 Rowena Place
Lafayette, CO 80226
(301) 541-1000
www.eyethinkcorp.com
TSP-6001 $39.00
Low-cost silver/silver chloride or Ag/AgCL pH
reference electrodes

Fat Quarters Software
24774 Shoshonee Drive
Murieta, CA 92562
(909) 698-7950
www.fatquarterssoftware.com
FM-1, FM-3 magnetometer

Figaro USA
3703 West Lake Ave, Suite 203
Glenview, IL 60026
www.figarosensor.com
TGS826 gas sensor

Freescale Semiconductor
7700 West Parmer Lane
Mailstop PL-02
Austin, TX 78729
Motorola MPX2100DP pressure sensor

General Eastern Instruments
500 Research Drive
Wilmington, MA 01887
(800) 334-8643
G-Cap2 capacitive humidity sensor

Glolab Corporation
307 Pine Ridge Drive
Wappingers Falls, NY 12590
(845) 297-9772 fax (for credit card orders)
Infrared motion detector kit and parts

Hamamatsu Corporation
PO Box 6910
Bridgewater, NJ 08807

Honeywell International
101 Columbia Road
Morristown, NJ 07962
http://content.honeywell.com/sensing/prodinfo/pressure/default.asp
SPX50DN Sensym pressure sensor (Syn SenSym Inc. division)
ASCX01DN Sensym pressure sensor (Syn SenSym Inc. division)

Images SI, Inc.
109 Woods of Arden Road
Staten Island, NY 10312
(718) 966-3694 phone
(718) 966-3695 fax
GCK-02 Geiger counter kit and parts

IRIS Headquarters
1200 New York Ave, NW, Suite 800
Washington, DC 20005
AS-1 seismograph available to qualified schools

Dr. Alan Jones
AmaSeis seismic software
www.binghamton.edu/faculty/jones/AmaSeis.html

Keithley Instruments
Keithley Instruments,Inc.
28775 Aurora Road
Cleveland, Ohio 44139

MicroCoatings
One Liberty Way
Westford, MA 01886
Interference filters

Micrometals
5615 E La Palma Ave, Anaheim, CA
Anaheim, CA
(714) 970-9400

Mouser Electronics
PO Box 699
Mansfield, TX 76063
(800) 346-6873

National Institute of Standards and Technology (NIST)
Washington, DC 20402
Publications available

National Semiconductor
2900 Semiconductor Drive
P.O. Box 58090
Santa Clara, CA 95052

Nemototech
http://www.nemototech.com/technology.htm
Pellistor combustible gas sensors

Newark Electronics
4801 North Ravenswood Avenue
Chicago, IL 60640-4496
(800) 342-1445
Honeywell SD3421-002 silicon photo diode

Nuclear Products Co.
PO Box 1178
El Monte, CA 91734
Ionizing radioactive cartridge 3-inch model 3Cl25 cartridge

Onset Computers
PO Box 3450
Pocasset, MA 02559
(800) 564-4377
www.onsetcomp.com
HOBO data loggers

Philips ECG Products
PO Box 967
1001 Snapps Ferry Road
Greeneville, TN 37744-0967
Photo transistor ECG-3031

Purdy Electronics
720 Palomar Drive
Sunnyvale, CA 94086
(408) 733-1287

Radio Jove Project
c/o Dr. James Thieman
Code 633
NASA's GSFC
Greenbelt, MD 20771
(301) 286-9790 phone
(301) 286-1771 fax

Sixth Sense
4 Stinsford Road
Poole Dorset
England BH17 ORZ
CAT16/25 pellistor gas sensor
www.sixthsense.com

Speake & CO Limited
Elvicta Estate
Crickhowell
Powys NP8 1DF
United Kingdom
FM-1, FM-3 magnetometer

Texas Instruments Incorporated
12500 TI Boulevard
Dallas, TX 75243-4136
(800) 336-5236
TI TLC271BCP light detector/amplifier chip

TOKO Coils
www.tokoam.com

Transtronics, Inc.
3209 West 9th Street
Lawrence, KS 66049
(785) 841-3089
(785) 841-0434
inform@xtronics.com
http://xtronics.com
Ultrasonic listener kit #207

Twardy Technology, Inc.
PO Box 2221
Darien, CT 06820
Interference filters

Vernier
8565 Southwest Beaverton Highway
Portland, OR, 97225
(503) 297-5317
www.vernier.com

Victoreen
Ohmite Mfg. Co.
3601 Howard Street
Skokie, IL 60076
(847) 675-2600 telephone
(847) 675-1505 fax
info@ohmite.com
www.ohmite.com

Vishay Intertechnology, Inc.
63 Lincoln Highway
Malvern, PA 19355-2143
http://www.vishay.com/
U401 Siliconix dual matched N-Channel FETs

Zetex
Sales Office
Zetex, Inc.
700 Veterans Memorial Highway, Suite 315
Hauppauge, NY 11788

Data Sheets

The following pages include data sheets from a
number of important products:

CAT-16

CAT-25

Dataq analog-to-digital converter

DP650 Acculex LCD

AND LCD

FM-3 magnetometer

Humid-1 and Humid-2—GE humidity sensors

ICL7136-1—National Semiconductor analog-to-
digital converter

MPX2100 Motorola pressure sensor

ZN414 mini AM radio chip

Honeywell pressure sensor

Catalytic Elements - CAT16

Innovation, Quality and Expertise for Gas Detection.

Operating Performance

Operating Principle	Constant current
Gas Detected	Most combustible gases & vapours
Measurement Range	0-100% LEL
Operating Voltage	2.7V ± 0.2V
Operating Current	200mA
Maximum Power Consumption	580mW
Maximum Methane Concentration	5% v/v
Expected Operating Life	Greater than five years in air
Output Sensitivity	>12mV/% methane
Temperature Range	-40°C to +50°C
Pressure Range	1 atm ± 10%
Humidity Range (non-condensing)	Continuous: 15-90% RH
	Intermittent: 0-99% RH
Response Time (T^5 90)	<10 seconds
Long Term Zero Drift	<±3% LEL methane per year
Long Term Span Drift	<±3% LEL methane per year
Linearity	±10% LEL up to 100% LEL

Poison Resistance

Hexamethyl-Disiloxane	Very high
Hydrogen Sulphide	Very high

Physical Specification

Can Type	Closed
Storage Life	6 months in sealed container
Storage Conditions	10-20°C, 45-75% RH in clean air
Orientation	Any
Warranty Period	12 months from date of despatch

Ordering Details

Part Number	2111B2016
Order From	Sixth Sense
	4 Stinsford Road, Poole, Dorset, England BH17 0RZ
	Tel: (44) 01202 645770
	Fax: (44) 01202 665331
	e.mail: sensors@sixth-sense.com
Code date: 06/00	www.sixthsense.com

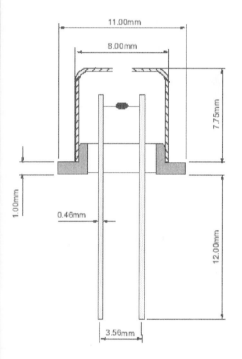

Innovation, Quality and Expertise for Gas Detection.

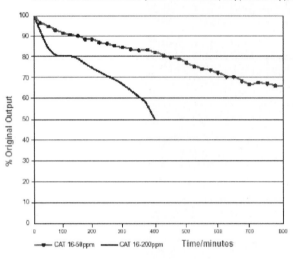

CAT16 Poison Resistance (2.5% v/v methane, 50ppm & 200ppm HMDS)

% Original Output

Time/minutes

—■— CAT 16-50ppm   —— CAT 16-200ppm

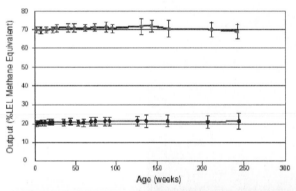

CAT16 Lifetests

Output (%LEL Methane Equivalent)

Age (weeks)

Relative Response Data*			
Gas/Vapour	**Relative Response%**	**Gas/Vapour**	**Relative Response%**
Methane	100	Ethanol	54
Hydrogen	121	Propan-2-ol	40
Ethane	70	Acetone	42
Propane	61	Butan-2-one (MEK)	40
Butane	49	MIBK	30
Pentane	42	Cyclohexane	37
Hexane	39	Di-Ethyl Ether	39
Heptane	35	Ethyl Acetate	37
Octane	32	Toluene	35
Ethylene	70	Xylene	26
Methanol	72	Acetylene	39

*Note: These figures are to be used as a guide only. For greatest accuracy, gas detectors should be calibrated with the target gas.
In the interest of product improvement Sixth Sense reserve the right to alter design features and specifications without notice.

Appendix B – Data Sheets

Catalytic Elements - CAT25

Operating Performance

Operating Principle	Constant voltage
Gas Detected	Most combustible gases & vapours
Measurement Range	0-100% LEL
Operating Voltage	3.3V ± 0.02V
Operating Current	70mA ± 5mA
Maximum Power Consumption	230mW
Maximum Methane Concentration	5% v/v
Expected Operating Life	Greater than two years in air
Output Sensitivity	> 25mV/% methane
Temperature Range	-40°C to +50°C
Pressure Range	1 atm ± 10%
Humidity Range (non-condensing)	Continuous: 0-90% RH
	Intermittent: 0-99% RH
Response Time (T^S 90)	< 10 seconds
Long Term Zero Drift	< ±5% LEL methane per year
Long Term Span Drift	< ±2% LEL methane per month
Linearity	± 10% LEL up to 100% LEL

Poison Resistance

Hexamethyl-Disiloxane	Some
Hydrogen Sulphide	Some

Physical Specification

Can Type	Open
Storage Life	6 months in sealed container
Storage Conditions	10-20°C, 45-75% RH in clean air
Orientation	Any
Warranty Period	12 months from date of despatch

Ordering Details

Part Number	2111B2125

Order From   Sixth Sense
4 Stinsford Road, Poole, Dorset,
England BH17 0RZ
Tel: (44) 01202 645770
Fax: (44) 01202 665331
e.mail: sensors@sixth-sense.com
Code date: 06/00   www.sixthsense.com

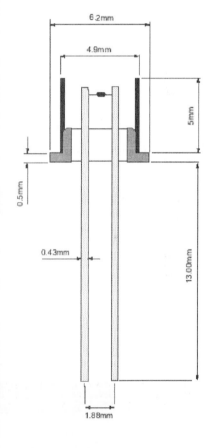

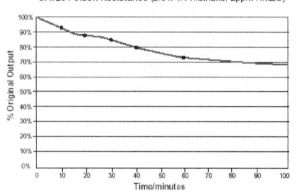

CAT25 Poison Resistance (2.5% v/v methane, 2ppm HMDS)

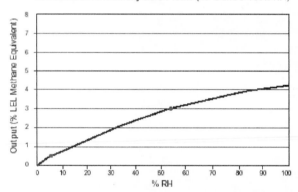

CAT25 Relative Humidity Zero Profile (40°C and 0-100% RH)

Relative Response Data*			
Gas/Vapour	**Relative Response%**	**Gas/Vapour**	**Relative Response%**
Methane	100	Ethanol	64
Hydrogen	107	Propan-2-ol	49
Ethane	82	Acetone	50
Propane	63	Butan-2-one (MEK)	48
Butane	51	MIBK	-
Pentane	50	Cyclohexane	-
Hexane	46	Di-Ethyl Ether	40
Heptane	44	Ethyl Acetate	46
Octane	38	Toluene	44
Ethylene	81	Xylene	31
Methanol	84	Acetylene	47

*Note: These figures are to be used as a guide only. For greatest accuracy, gas detectors should be calibrated with the target gas.
In the interest of product improvement Sixth Sense reserve the right to alter design features and specifications without notice.

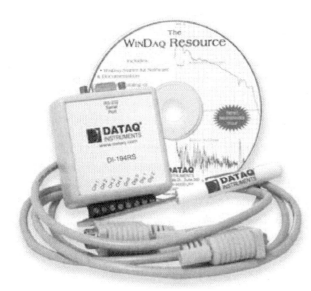

Inexpensive Analog to Digital Converter

Chart Recorder Starter Kit for $24.95
Chart recorder performance without messy and expensive paper.

More compact and a fraction of a traditional chart recorder price.

4-Channel, 10-Bit, ±10V ADC.

Provided with a Serial Port Interface Cable.

Two Digital Inputs For Remote Start/Stop and Remote Event Marker Control.

Includes WinDaq/Lite Chart Recorder Software, WinDaq Waveform Browser Playback and Analysis Software, and Documentation.

Provided ActiveX Controls allow you to program the DI-194 from any Windows programming environment.

NEW DataqSDK 1.0 (beta) Linux Package. Free C++ Software Developers Kit for use with the DI-194RS and DI-154RS Starter Kit devices. Click here to learn more...
Download our Starter Kit Datasheet

Download a PDF of the DI-194RS Starter Kit Manual
starterkit_pdf.jpg (6666 bytes)

Questions? Check out our Starter Kit Frequently Asked Questions Page.
Need more power? Take a look at our new starter kits and other data acquisition units:
NEW DI-148 and DI-158 Series Data Acquisition Starter Kits. With up to 14,400 Hz sample rates, USB interface, and up to ±64V FS range.
NEW DI-710 Data Acquisition System with stand-alone capability. Low-cost yet powerful data acquisition system with stand-alone option allowing you to record data directly to memory card.

View our full spectrum of products: Click here to check out our selection guide

Why order through our online store?
Click here to learn the advantages of purchasing your DATAQ Starter Kit online.

Part Number   Description   Price
(US Dollars)   DATAQ Store
DI-194RS   4-channel, 10-bit serial port instrument with cable. Includes WinDaq/Lite chart recorder software and WinDaq Waveform Browser playback and analysis software. $ 24.95

Quantity:
Note: We do NOT accept Purchase Orders for the DI-194RS. If your order is for this product only, you must pay with a credit card or PayPal. Online Ordering Frequently Asked Questions.

Product Highlights
The DI-194RS is a four-channel data acquisition module designed to familiarize you with WinDaq chart recorder software. It provides 10-bit measurement accuracy, a ±10V analog measurement range, up to 240 samples/second throughput, and four analog input channels. It features a serial port interface, making installation a snap. It also features two digital inputs that can be used for remote start/stop and remote event marker control with WinDaq/Lite software.

Our DI-194RS starter kit provides a taste of the exceptional power and speed possible with WinDaq software. A free CD demonstrates the WinDaq Waveform Browser, our playback and analysis software (just call and ask), but to get a hands-on illustration of the chart recorder and display capabilities of WinDaq/Lite, you need the DI-194RS starter kit. When connected to your PCs serial port, the DI-194RS starter kit allows you to record, chart, and analyze data using your own signals. The DI-194RS starter kit consists of a portable, four-channel, 2.5 x 2.5 x 1.25 inch A/D module that can be directly connected to the serial port of any PC. The module features four, single-ended, bipolar analog inputs (maximum measurement range ±10 volts) and two digital input ports for remote stop/start or remote event marker control. It also features an onboard waveform generator that can be used as a quick, convenient input signal. The kit ships with WinDaq/Lite chart recorder software and WinDaq Waveform Browser playback and analysis software. Data acquisition rates up to 240 samples per second are supported for Windows 95, 98, NT, ME, 2000, and XP.

Self-Powered Advantage
The DI-194RS derives its power directly from the RS-232 serial port to which it is connected — no batteries to replace or external power supplies to connect.

WINDAQ Software Included
The DI-194RS starter kit includes WINDAQ/Lite recording software and WINDAQ Waveform Browser playback and analysis software.

See webpage --- http://www.dataq.com/194.htm

DP-650 Series

Miniature LCD Display Digital Panel Meters

Description

The DP-650 Series is an ultra-compact, low-power LCD meter requiring only 15mW of power. This meter is ideal for applications requiring a highly reliable LCD display in a small package. The snap-in bezel packaging enables easy and rapid installation of the meter. The simplicity of design, overvoltage protection, and low power consumption all contribute to the high reliability of this device.

Three input voltage ranges are available: ±200mV, ±2VDC, and ±20VDC. A 4-20mA current input adapter (DP-670) is also offered. Automatic polarity, overrange indication, adjustable decimal point, and display HOLD are standard features. The DP-650 Series employs a dual slope integrating A/D converter, and operates from an internal 100mV reference via a precision DC-to-DC converter. Typical accuracy is ±0.1%.

Features

- Convenient snap-in bezel package
- Very low power consumption (5VDC @ 3mA)
- Differential input, 3 voltage ranges
- 0.1% typical accuracy
- User-selectable decimal point
- Hold-display function
- Overvoltage protection
- Quantity discounts

Technical Notes

1. The connection shown utilizes differential inputs. For single-ended operation, INPUT(-) is strapped to GND. If the signal is isolated from the power source, connect INPUT(-) to GND either directly or through a 10K resistor network as shown to reduce common mode errors.

2. Display holds reading when HOLD is high (+5V). Readings are updated when HOLD is low or floating.

3. The decimal point is user selectable by connecting the decimal point pins (4, 5, or 6) to DP COM (pin 3).

4. Full-scale input span is adjusted prior to shipment. Further calibration is possible by adjusting the trimpot mounted on the back of the DPM.

5. Unused pins should be left open.

Applications

- Portable/mobile equipment
- Medical instruments
- Avionics/aerospace systems
- Office equipment

Ordering Information

Model	Description
DP-650	Miniature DPM LCD ±200mV
DP-652	Miniature DPM LCD ±2V
DP-654	Miniature DPM LCD ±20V
APS-5	120V AC to 5V DC Power Supply
C-10	Connector - 10 pin

6 **ACCULEX** · 1-888-472-6676 · Fax 1-888-210-0119 · acculex.com

DP-650 Series

Dimensions

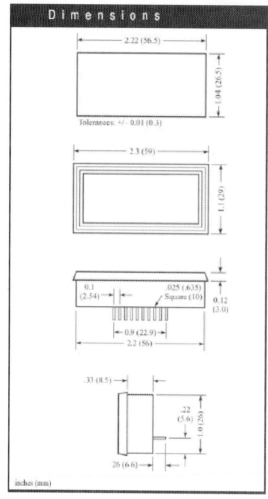

2.22 (56.5)

1.04 (26.5)

Tolerances: +/- 0.01 (0.3)

2.3 (59)

1.1 (29)

0.1 (2.54)   .025 (.635) Square (10)   0.12 (3.0)

0.9 (22.9)

2.2 (56)

.33 (8.5)

.22 (5.6)

1.0 (26)

26 (6.6)

inches (mm)

Connections

Pin No.	Pin Name	Description
1	V+	+5V power supply
2	GND	Power supply ground
3	DP COM	Decimal point return
4	DP1	1XX.X (connect to DP COM to turn on)
5	DP2	1X.XX (connect to DP COM to turn on)
6	DP3	1.XXX (connect to DP COM to turn on)
7	INPUT(+)	Positive input signal
8	INPUT(-)	Negative input signal (see Technical Note 1)
9	NC	No connection required
10	HOLD	Hold last display (see Technical Note 2)

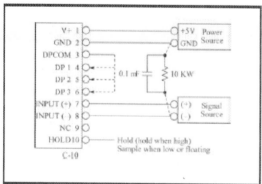

V+ 1
GND 2
DPCOM 3
DP 1 4
DP 2 5
DP 3 6
INPUT (+) 7
INPUT (-) 8
NC 9
HOLD 10

+5V Power Source
GND

0.1 mF   10 KW

(+) Signal Source
(-)

Hold (hold when high)
Sample when low or floating

C-10

Specifications

Display	Digits	3½ (±1999 count)
	Type	7-segment LCD
	Digit Height	.39in (10mm)
	Polarity Indication	Automatic - for neg input
	Decimal Point	3 position selectable
	Overrange Indication	1--- for positive, -1--- for negative
	Other Features/Options	Display hold function
Signal Inputs	Configuration	Bipolar, differential
	Full-Scale Input	±200mV, ±2V, ±20V
	Input Offset Adjustment	Auto zero
	Input Impedance	10MΩ (±200mV), 1MΩ (other ranges)
	Common Mode Range	±1VDC
	Common Mode Rejection	>86dB
	Overrange Protection	±350VDC (±100 VDC for 200mV range)
	Input Bias Current	1pA typical, 100pA max
	Control Inputs	Decimal point select, hold function
Performance	Sampling Rate	3.3 readings/s
	Accuracy	±(0.1% + 2 counts) typ, ±(0.2% + 2 counts) max
	Warmup, typical	10 min
	Temperature Drift, typical	100 ppm/°C
Power Supply Requirements		
	Supply Voltage	+5VDC ±5%
	Supply Current, typical	3mA
Physical	Package Style	Snap-in bezel mount
	Dimensions	2.30 x 1.10 x .71
	Panel Cutout	2.22 x 1.04
	Weight	0.7oz (20g)
	Connector	C-10 (10-pin) optional
	Bezel	Snap-in bezel included
Environmental Requirements		
	Operating Temperature	0 to 50°C
	Storage Temperature	-10 to 60°C
	Relative Humidity	0 to 95% non-condensing

FE0203

3 1/2 Digits, 0.5 Inch (12.7mm) Character Height

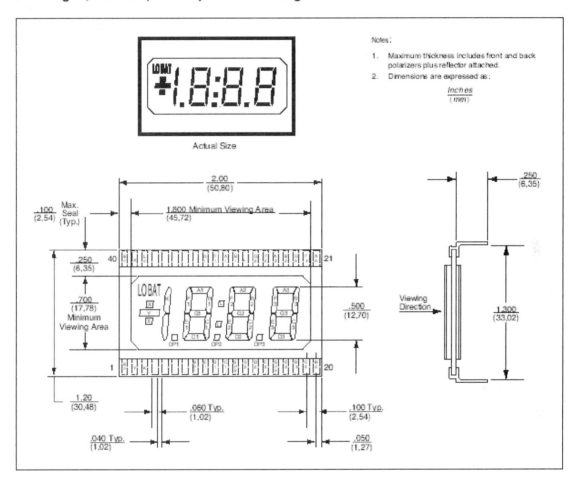

Notes:

1. Maximum thickness includes front and back polarizers plus reflector attached.

2. Dimensions are expressed as:

$$\frac{inches}{(mm)}$$

Pin Assignments

Pin #	Seg.	Pin #	Seg.	Pin #	Seg.	Pin #	Seg.	Pin #	Seg.	Pin #	Seg.	Pin #	Seg.	Pin #	Seg.	Pin #	Seg.	Pin #	Seg.
1	BP	5	NC	9	E1	13	E2	17	E3	21	A3	25	A2	29	B1	33	NC	37	NC
2	Y	6	NC	10	D1	14	D2	18	D3	22	F3	26	F2	30	A1	34	NC	38	LO
3	K	7	NC	11	C1	15	C2	19	C3	23	G3	27	G2	31	F1	35	NC	39	X
4	NC	8	DP1	12	DP2	16	DP3	20	B3	24	B2	28	L	32	G1	36	NC	40	BP

Product specifications contained herein may be changed without prior notice.
It is therefore advisable to contact Purdy Electronics before proceeding with the design of equipment incorporating this product.

Purdy Electronics Corporation • 720 Palomar Avenue • Sunnyvale, CA 94085
4/2/05   Tel: 408.523.8200 • Fax: 408.733.1287 • email@purdyelectronics.com • www.purdyelectronics.com   1

FGM-series
Magnetic Field Sensors

PHYSICAL CHARACTERISTICS

Sensor Outline - FGM-3

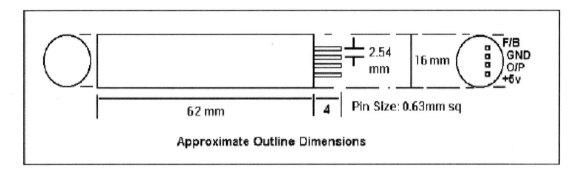

2.54 mm

16 mm

F/B
GND
O/P
+5v

62 mm

4

Pin Size: 0.63mm sq

Approximate Outline Dimensions

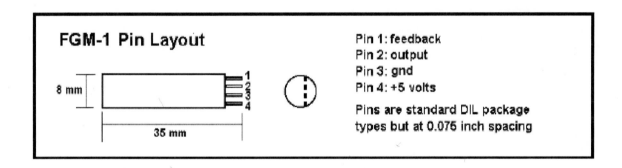

FGM-1 Pin Layout

8 mm

35 mm

1
2
3
4

Pin 1: feedback
Pin 2: output
Pin 3: gnd
Pin 4: +5 volts

Pins are standard DIL package
types but at 0.075 inch spacing

GE Infrastructure
Sensing

Performance

The unique properties of the G-CAP2's humidity-sensing polymer film allows the sensor to operate and survive when exposed to 100% RH condensing environments. These environments will often affect other sensors, leading to an increased drift rate and potential sensor failure. The G-CAP2 sensor offers low drift performance, typically less than 1% per year. As a result, Original Equipment Manufacturers (OEMs) can anticipate fewer failures and lower costs associated with replacement sensors and calibration services.

The G-CAP2 sensor also features a negligible temperature coefficient of less than 0.05% RH/°F (°C), ensuring accurate and reliable operation in applications where the final product will be exposed to wide temperature ranges.

Applications

Ideally suited for high-volume OEM applications, the G-CAP2 sensor can be used in automotive applications, data loggers, consumer appliances, HVAC controls, and medical instrumentation.

Sensing Element

The sensing element consists of a proprietary electrode metallization deposited over the humidity-sensitive polymer. This allows for rapid diffusion of water vapor, which accelerates desorption and makes recalibration easy. The properties of the film allow the sensor to survive total immersion in water with no loss of accuracy.

Sensor Enclosure

Unlike competitive designs where electrical contacts are made on active sensing surfaces, the G-CAP2's contacts are terminated on nonactive electrodes. This feature ensures continued performance and reliability without degradation of active sensor electrodes.

Assured Accuracy

Every G-CAP2 sensor is measured and verified in a controlled, NIST-traceable RH environment. Calibration chambers, equipment, facilities and processes are in compliance with MIL-STD 45662/A, Nuclear Regulation 10CFR Part 21 and ISO9001.

Custom-Designed OEM Applications

Depending on your requirements, GE Infrastructure Sensing can provide the G-CAP2 RH sensor with custom-designed, calibrated signal conditioning electronics to meet your specific application.

Evaluation Kit

GE offers a G-CAP2 relative humidity sensor evaluation kit. This kit includes five G-CAP2 sensors and a PC board with a 0 to 5 volt output. The PC board features plug-in connectors that allow an engineer to evaluate each G-CAP2 sensor without the need to design and build special test circuits. The kit includes full documentation that offers helpful hints on how to devise some simple tests.

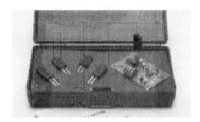

Performance

The unique properties of the G-CAP2's humidity-sensing polymer film allows the sensor to operate and survive when exposed to 100% RH condensing environments. These environments will often affect other sensors, leading to an increased drift rate and potential sensor failure. The G-CAP2 sensor offers low drift performance, typically less than 1% per year. As a result, Original Equipment Manufacturers (OEMs) can anticipate fewer failures and lower costs associated with replacement sensors and calibration services.

The G-CAP2 sensor also features a negligible temperature coefficient of less than 0.05% RH/°F (°C), ensuring accurate and reliable operation in applications where the final product will be exposed to wide temperature ranges.

Applications

Ideally suited for high-volume OEM applications, the G-CAP2 sensor can be used in automotive applications, data loggers, consumer appliances, HVAC controls, and medical instrumentation.

Sensing Element

The sensing element consists of a proprietary electrode metallization deposited over the humidity-sensitive polymer. This allows for rapid diffusion of water vapor, which accelerates desorption and makes recalibration easy. The properties of the film allow the sensor to survive total immersion in water with no loss of accuracy.

Sensor Enclosure

Unlike competitive designs where electrical contacts are made on active sensing surfaces, the G-CAP2's contacts are terminated on nonactive electrodes. This feature ensures continued performance and reliability without degradation of active sensor electrodes.

Assured Accuracy

Every G-CAP2 sensor is measured and verified in a controlled, NIST-traceable RH environment. Calibration chambers, equipment, facilities and processes are in compliance with MIL-STD 45662/A, Nuclear Regulation 10CFR Part 21 and ISO9001.

Custom-Designed OEM Applications

Depending on your requirements, GE Infrastructure Sensing can provide the G-CAP2 RH sensor with custom-designed, calibrated signal conditioning electronics to meet your specific application.

Evaluation Kit

GE offers a G-CAP2 relative humidity sensor evaluation kit. This kit includes five G-CAP2 sensors and a PC board with a 0 to 5 volt output. The PC board features plug-in connectors that allow an engineer to evaluate each G-CAP2 sensor without the need to design and build special test circuits. The kit includes full documentation that offers helpful hints on how to devise some simple tests.

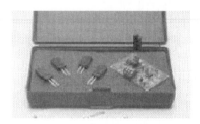

| Data Sheet | October 25, 2004 | FN3086.5 |

31/2 Digit LCD, Low Power Display, A/D Converter with Overrange Recovery

The Intersil ICL7136 is a high performance, low power $3\frac{1}{2}$ digit, A/D converter. Included are seven segment decoders, display drivers, a reference, and a clock. The ICL7136 is designed to interface with a liquid crystal display (LCD) and includes a multiplexed backplane drive.

The ICL7136 brings together a combination of high accuracy, versatility, and true economy. It features auto-zero to less than 10μV, zero drift of less than 1μV/ºC, input bias current of 10pA (Max), and rollover error of less than one count. True differential inputs and reference are useful in all systems, but give the designer an uncommon advantage when measuring load cells, strain gauges and other bridge type transducers. Finally, the true economy of single power supply operation, enables a high performance panel meter to be built with the addition of only 10 passive components and a display.

The ICL7136 is an improved version of the ICL7126, eliminating the overrange hangover and hysteresis effects, and should be used in its place in all applications. It can also be used as a plug-in replacement for the ICL7106 in a wide variety of applications, changing only the passive components.

Features

- First Reading Overrange Recovery in One Conversion Period
- Guaranteed Zero Reading for 0V Input on All Scales
- True Polarity at Zero for Precise Null Detection
- 1pA Typical Input Current
- True Differential Input and Reference, Direct Display Drive
  - LCD ICL7136
- Low Noise - Less Than 15μV$_{P-P}$
- On Chip Clock and Reference
- No Additional Active Circuits Required
- Low Power - Less Than 1mW
- Surface Mount Package Available
- Drop-In Replacement for ICL7126, No Changes Needed
- Pb-Free Available (RoHS Compliant)

Ordering Information

PART NUMBER	TEMP. RANGE (°C)	PACKAGE	PKG. DWG. #
ICL7136CPL	0 to 70	40 Ld PDIP	E40.6
ICL7136CPLZ (Note 1)	0 to 70	40 Ld PDIP (Pb-free) (Note 2)	E40.6
ICL7136CM44	0 to 70	44 Ld MQFP	Q44.10x10

NOTES:

1. Intersil Pb-free products employ special Pb-free material sets; molding compounds/die attach materials and 100% matte tin plate termination finish, which are RoHS compliant and compatible with both SnPb and Pb-free soldering operations. Intersil Pb-free products are MSL classified at Pb-free peak reflow temperatures that meet or exceed the Pb-free requirements of IPC/JEDEC J STD-020C.

2. Pb-free PDIPs can be used for through hole wave solder processing only. They are not intended for use in Reflow solder processing applications.

Pinouts

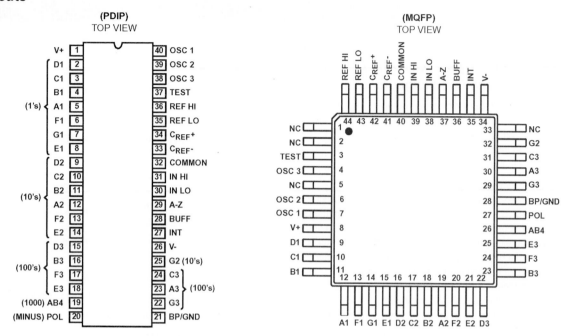

(PDIP)
TOP VIEW

V+	1	40	OSC 1
D1	2	39	OSC 2
C1	3	38	OSC 3
B1	4	37	TEST
A1	5	36	REF HI
F1	6	35	REF LO
G1	7	34	C_{REF}^+
E1	8	33	C_{REF}^-
D2	9	32	COMMON
C2	10	31	IN HI
B2	11	30	IN LO
A2	12	29	A-Z
F2	13	28	BUFF
E2	14	27	INT
D3	15	26	V-
B3	16	25	G2 (10's)
F3	17	24	C3
E3	18	23	A3
(1000) AB4	19	22	G3
(MINUS) POL	20	21	BP/GND

(1's) — A1, F1, G1, E1
(10's) — D2, C2, B2, A2, F2, E2
(100's) — D3, B3, F3, E3
A3, G3, C3 (100's)

(MQFP)
TOP VIEW

Top pins: REF HI, REF LO, C_{REF}^+, C_{REF}^-, COMMON, IN HI, IN LO, A-Z, BUFF, INT, V-

Left pins: NC (1), NC (2), TEST (3), OSC 3 (4), NC (5), OSC 2 (6), OSC 1 (7), V+ (8), D1 (9), C1 (10), B1 (11)

Right pins: NC (33), G2 (32), C3 (31), A3 (30), G3 (29), BP/GND (28), POL (27), AB4 (26), E3 (25), F3 (24), B3 (23)

Bottom pins: A1, F1, G1, E1, D2, C2, B2, A2, F2, E2, D3

Pin numbers: 44 43 42 41 40 39 38 37 36 35 34 (top); 12 13 14 15 16 17 18 19 20 21 22 (bottom)

Absolute Maximum Ratings

Supply Voltage
  ICL7136, V+ to V-. 15V
Analog Input Voltage (Either Input) (Note 1) . . . . V+ to V-
Reference Input Voltage (Either Input). . . . . V+ to V-
Clock Input
  ICL7136 . . . . TEST to V+

Operating Conditions

Temperature Range. . . . . 0°C to 70°C

Thermal Information

Thermal Resistance (Typical, Note 2) θ_{JA} (°C/W)
  PDIP Package   50
  MQFP Package   75
Maximum Junction Temperature . . . . .150°C
Maximum Storage Temperature Range. . . . . -65°C to 150°C
Maximum Lead Temperature (Soldering 10s) . . . . .300°C
  (MQFP - Lead Tips Only)

NOTE: Pb-free PDIPs can be used for through hole wave solder processing only. They are not intended for use in Reflow solder processing applications.

CAUTION: Stresses above those listed in "Absolute Maximum Ratings" may cause permanent damage to the device. This is a stress only rating and operation of the device at these or any other conditions above those indicated in the operational sections of this specification is not implied.

NOTES:

1. Input voltages may exceed the supply voltages provided the input current is limited to ±100μA.

2. θ_{JA} is measured with the component mounted on a low effective thermal conductivity test board in free air. See Tech Brief TB379 for details.

Electrical Specifications   (Note 3)

PARAMETER	TEST CONDITIONS	MIN	TYP	MAX	UNITS
SYSTEM PERFORMANCE					
Zero Input Reading	V_{IN} = 0V, Full Scale = 200mV	-000.0	±000.0	+000.0	Digital Reading
Ratiometric Reading	V_{IN} = V_{REF}, V_{REF} = 100mV	999	999/1000	1000	Digital Reading
Rollover Error	$-V_{IN}$ = $+V_{IN}$ ≅ 200mV Difference in Reading for Equal Positive and Negative Inputs Near Full Scale	-	±0.2	±1	Counts
Linearity	Full Scale = 200mV or Full Scale = 2V Maximum Deviation from Best Straight Line Fit (Note 5)	-	±0.2	±1	Counts
Common Mode Rejection Ratio	V_{CM} = ±1V, V_{IN} = 0V, Full Scale = 200mV (Note 5)	-	50	-	μV/V
Noise	V_{IN} = 0V, Full Scale = 200mV (Peak-To-Peak Value Not Exceeded 95% of Time) (Note 5)	-	15	-	μV
Leakage Current Input	V_{IN} = 0V (Note 5)	-	1	10	pA
Zero Reading Drift	V_{IN} = 0V, 0°C To 70°C (Note 5)	-	0.2	1	μV/°C
Scale Factor Temperature Coefficient	V_{IN} = 199mV, 0°C To 70°C, (Ext. Ref. 0ppm/x°C) (Note 5)	-	1	5	ppm/°C
COMMON Pin Analog Common Voltage	25kΩ Between Common and Positive Supply (With Respect to + Supply)	2.4	3.0	3.2	V
Temperature Coefficient of Analog Common	25kΩ Between Common and Positive Supply (With Respect to + Supply) (Note 5)	-	150	-	ppm/°C
SUPPLY CURRENT					
V+ Supply Current	V_{IN} = 0 (Does Not Include Common Current) 16kHz Oscillator (Note 6)	-	70	100	μA
DISPLAY DRIVER					
Peak-To-Peak Segment Drive Voltage and Peak-To-Peak Backplane Drive Voltage	V+ to V- = 9V (Note 4)	4	5.5	6	V

NOTES:

3. Unless otherwise noted, specifications apply to the ICL7136 at T_A = 25°C, f_{CLOCK} = 48kHz. ICL7136 is tested in the circuit of Figure 1.

4. Back plane drive is in phase with segment drive for "off" segment, 180 degrees out of phase for "on" segment. Frequency is 20 times conversion rate. Average DC component is less than 50mV.

5. Not tested, guaranteed by design.

6. 48kHz oscillator increases current by 20μA (Typ).

Typical Applications and Test Circuits

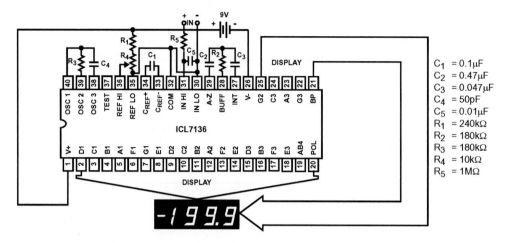

$C_1 = 0.1\mu F$
$C_2 = 0.47\mu F$
$C_3 = 0.047\mu F$
$C_4 = 50pF$
$C_5 = 0.01\mu F$
$R_1 = 240k\Omega$
$R_2 = 180k\Omega$
$R_3 = 180k\Omega$
$R_4 = 10k\Omega$
$R_5 = 1M\Omega$

FIGURE 1. ICL7136 TEST CIRCUIT AND TYPICAL APPLICATION WITH LCD DISPLAY COMPONENTS SELECTED FOR 200mV FULL SCALE

4 | **intersil**

FN3086.5

Design Information Summary Sheet

- **OSCILLATOR FREQUENCY**

  $f_{OSC} = 0.45/RC$
  $C_{OSC} > 50pF$; $R_{OSC} > 50k\Omega$
  f_{OSC} (Typ) = 48kHz

- **OSCILLATOR PERIOD**

  $t_{OSC} = RC/0.45$

- **INTEGRATION CLOCK FREQUENCY**

  $f_{CLOCK} = f_{OSC}/4$

- **INTEGRATION PERIOD**

  $t_{INT} = 1000 \times (4/f_{OSC})$

- **60/50Hz REJECTION CRITERION**

  t_{INT}/t_{60Hz} or t_{INT}/t_{50Hz} = Integer

- **OPTIMUM INTEGRATION CURRENT**

  $I_{INT} = 1\mu A$

- **FULL SCALE ANALOG INPUT VOLTAGE**

  V_{INFS} (Typ) = 200mV or 2V

- **INTEGRATE RESISTOR**

  $R_{INT} = \dfrac{V_{INFS}}{I_{INT}}$

- **INTEGRATE CAPACITOR**

  $C_{INT} = \dfrac{(t_{INT})(I_{INT})}{V_{INT}}$

- **INTEGRATOR OUTPUT VOLTAGE SWING**

  $V_{INT} = \dfrac{(t_{INT})(I_{INT})}{C_{INT}}$

- **V_{INT} MAXIMUM SWING:**

  $(V- + 0.5V) < V_{INT} < (V+ - 0.5V)$, V_{INT} (Typ) = 2V

- **DISPLAY COUNT**

  $COUNT = 1000 \times \dfrac{V_{IN}}{V_{REF}}$

- **CONVERSION CYCLE**

  $t_{CYC} = t_{CLOCK} \times 4000$
  $t_{CYC} = t_{OSC} \times 16,000$
  when $f_{OSC} = 48kHz$; $t_{CYC} = 333ms$

- **COMMON MODE INPUT VOLTAGE**

  $(V- + 1V) < V_{IN} < (V+ - 0.5V)$

- **AUTO-ZERO CAPACITOR**

  $0.01\mu F < C_{AZ} < 1\mu F$

- **REFERENCE CAPACITOR**

  $0.1\mu F < C_{REF} < 1\mu F$

- **V_{COM}**

  Biased between V+ and V-.

- **$V_{COM} \cong V+ - 2.8V$**

  Regulation lost when V+ to V- < $\cong 6.8V$.
  If V_{COM} is externally pulled down to (V+ to V-)/2, the V_{COM} circuit will turn off.

- **POWER SUPPLY: SINGLE 9V**

  V+ - V- = 9V
  Digital supply is generated internally
  $V_{TEST} \cong V+ - 4.5V$

- **DISPLAY: LCD**

  Type: Direct drive with digital logic supply amplitude.

Typical Integrator Amplifier Output Waveform (INT Pin)

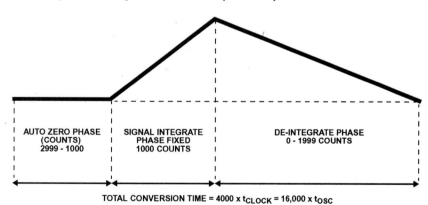

AUTO ZERO PHASE (COUNTS) 2999 - 1000 | SIGNAL INTEGRATE PHASE FIXED 1000 COUNTS | DE-INTEGRATE PHASE 0 - 1999 COUNTS

TOTAL CONVERSION TIME = 4000 x t_{CLOCK} = 16,000 x t_{OSC}

Pin Descriptions

PIN NUMBER		NAME	FUNCTION	DESCRIPTION
40 PIN DIP	44 PIN FLATPACK			
1	8	V+	Supply	Power Supply.
2	9	D1	Output	Driver Pin for Segment "D" of the display units digit.
3	10	C1	Output	Driver Pin for Segment "C" of the display units digit.
4	11	B1	Output	Driver Pin for Segment "B" of the display units digit.
5	12	A1	Output	Driver Pin for Segment "A" of the display units digit.
6	13	F1	Output	Driver Pin for Segment "F" of the display units digit.
7	14	G1	Output	Driver Pin for Segment "G" of the display units digit.
8	15	E1	Output	Driver Pin for Segment "E" of the display units digit.
9	16	D2	Output	Driver Pin for Segment "D" of the display tens digit.
10	17	C2	Output	Driver Pin for Segment "C" of the display tens digit.
11	18	B2	Output	Driver Pin for Segment "B" of the display tens digit.
12	19	A2	Output	Driver Pin for Segment "A" of the display tens digit.
13	20	F2	Output	Driver Pin for Segment "F" of the display tens digit.
14	21	E2	Output	Driver Pin for Segment "E" of the display tens digit.
15	22	D3	Output	Driver pin for segment "D" of the display hundreds digit.
16	23	B3	Output	Driver pin for segment "B" of the display hundreds digit.
17	24	F3	Output	Driver pin for segment "F" of the display hundreds digit.
18	25	E3	Output	Driver pin for segment "E" of the display hundreds digit.
19	26	AB4	Output	Driver pin for both "A" and "B" segments of the display thousands digit.
20	27	POL	Output	Driver pin for the negative sign of the display.
21	28	BP/GND	Output	Driver pin for the LCD backplane/Power Supply Ground.
22	29	G3	Output	Driver pin for segment "G" of the display hundreds digit.
23	30	A3	Output	Driver pin for segment "A" of the display hundreds digit.
24	31	C3	Output	Driver pin for segment "C" of the display hundreds digit.
25	32	G2	Output	Driver pin for segment "G" of the display tens digit.
26	34	V⁻	Supply	Negative power supply.
27	35	INT	Output	Integrator amplifier output. To be connected to integrating capacitor.
28	36	BUFF	Output	Input buffer amplifier output. To be connected to integrating resistor.
29	37	A-Z	Input	Integrator amplifier input. To be connected to auto-zero capacitor.
30 / 31	38 / 39	IN LO / IN HI	Input	Differential inputs. To be connected to input voltage to be measured. LO and HI designators are for reference and do not imply that LO should be connected to lower potential, e.g., for negative inputs IN LO has a higher potential than IN HI.
32	40	COMMON	Supply/ Output	Internal voltage reference output.
33 / 34	41 / 42	C_REF- / C_REF+		Connection pins for reference capacitor.
35 / 36	43 / 44	REF LO / REF HI	Input	Input pins for reference voltage to the device. REF HI should be positive reference to REF LO.
37	3	TEST	Input	Display test. Turns on all segments when tied to V+.
38 / 39 / 40	4 / 6 / 7	OSC3 / OSC2 / OSC1	Output / Output / Input	Device clock generator circuit connection pins.

Detailed Description

Analog Section

Figure 2 shows the Analog Section for the ICL7136. Each measurement cycle is divided into four phases. They are (1) auto-zero (A-Z), (2) signal integrate (INT) and (3) de-integrate (DE), (4) zero integrate (ZI).

Auto-Zero Phase

During auto-zero three things happen. First, input high and low are disconnected from the pins and internally shorted to analog COMMON. Second, the reference capacitor is charged to the reference voltage. Third, a feedback loop is closed around the system to charge the auto-zero capacitor C_{AZ} to compensate for offset voltages in the buffer amplifier, integrator, and comparator. Since the comparator is included in the loop, the A-Z accuracy is limited only by the noise of the system. In any case, the offset referred to the input is less than $10\mu V$.

Signal Integrate Phase

During signal integrate, the auto-zero loop is opened, the internal short is removed, and the internal input high and low are connected to the external pins. The converter then integrates the differential voltage between IN HI and IN LO for a fixed time. This differential voltage can be within a wide common mode range: up to 1V from either supply. If, on the other hand, the input signal has no return with respect to the converter power supply, IN LO can be tied to analog COMMON to establish the correct common mode voltage. At the end of this phase, the polarity of the integrated signal is determined.

De-Integrate Phase

The final phase is de-integrate, or reference integrate. Input low is internally connected to analog COMMON and input high is connected across the previously charged reference capacitor. Circuitry within the chip ensures that the capacitor will be connected with the correct polarity to cause the integrator output to return to zero. The time required for the output to return to zero is proportional to the input signal. Specifically the digital reading displayed is:

$$\text{DISPLAY READING} = 1000\left(\frac{V_{IN}}{V_{REF}}\right).$$

Zero Integrator Phase

The final phase is zero integrator. First, input low is shorted to analog COMMON. Second, the reference capacitor is charged to the reference voltage. Finally, a feedback loop is closed around the system to IN HI to cause the integrator output to return to zero. Under normal conditions, this phase lasts for between 11 to 140 clock pulses, but after a "heavy" overrange conversion, it is extended to 740 clock pulses.

Differential Input

The input can accept differential voltages anywhere within the common mode range of the input amplifier, or specifically from 0.5V below the positive supply to 1V above the negative supply. In this range, the system has a CMRR of 86dB typical. However, care must be exercised to assure the integrator output does not saturate. A worst case condition would be a large positive common mode voltage with a near full scale negative differential input voltage. The negative input signal drives the integrator positive when most of its swing has been used up by the positive common mode voltage. For these critical applications the integrator output swing can be reduced to less than the recommended 2V full scale swing with little loss of accuracy. The integrator output can swing to within 0.3V of either supply without loss of linearity.

Differential Reference

The reference voltage can be generated anywhere within the power supply voltage of the converter. The main source of common mode error is a roll-over voltage caused by the reference capacitor losing or gaining charge to stray capacity on its nodes. If there is a large common mode voltage, the reference capacitor can gain charge (increase voltage) when called up to de-integrate a positive signal but lose charge (decrease voltage) when called up to de-integrate a negative input signal. This difference in reference for positive or negative input voltage will give a roll-over error. However, by selecting the reference capacitor such that it is large enough in comparison to the stray capacitance, this error can be held to less than 0.5 count worst case. (See Component Value Selection.)

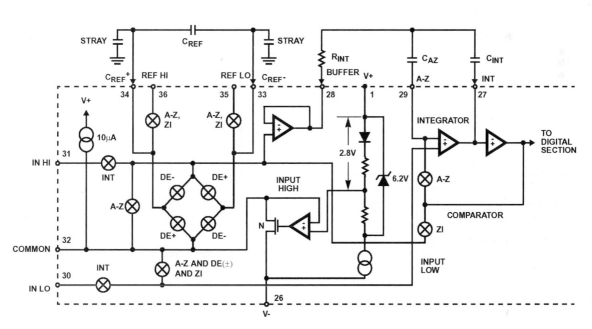

FIGURE 2. ANALOG SECTION OF ICL7136

Analog COMMON

This pin is included primarily to set the common mode voltage for battery operation or for any system where the input signals are floating with respect to the power supply. The COMMON pin sets a voltage that is approximately 2.8V more negative than the positive supply. This is selected to give a minimum end-of-life battery voltage of about 6.8V. However, analog COMMON has some of the attributes of a reference voltage. When the total supply voltage is large enough to cause the zener to regulate (>7V), the COMMON voltage will have a low voltage coefficient (0.001%/V), low output impedance ($\cong$15Ω), and a temperature coefficient typically less than 150ppm/^{O}C.

The limitations of the on chip reference should also be recognized, however. Due to their higher thermal resistance, plastic parts are poorer in this respect than ceramic. The combination of reference Temperature Coefficient (TC), internal chip dissipation, and package thermal resistance can increase noise near full scale from 25μV to 80μV$_{P-P}$. Also the linearity in going from a high dissipation count such as 1000 (20 segments on) to a low dissipation count such as 1111 (8 segments on) can suffer by a count or more. Devices with a positive TC reference may require several counts to pull out of an over range condition. This is because over-range is a low dissipation mode, with the three least significant digits blanked. Similarly, units with a negative TC may cycle between over range and a non-over range count as the die alternately heats and cools. All these problems are of course eliminated if an external reference is used.

The ICL7136, with its negligible dissipation, suffers from none of these problems. In either case, an external reference can easily be added, as shown in Figure 3.

Analog COMMON is also used as the input low return during auto-zero and de-integrate. If IN LO is different from analog COMMON, a common mode voltage exists in the system and is taken care of by the excellent CMRR of the converter. However, in some applications IN LO will be set at a fixed known voltage (power supply common for instance). In this application, analog COMMON should be tied to the same point, thus removing the common mode voltage from the converter. The same holds true for the reference voltage. If reference can be conveniently tied to analog COMMON, it should be since this removes the common mode voltage from the reference system.

Within the IC, analog COMMON is tied to an N-Channel FET that can sink approximately 3mA of current to hold the voltage 2.8V below the positive supply (when a load is trying to pull the common line positive). However, there is only 10μA of source current, so COMMON may easily be tied to a more negative voltage thus overriding the internal reference.

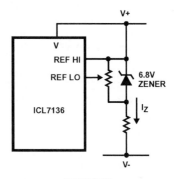

FIGURE 3A.

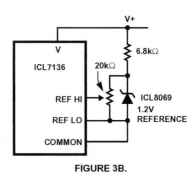

FIGURE 3B.

FIGURE 3. USING AN EXTERNAL REFERENCE

TEST

The TEST pin serves two functions. On the ICL7136 it is coupled to the internally generated digital supply through a 500Ω resistor. Thus it can be used as the negative supply for externally generated segment drivers such as decimal points or any other presentation the user may want to include on the LCD display. Figures 4 and 5 show such an application. No more than a 1mA load should be applied.

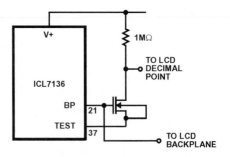

FIGURE 4. SIMPLE INVERTER FOR FIXED DECIMAL POINT

The second function is a "lamp test". When TEST is pulled high (to V+) all segments will be turned on and the display should read "-1888". The TEST pin will sink about 5mA under these conditions.

CAUTION: On the ICL7136, in the lamp test mode, the segments have a constant DC voltage (no square-wave) and may burn the LCD display if left in this mode for several minutes.

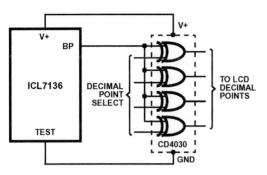

FIGURE 5. EXCLUSIVE "OR" GATE FOR DECIMAL POINT DRIVE

Digital Section

Figures 6 shows the digital section for the ICL7136. In the ICL7136, an internal digital ground is generated from a 6V Zener diode and a large P-Channel source follower. This supply is made stiff to absorb the relatively large capacitive currents when the back plane (BP) voltage is switched. The BP frequency is the clock frequency divided by 800. For three readings/second this is a 60Hz square wave with a nominal amplitude of 5V. The segments are driven at the same frequency and amplitude and are in phase with BP when OFF, but out of phase when ON. In all cases negligible DC voltage exists across the segments.

The polarity indication is "on" for negative analog inputs. If IN LO and IN HI are reversed, this indication can be reversed also, if desired.

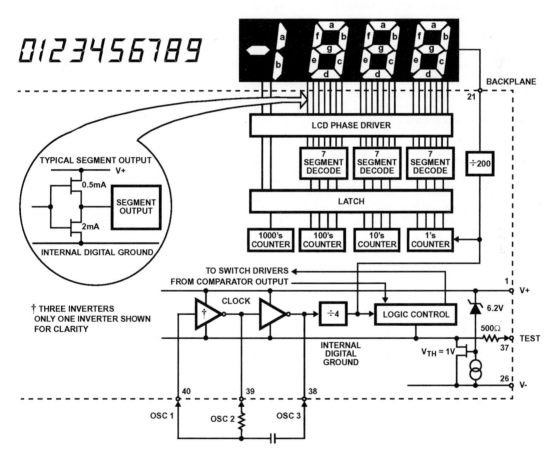

FIGURE 6. ICL7136 DIGITAL SECTION

System Timing

Figure 7 shows the clocking arrangement used in the ICL7136. Two basic clocking arrangements can be used:

1. Figure 9A, an external oscillator connected to DIP pin 40.
2. Figure 9B, an R-C oscillator using all three pins.

The oscillator frequency is divided by four before it clocks the decade counters. It is then further divided to form the three convert-cycle phases. These are signal integrate (1000 counts), reference de-integrate (0 to 2000 counts) and auto-zero (1000 to 3000 counts). For signals less than full scale, auto-zero gets the unused portion of reference de-integrate. This makes a complete measure cycle of 4,000 counts (16,000 clock pulses) independent of input voltage. For three readings/second, an oscillator frequency of 48kHz would be used.

To achieve maximum rejection of 60Hz pickup, the signal integrate cycle should be a multiple of 60Hz. Oscillator frequencies of 240kHz, 120kHz, 80kHz, 60kHz, 48kHz, 40kHz, $33^1/_3$kHz, etc., should be selected. For 50Hz rejection, Oscillator frequencies of 200kHz, 100kHz, $66^2/_3$kHz, 50kHz, 40kHz, etc. would be suitable. Note that 40kHz (2.5 readings/sec.) will reject both 50Hz and 60Hz (also 400Hz and 440Hz).

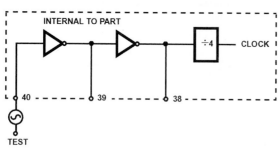

FIGURE 7A. EXTERNAL OSCILLATOR

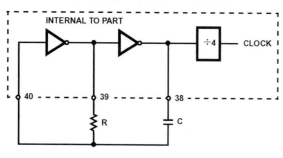

FIGURE 7B. RC OSCILLATOR

FIGURE 7. CLOCK CIRCUITS

Component Value Selection

Integrating Resistor

Both the buffer amplifier and the integrator have a class A output stage with 100µA of quiescent current. They can supply 1µA of drive current with negligible nonlinearity. The integrating resistor should be large enough to remain in this very linear region over the input voltage range, but small enough that undue leakage requirements are not placed on the PC board. For 2V full scale, 1.8MΩ is near optimum and similarly a 180kΩ for a 200mV scale.

Integrating Capacitor

The integrating capacitor should be selected to give the maximum voltage swing that ensures tolerance buildup will not saturate the integrator swing (approximately 0.3V from either supply). In the ICL7136, when the analog COMMON is used as a reference, a nominal +2V full-scale integrator swing is fine. For three readings/second (48kHz clock) nominal values for C_{INT} are 0.047µF and 0.5µF, respectively. Of course, if different oscillator frequencies are used, these values should be changed in inverse proportion to maintain the same output swing.

An additional requirement of the integrating capacitor is that it must have a low dielectric absorption to prevent roll-over errors. While other types of capacitors are adequate for this application, polypropylene capacitors give undetectable errors at reasonable cost.

Auto-Zero Capacitor

The size of the auto-zero capacitor has some influence on the noise of the system. For 200mV full scale where noise is very important, a 0.47µF capacitor is recommended. On the 2V scale, a 0.047µF capacitor increases the speed of recovery from overload and is adequate for noise on this scale.

Reference Capacitor

A 0.1µF capacitor gives good results in most applications. However, where a large common mode voltage exists (i.e., the REF LO pin is not at analog COMMON) and a 200mV scale is used, a larger value is required to prevent roll-over error. Generally 1µF will hold the roll-over error to 0.5 count in this instance.

Oscillator Components

For all ranges of frequency a 180kΩ resistor is recommended and the capacitor is selected from the equation:

$$f = \frac{0.45}{RC} \quad \text{For 48kHz Clock (3 Readings/s.),}$$

$$C = 50\text{pF}.$$

Reference Voltage

The analog input required to generate full scale output (2000 counts) is: $V_{IN} = 2V_{REF}$. Thus, for the 200mV and 2V scale, V_{REF} should equal 100mV and 1V, respectively. However, in many applications where the A/D is connected to a transducer, there will exist a scale factor other than unity between the input voltage and the digital reading. For instance, in a weighing system, the designer might like to have a full scale reading when the voltage from the transducer is 0.662V. Instead of dividing the input down to 200mV, the designer should use the input voltage directly and select $V_{REF} = 0.341V$. Suitable values for integrating resistor and capacitor would be 330kΩ and 0.047μF. This makes the system slightly quieter and also avoids a divider network on the input. Another advantage of this system occurs when a digital reading of zero is desired for $V_{IN} \neq 0$. Temperature and weighing systems with a variable fare are examples. This offset reading can be conveniently generated by connecting the voltage transducer between IN HI and COMMON and the variable (or fixed) offset voltage between COMMON and IN LO.

Typical Applications

The ICL7136 may be used in a wide variety of configurations. The circuits which follow show some of the possibilities, and serve to illustrate the exceptional versatility of these A/D converters.

The following application notes contain very useful information on understanding and applying this part and are available from Intersil.

Application Notes

NOTE #	DESCRIPTION
AN016	"Selecting A/D Converters"
AN017	"The Integrating A/D Converter"
AN018	"Do's and Don'ts of Applying A/D Converters"
AN023	"Low Cost Digital Panel Meter Designs"
AN032	"Understanding the Auto-Zero and Common Mode Performance of the ICL7136/7/9 Family"
AN046	"Building a Battery-Operated Auto Ranging DVM with the ICL7106"
AN052	"Tips for Using Single Chip $3\frac{1}{2}$ Digit A/D Converters"

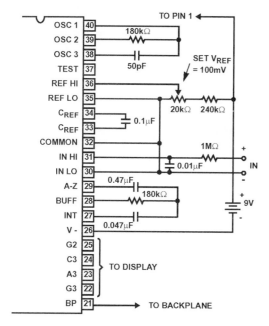

Values shown are for 200mV full scale, 3 readings/sec., floating supply voltage (9V battery).

FIGURE 8. ICL7136 USING THE INTERNAL REFERENCE

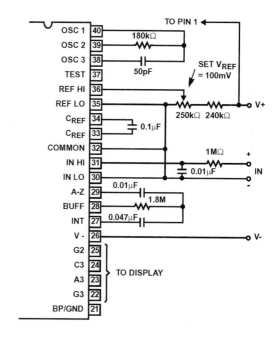

FIGURE 9. RECOMMENDED COMPONENT VALUES FOR 2V FULL SCALE

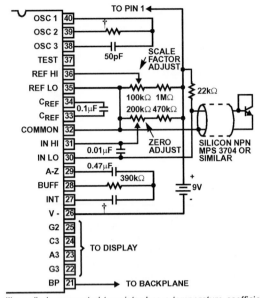

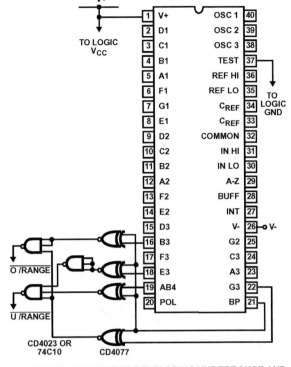

A silicon diode-connected transistor has a temperature coefficient of about -2mV/°C. Calibration is achieved by placing the sensing transistor in ice water and adjusting the zeroing potentiometer for a 000.0 reading. The sensor should then be placed in boiling water and the scale-factor potentiometer adjusted for a 100.0 reading.
† Value depends on clock frequency.

FIGURE 10. ICL7136 USED AS A DIGITAL CENTIGRADE THERMOMETER

FIGURE 11. CIRCUIT FOR DEVELOPING UNDERRANGE AND OVERRANGE SIGNAL FROM ICL7136 OUTPUTS

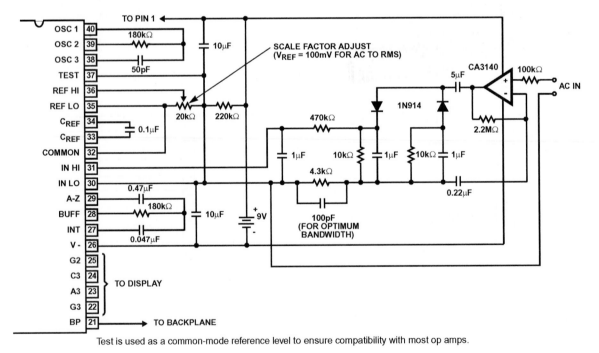

Test is used as a common-mode reference level to ensure compatibility with most op amps.

FIGURE 12. AC TO DC CONVERTER WITH ICL7136

Die Characteristics

DIE DIMENSIONS:

    127 mils x 149 mils

METALLIZATION:

    Type: Al
    Thickness: 10kÅ ±1kÅ

PASSIVATION:

    Type: PSG Nitride
    Thickness: 15kÅ ±3kÅ

Metallization Mask Layout

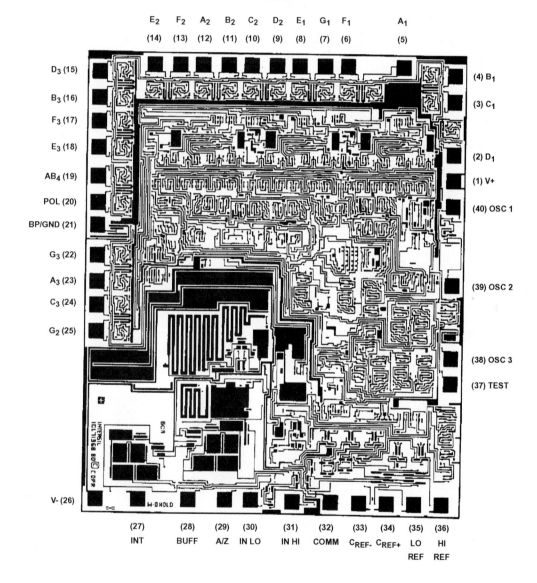

ICL7136

Dual-In-Line Plastic Packages (PDIP)

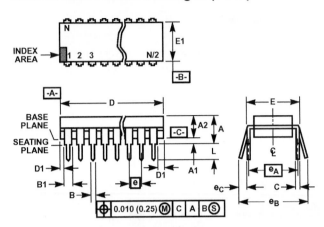

E40.6 (JEDEC MS-011-AC ISSUE B)
40 LEAD DUAL-IN-LINE PLASTIC PACKAGE

SYMBOL	INCHES		MILLIMETERS		NOTES
	MIN	MAX	MIN	MAX	
A	-	0.250	-	6.35	4
A1	0.015	-	0.39	-	4
A2	0.125	0.195	3.18	4.95	-
B	0.014	0.022	0.356	0.558	-
B1	0.030	0.070	0.77	1.77	8
C	0.008	0.015	0.204	0.381	-
D	1.980	2.095	50.3	53.2	5
D1	0.005	-	0.13	-	5
E	0.600	0.625	15.24	15.87	6
E1	0.485	0.580	12.32	14.73	5
e	0.100 BSC		2.54 BSC		-
e_A	0.600 BSC		15.24 BSC		6
e_B	-	0.700	-	17.78	7
L	0.115	0.200	2.93	5.08	4
N	40		40		9

Rev. 0 12/93

NOTES:

1. Controlling Dimensions: INCH. In case of conflict between English and Metric dimensions, the inch dimensions control.

2. Dimensioning and tolerancing per ANSI Y14.5M-1982.

3. Symbols are defined in the "MO Series Symbol List" in Section 2.2 of Publication No. 95.

4. Dimensions A, A1 and L are measured with the package seated in JEDEC seating plane gauge GS-3.

5. D, D1, and E1 dimensions do not include mold flash or protrusions. Mold flash or protrusions shall not exceed 0.010 inch (0.25mm).

6. E and e_A are measured with the leads constrained to be perpendicular to datum -C- .

7. e_B and e_C are measured at the lead tips with the leads unconstrained. e_C must be zero or greater.

8. B1 maximum dimensions do not include dambar protrusions. Dambar protrusions shall not exceed 0.010 inch (0.25mm).

9. N is the maximum number of terminal positions.

10. Corner leads (1, N, N/2 and N/2 + 1) for E8.3, E16.3, E18.3, E28.3, E42.6 will have a B1 dimension of 0.030 - 0.045 inch (0.76 - 1.14mm).

Metric Plastic Quad Flatpack Packages (MQFP)

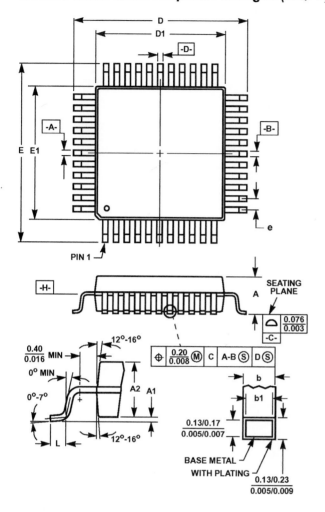

Q44.10x10 (JEDEC MS-022AB ISSUE B)
44 LEAD METRIC PLASTIC QUAD FLATPACK PACKAGE

SYMBOL	INCHES		MILLIMETERS		NOTES
	MIN	MAX	MIN	MAX	
A	-	0.096	-	2.45	-
A1	0.004	0.010	0.10	0.25	-
A2	0.077	0.083	1.95	2.10	-
b	0.012	0.018	0.30	0.45	6
b1	0.012	0.016	0.30	0.40	-
D	0.515	0.524	13.08	13.32	3
D1	0.389	0.399	9.88	10.12	4, 5
E	0.516	0.523	13.10	13.30	3
E1	0.390	0.398	9.90	10.10	4, 5
L	0.029	0.040	0.73	1.03	-
N	44		44		7
e	0.032 BSC		0.80 BSC		

Rev. 2 4/99

NOTES:

1. Controlling dimension: MILLIMETER. Converted inch dimensions are not necessarily exact.

2. All dimensions and tolerances per ANSI Y14.5M-1982.

3. Dimensions D and E to be determined at seating plane -C-.

4. Dimensions D1 and E1 to be determined at datum plane -H-.

5. Dimensions D1 and E1 do not include mold protrusion. Allowable protrusion is 0.25mm (0.010 inch) per side.

6. Dimension b does not include dambar protrusion. Allowable dambar protrusion shall be 0.08mm (0.003 inch) total.

7. "N" is the number of terminal positions.

For information regarding Intersil Corporation and its products, see www.intersil.com

15 **intersil**

MOTOROLA — **Freescale Semiconductor, Inc.**

SEMICONDUCTOR TECHNICAL DATA

100 kPa On-Chip Temperature Compensated & Calibrated Silicon Pressure Sensors

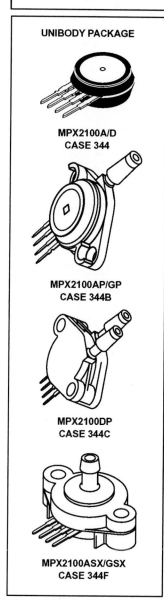

MPX2100 SERIES

0 to 100 kPa (0 to 14.5 psi)
40 mV FULL SCALE SPAN
(TYPICAL)

UNIBODY PACKAGE

MPX2100A/D CASE 344

MPX2100AP/GP CASE 344B

MPX2100DP CASE 344C

MPX2100ASX/GSX CASE 344F

The MPX2100 series device is a silicon piezoresistive pressure sensor providing a highly accurate and linear voltage output — directly proportional to the applied pressure. The sensor is a single, monolithic silicon diaphragm with the strain gauge and a thin–film resistor network integrated on–chip. The chip is laser trimmed for precise span and offset calibration and temperature compensation.

Features

- Temperature Compensated Over 0°C to +85°C
- Easy–to–Use Chip Carrier Package Options
- Available in Absolute, Differential and Gauge Configurations
- Ratiometric to Supply Voltage
- ±0.25% Linearity (MPX2100D)

Application Examples

- Pump/Motor Controllers
- Robotics
- Level Indicators
- Medical Diagnostics
- Pressure Switching
- Barometers
- Altimeters

Figure 1 illustrates a block diagram of the internal circuitry on the stand–alone pressure sensor chip.

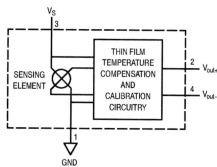

Figure 1. Temperature Compensated Pressure Sensor Schematic

VOLTAGE OUTPUT versus APPLIED DIFFERENTIAL PRESSURE

The differential voltage output of the sensor is directly proportional to the differential pressure applied.

The absolute sensor has a built–in reference vacuum. The output voltage will decrease as vacuum, relative to ambient, is drawn on the pressure (P1) side.

The output voltage of the differential or gauge sensor increases with increasing pressure applied to the pressure (P1) side relative to the vacuum (P2) side. Similarly, output voltage increases as increasing vacuum is applied to the vacuum (P2) side relative to the pressure (P1) side.

PIN NUMBER			
1	Gnd	3	V_S
2	+V_{out}	4	−V_{out}

NOTE: Pin 1 is noted by the notch in the lead.

Freescale Semiconductor, Inc.

MAXIMUM RATINGS(NOTE)

Rating	Symbol	Value	Unit
Maximum Pressure (P1 > P2)	P_{max}	400	kPa
Storage Temperature	T_{stg}	−40 to +125	°C
Operating Temperature	T_A	−40 to +125	°C

NOTE: Exposure beyond the specified limits may cause permanent damage or degradation to the device.

OPERATING CHARACTERISTICS (V_S = 10 Vdc, T_A = 25°C unless otherwise noted, P1 > P2)

Characteristic		Symbol	Min	Typ	Max	Unit
Pressure Range[1]		P_{OP}	0	—	100	kPa
Supply Voltage[2]		V_S	—	10	16	Vdc
Supply Current		I_o	—	6.0	—	mAdc
Full Scale Span[3]	MPX2100A, MPX2100D	V_{FSS}	38.5	40	41.5	mV
Offset[4]	MPX2100D MPX2100A Series	V_{off}	−1.0 −2.0	—	1.0 2.0	mV
Sensitivity		$\Delta V/\Delta P$	—	0.4	—	mV/kPa
Linearity[5]	MPX2100D Series MPX2100A Series		— —	−0.25 −1.0	— —	0.25 1.0	%V_{FSS}
Pressure Hysteresis[5] (0 to 100 kPa)			—	±0.1	—	%V_{FSS}
Temperature Hysteresis[5] (−40°C to +125°C)			—	±0.5	—	%V_{FSS}
Temperature Effect on Full Scale Span[5]		TCV_{FSS}	−1.0	—	1.0	%V_{FSS}
Temperature Effect on Offset[5]		TCV_{off}	−1.0	—	1.0	mV
Input Impedance		Z_{in}	1000	—	2500	Ω
Output Impedance		Z_{out}	1400	—	3000	Ω
Response Time[6] (10% to 90%)		t_R	—	1.0	—	ms
Warm–Up			—	20	—	ms
Offset Stability[7]			—	±0.5	—	%V_{FSS}

NOTES:
1. 1.0 kPa (kiloPascal) equals 0.145 psi.
2. Device is ratiometric within this specified excitation range. Operating the device above the specified excitation range may induce additional error due to device self–heating.
3. Full Scale Span (V_{FSS}) is defined as the algebraic difference between the output voltage at full rated pressure and the output voltage at the minimum rated pressure.
4. Offset (V_{off}) is defined as the output voltage at the minimum rated pressure.
5. Accuracy (error budget) consists of the following:
   - Linearity: Output deviation from a straight line relationship with pressure, using end point method, over the specified pressure range.
   - Temperature Hysteresis: Output deviation at any temperature within the operating temperature range, after the temperature is cycled to and from the minimum or maximum operating temperature points, with zero differential pressure applied.
   - Pressure Hysteresis: Output deviation at any pressure within the specified range, when this pressure is cycled to and from the minimum or maximum rated pressure, at 25°C.
   - TcSpan: Output deviation at full rated pressure over the temperature range of 0 to 85°C, relative to 25°C.
   - TcOffset: Output deviation with minimum rated pressure applied, over the temperature range of 0 to 85°C, relative to 25°C.
6. Response Time is defined as the time for the incremental change in the output to go from 10% to 90% of its final value when subjected to a specified step change in pressure.
7. Offset stability is the product's output deviation when subjected to 1000 hours of Pulsed Pressure, Temperature Cycling with Bias Test.

LINEARITY

Linearity refers to how well a transducer's output follows the equation: $V_{out} = V_{off}$ + sensitivity x P over the operating pressure range. There are two basic methods for calculating nonlinearity: (1) end point straight line fit (see Figure 2) or (2) a least squares best line fit. While a least squares fit gives the "best case" linearity error (lower numerical value), the calculations required are burdensome.

Conversely, an end point fit will give the "worst case" error (often more desirable in error budget calculations) and the calculations are more straightforward for the user. Motorola's specified pressure sensor linearities are based on the end point straight line method measured at the midrange pressure.

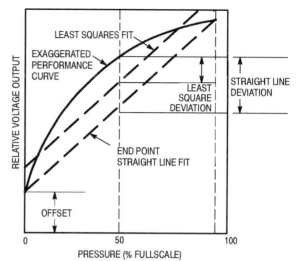

Figure 2. Linearity Specification Comparison

ON–CHIP TEMPERATURE COMPENSATION and CALIBRATION

Figure 3 shows the output characteristics of the MPX2100 series at 25°C. The output is directly proportional to the differential pressure and is essentially a straight line.

The effects of temperature on Full Scale Span and Offset are very small and are shown under Operating Characteristics.

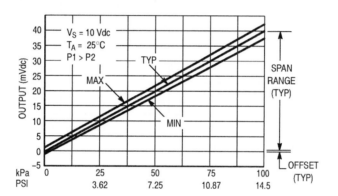

Figure 3. Output versus Pressure Differential

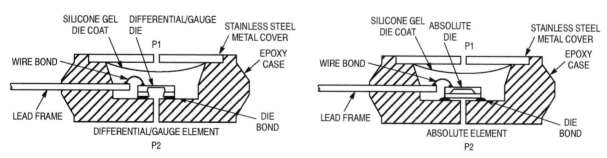

Figure 4. Cross–Sectional Diagrams (Not to Scale)

Figure 4 illustrates the absolute sensing configuration (right) and the differential or gauge configuration in the basic chip carrier (Case 344). A silicone gel isolates the die surface and wire bonds from the environment, while allowing the pressure signal to be transmitted to the silicon diaphragm.

The MPX2100 series pressure sensor operating characteristics and internal reliability and qualification tests are based on use of dry air as the pressure media. Media other than dry air may have adverse effects on sensor performance and long term reliability. Contact the factory for information regarding media compatibility in your application.

Freescale Semiconductor, Inc.

PRESSURE (P1)/VACUUM (P2) SIDE IDENTIFICATION TABLE

Motorola designates the two sides of the pressure sensor as the Pressure (P1) side and the Vacuum (P2) side. The Pressure (P1) side is the side containing the silicone gel which isolates the die. The differential or gauge sensor is designed to operate with positive differential pressure applied, P1 > P2. The absolute sensor is designed for vacuum applied to P1 side.

The Pressure (P1) side may be identified by using the table below:

Part Number		Case Type	Pressure (P1) Side Identifier
MPX2100A	MPX2100D	344	Stainless Steel Cap
MPX2100DP		344C	Side with Part Marking
MPX2100AP	MPX2100GP	344B	Side with Port Attached
MPX2100ASX	MPX2100GSX	344F	Side with Port Attached

ORDERING INFORMATION

MPX2100 series pressure sensors are available in absolute, differential and gauge configurations. Devices are available in the basic element package or with pressure port fittings which provide printed circuit board mounting ease and barbed hose pressure connections.

Device Type	Options	Case Type	MPX Series	
			Order Number	Device Marking
Basic Element	Absolute, Differential	344	MPX2100A MPX2100D	MPX2100A MPX2100D
Ported Elements	Differential, Dual Port	344C	MPX2100DP	MPX2100DP
	Absolute, Gauge	344B	MPX2100AP MPX2100GP	MPX2100AP MPX2100GP
	Absolute, Gauge Axial	344F	MPX2100ASX MPX2100GSX	MPX2100A MPX2100D

PACKAGE DIMENSIONS

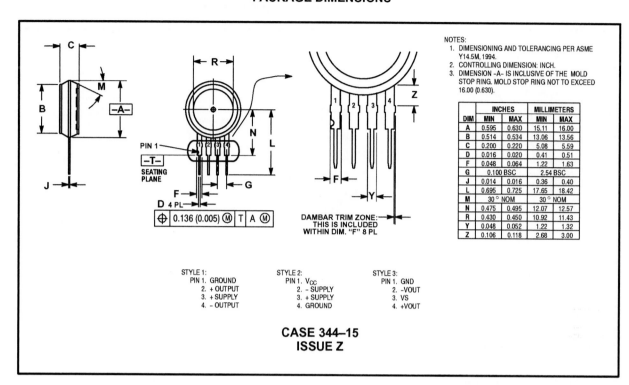

NOTES:
1. DIMENSIONING AND TOLERANCING PER ASME Y14.5M, 1994.
2. CONTROLLING DIMENSION: INCH.
3. DIMENSION –A– IS INCLUSIVE OF THE MOLD STOP RING. MOLD STOP RING NOT TO EXCEED 16.00 (0.630).

DIM	INCHES		MILLIMETERS	
	MIN	MAX	MIN	MAX
A	0.595	0.630	15.11	16.00
B	0.514	0.534	13.06	13.56
C	0.200	0.220	5.08	5.59
D	0.016	0.020	0.41	0.51
F	0.048	0.064	1.22	1.63
G	0.100 BSC		2.54 BSC	
J	0.014	0.016	0.36	0.40
L	0.695	0.725	17.65	18.42
M	30° NOM		30° NOM	
N	0.475	0.495	12.07	12.57
R	0.430	0.450	10.92	11.43
Y	0.048	0.052	1.22	1.32
Z	0.106	0.118	2.68	3.00

DAMBAR TRIM ZONE:
THIS IS INCLUDED
WITHIN DIM. "F" 8 PL

STYLE 1:
PIN 1. GROUND
2. + OUTPUT
3. + SUPPLY
4. – OUTPUT

STYLE 2:
PIN 1. V$_{CC}$
2. – SUPPLY
3. + SUPPLY
4. GROUND

STYLE 3:
PIN 1. GND
2. –VOUT
3. VS
4. +VOUT

CASE 344–15
ISSUE Z

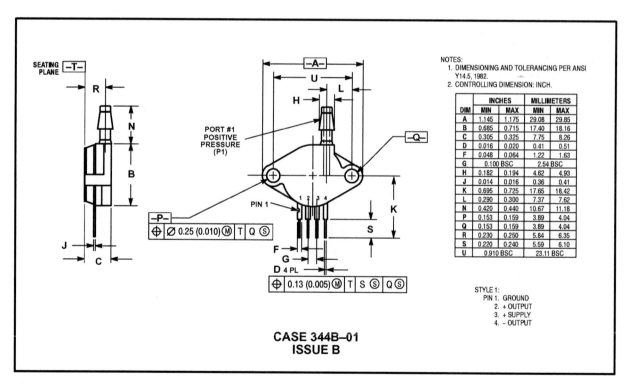

NOTES:
1. DIMENSIONING AND TOLERANCING PER ANSI Y14.5, 1982.
2. CONTROLLING DIMENSION: INCH.

DIM	INCHES		MILLIMETERS	
	MIN	MAX	MIN	MAX
A	1.145	1.175	29.08	29.85
B	0.685	0.715	17.40	18.16
C	0.305	0.325	7.75	8.26
D	0.016	0.020	0.41	0.51
F	0.048	0.064	1.22	1.63
G	0.100 BSC		2.54 BSC	
H	0.182	0.194	4.62	4.93
J	0.014	0.016	0.36	0.41
K	0.695	0.725	17.65	18.42
L	0.290	0.300	7.37	7.62
N	0.420	0.440	10.67	11.18
P	0.153	0.159	3.89	4.04
Q	0.153	0.159	3.89	4.04
R	0.230	0.250	5.84	6.35
S	0.220	0.240	5.59	6.10
U	0.910 BSC		23.11 BSC	

PORT #1
POSITIVE
PRESSURE
(P1)

STYLE 1:
PIN 1. GROUND
2. + OUTPUT
3. + SUPPLY
4. – OUTPUT

CASE 344B–01
ISSUE B

Freescale Semiconductor, Inc.

PACKAGE DIMENSIONS — CONTINUED

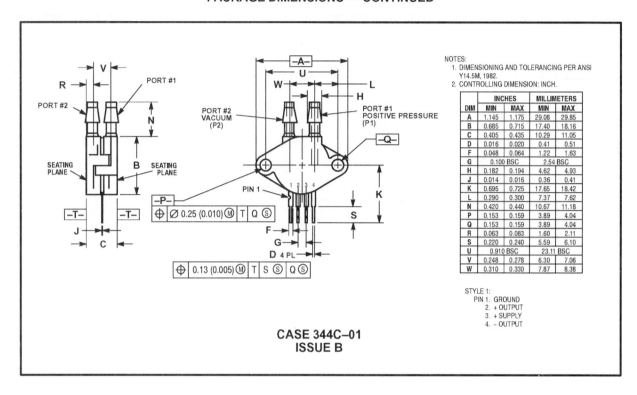

NOTES:
1. DIMENSIONING AND TOLERANCING PER ANSI Y14.5M, 1982.
2. CONTROLLING DIMENSION: INCH.

DIM	INCHES		MILLIMETERS	
	MIN	MAX	MIN	MAX
A	1.145	1.175	29.08	29.85
B	0.685	0.715	17.40	18.16
C	0.405	0.435	10.29	11.05
D	0.016	0.020	0.41	0.51
F	0.048	0.064	1.22	1.63
G	0.100 BSC		2.54 BSC	
H	0.182	0.194	4.62	4.93
J	0.014	0.016	0.36	0.41
K	0.695	0.725	17.65	18.42
L	0.290	0.300	7.37	7.62
N	0.420	0.440	10.67	11.18
P	0.153	0.159	3.89	4.04
Q	0.153	0.159	3.89	4.04
R	0.063	0.083	1.60	2.11
S	0.220	0.240	5.59	6.10
U	0.910 BSC		23.11 BSC	
V	0.248	0.278	6.30	7.06
W	0.310	0.330	7.87	8.38

STYLE 1:
PIN 1. GROUND
2. + OUTPUT
3. + SUPPLY
4. – OUTPUT

CASE 344C–01
ISSUE B

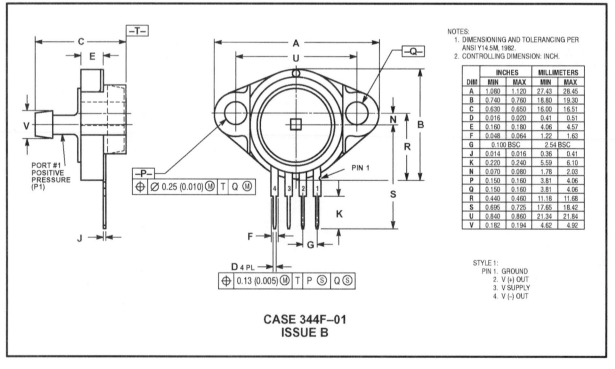

NOTES:
1. DIMENSIONING AND TOLERANCING PER ANSI Y14.5M, 1982.
2. CONTROLLING DIMENSION: INCH.

DIM	INCHES		MILLIMETERS	
	MIN	MAX	MIN	MAX
A	1.080	1.120	27.43	28.45
B	0.740	0.760	18.80	19.30
C	0.630	0.650	16.00	16.51
D	0.016	0.020	0.41	0.51
E	0.160	0.180	4.06	4.57
F	0.048	0.064	1.22	1.63
G	0.100 BSC		2.54 BSC	
J	0.014	0.016	0.36	0.41
K	0.220	0.240	5.59	6.10
N	0.070	0.080	1.78	2.03
P	0.150	0.160	3.81	4.06
Q	0.150	0.160	3.81	4.06
R	0.440	0.460	11.18	11.68
S	0.695	0.725	17.65	18.42
U	0.840	0.860	21.34	21.84
V	0.182	0.194	4.62	4.92

STYLE 1:
PIN 1. GROUND
2. V (+) OUT
3. V SUPPLY
4. V (–) OUT

CASE 344F–01
ISSUE B

Appendix B — Data Sheets

GEC PLESSEY
SEMICONDUCTORS

UR70

ZN414Z, ZN415E, ZN416E
AM RADIO RECEIVERS

FEATURES

- Single cell operation (1.1 to 1.6 volt operating range)
- Low current consumption
- 150kHz to 3MHz frequency range (i.e. full coverage of medium and long wavebands)
- Easy to assemble, no alignment necessary
- Simple and effective AGC action
- Will drive crystal earphone direct (ZN414Z)
- Will drive headphones direct (ZN415E and ZN416E)
- Excellent audio quality
- Typical power gain of 72dB (ZN414Z)
- Minimum of external components required

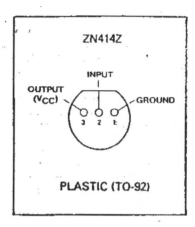

PLASTIC (TO-92)

DP8

GENERAL DESCRIPTION

The ZN414Z is a 10 transistor tuned radio frequency (TRF) circuit packaged in a 3-pin TO-92 plastic package for simplicity and space economy.

The circuit provides a complete R.F. amplifier, detector and AGC circuit which requires only six external components to give a high quality A.M. tuner. Effective AGC action is available and is simply adjusted by selecting one external resistor value. Excellent audio quality can be achieved, and current consumption is extremely low. No setting-up or alignment is required and the circuit is completely stable in use.

The ZN415E retains all the features of the ZN414Z but also incorporates a buffer stage giving sufficient output to drive headphones directly from the 8 pin DIL.

Similarly the ZN416E is a buffered output version of the ZN414Z giving typically 120mV (r.m.s.) output into a 64Ω load. The same package and pinning is used for the ZN416E as the ZN415E.

EVICE SPECIFICATIONS $T_{amb} = 25°C$, $V_{CC} = 1.4V$. Parameters apply to all types unless therwise stated.

Parameter		Min.	Typ	Max.	Units
Supply voltage, V_{CC}		1.1	1.3	1.6	volts
Supply current, I_S with 64Ω headphones	ZN414Z	-	0.3	0.5	mA
	ZN415E		2.3	3	
	ZN416E		4	5	
Input frequency range		0.15	-	3.0	MHz
Input resistance		-	4.0	-	MΩ
Threshold sensitivity (Dependant on Q of coil)			50		μV
Selectivity		-	4.0	-	kHz
Total harmonic distortion		-	3.0	-	%
AGC range		-	20	-	dB
Power gain (ZN414Z)			72		dB
Voltage gain of output stage	ZN415E	-	6	-	dB
	ZN416E		18		
Output voltage into 64Ω load before clipping	ZN414Z	-	60	-	mVpp
	ZN415E		120		
	ZN416E		340		
Upper cut-off frequency of output stage, No capacitor, (ZN415E and ZN416E)		20	-	-	kHz
With 0.01μF between pin 7 and 0V (ZN415E)		-	6	-	kHz
With 0.01μF between pin 7 and 0V (ZN416E)		-	10	-	kHz
Lower cut-off frequency of output stage 0.1μF between pins 2 and 3 for ZN415E 0.47μF between pins 2 and 3 for ZN416E		-	50	-	Hz
Quiescent output voltage	ZN414Z	-	40	-	mV
	ZN415E		80		
	ZN416E		200		
Operating temperature range		0	-	70	°C
Maximum storage temperature		-65	-	125	°C

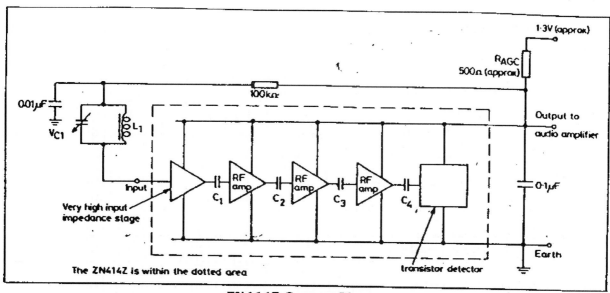

ZN414Z System Diagram

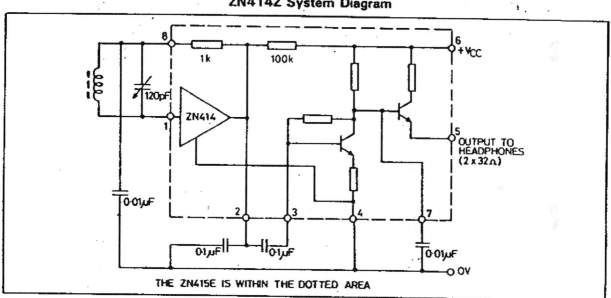

ZN415E System Diagram

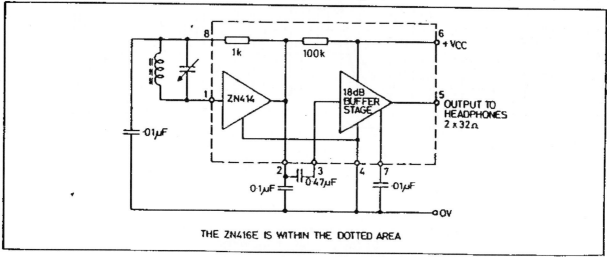

ZN416E System Diagram

ZN414Z CHARACTERISTICS — All measurements performed with 30% modulation, $F_M = 400Hz$

Gain and AGC characteristics

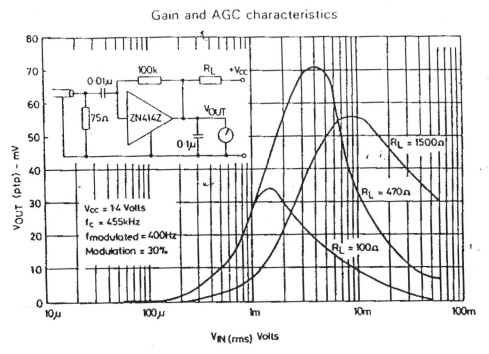

See operating notes for explanation of AGC action.

Frequency response of the ZN414Z

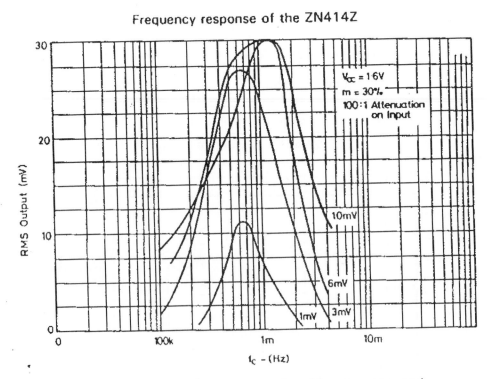

Note that this graph represents the chip response, and
not the receiver bandwidth.

ZN414Z CHARACTERISTICS – *(Continued)*

Gain variation with supply volts.

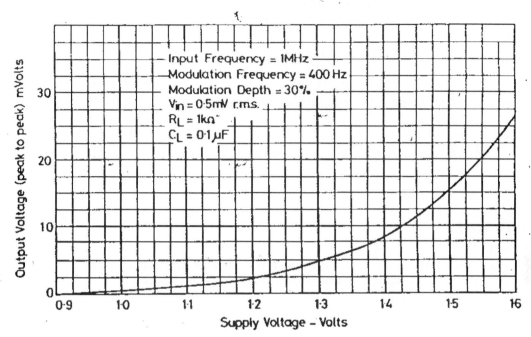

D.C. level at output

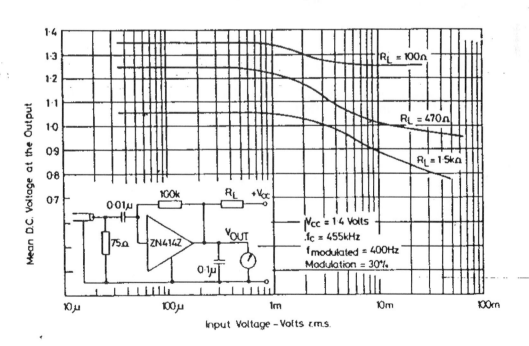

LAYOUT REQUIREMENTS

As with any high gain R.F. device, certain basic layout rules must be adhered to if stable and reliable operation is to be obtained. These are listed below:

1. The output decoupling capacitor should be soldered as near as possible to the output and earth leads of the ZN414Z. Furthermore, its value together with the AGC resistor (R_{AGC}) should be calculated at $\approx 4kHz$, i.e.:

$$C \text{ (farads)} = \frac{1}{2\pi \cdot R_{AGC} \cdot 4 \cdot 10^3}$$

2. All leads should be kept as short as possible, especially those in close proximity to the ZN414Z.

3. The tuning assembly should be some distance from the battery, loudspeaker and their associated leads.

4. The 'earthy' side of the tuning capacitor should be connected to the junction of the 100kΩ resistor and the 0.01μF capacitor.

OPERATING NOTES

(a) Selectivity

To obtain good selectivity, essential with any T.R.F. device, the ZN414Z must be fed from an efficient, high 'Q' coil and capacitor tuning network. With suitable components the selectivity is comparable to superhet designs, except that a very strong signal in proximity to the receiver may swamp the device unless the ferrite rod aerial is rotated to "null-out" the strong signal.

Two other factors affect the apparent selectivity of the device. Firstly, the gain of the ZN414Z is voltage sensitive (see previous page) so that, in strong signal areas, less supply voltage will be needed to obtain correct AGC action. Incorrect adjustment of the AGC causes a strong station to occupy a much wider bandwidth than necessary and in extreme cases can cause the RF stages to saturate before the AGC can limit RF gain. This gives the effect of swamping together with reduced AF output. All the above factors have to be considered if optimum performance is to be obtained.

(b) Ferrite aerial size

Because of the gain variation available by altering supply voltage, the size of the ferrite rod is relatively unimportant. However, the ratio of aerial rod length to diameter should ideally be large to give the receiver better directional properties. Successful receivers have been constructed with ferrite rod aerials of 4cm (1.5″) and up to 20cm (8″).

DRIVE CIRCUITS

Three types of drive circuit are shown, each has been used successfully. The choice is largely an economic one, but circuit 3 is recommended wherever possible, having several advantages over the other circuits. Values for 9V supplies are shown, simple calculations will give values for other supplies.

1. Resistive Divider (ZN414Z)

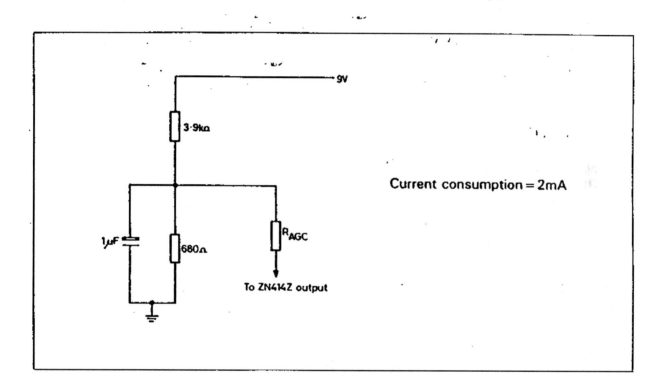

Current consumption = 2mA

Note: Replacing the 680Ω resistor with a 500Ω resistor and a 250Ω preset, sensitivity may be adjusted and will enable optimum reception to be realised under most conditions.

2. Diode Drive (ZN414Z)

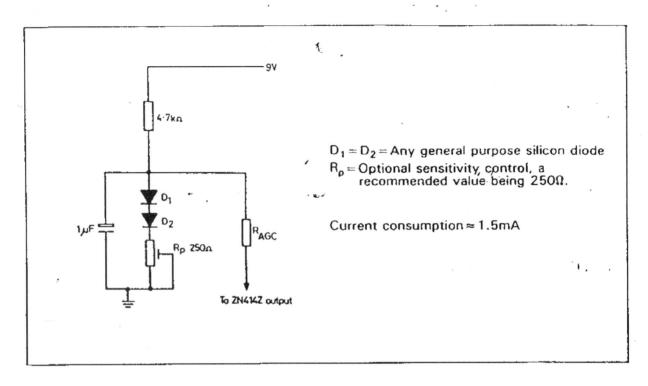

$D_1 = D_2 =$ Any general purpose silicon diode
$R_p =$ Optional sensitivity control, a recommended value being 250Ω.

Current consumption ≈ 1.5mA

3. Transistor Drive (ZN414Z and ZN415E)

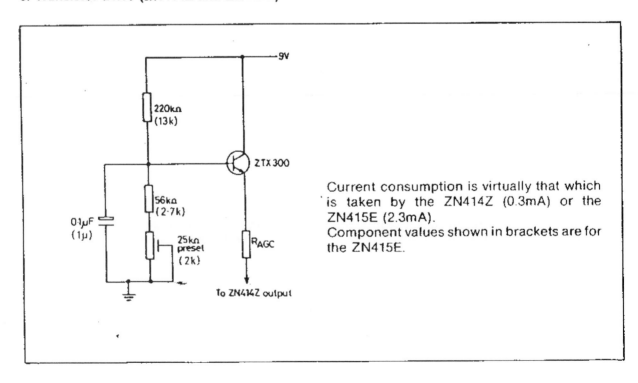

Current consumption is virtually that which is taken by the ZN414Z (0.3mA) or the ZN415E (2.3mA).
Component values shown in brackets are for the ZN415E.

RECOMMENDED CIRCUITS

(a) Earphone radio

The ZN414Z will drive a sensitive earpiece directly. In this case, an earpiece of equivalent impedance to R_{AGC} is substituted for R_{AGC} in the basic tuner circuit. Unfortunately, the cost of a sensitive earpiece is high, and unless an ultra-miniature radio is wanted, it is considerably cheaper to use a low cost crystal earpiece and add a single gain stage. One further advantage of this technique is that provision for a volume control can be made. A suitable circuit is shown below.

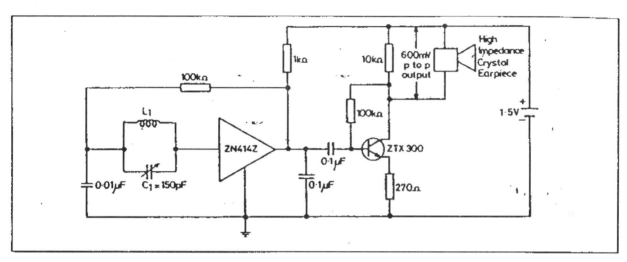

$L_1 \approx 80$ turns of 0.3mm dia. enamelled copper wire on a 5cm or 7.5cm long ferrite rod. Do not expect to adhere rigidly to the coil-capacitor details given. Any value of L_1 and C_1 which will give a high 'Q' at the desired frequency may be used.

Volume Control: a 250Ω potentiometer in series with a 100Ω fixed resistor substituted for the 270Ω emitter resistor provides an effective volume control.

(b) Domestic portable receiver

The circuit shown is capable of excellent quality, and its cost relative to conventional designs is much lower.

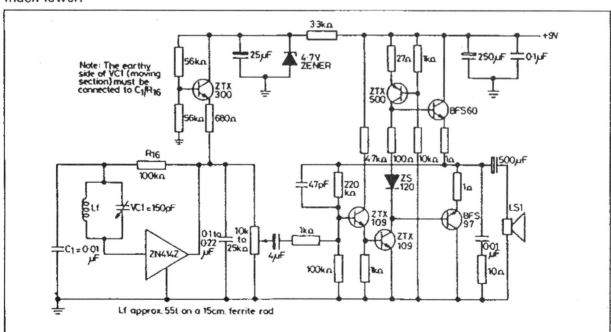

The complete circuit diagram of the Triffid receiver

Honeywell

Microstructure Pressure Sensors
Signal Conditioned
0 psi to 1 psi up to 0 psi to 150 psi

ASCX Series

FEATURES
- 5 Vdc Supply
- High Level Voltage Output
- Field Interchangeable
- Calibrated and Temperature Compensated
- Small Form Factor
- Low Power
- Offset Adjust

TYPICAL APPLICATIONS
- Medical Equipment
- Industrial Controls
- Pneumatic Controls

This series is a signal-conditioned version of Honeywell's proven performer and industry leading SCX series sensor.

This amplified ASCX device is in a package the same as the SCX but it offers a high level (4.5 V span) output on a very cost-effective basis. This family is fully calibrated and temperature compensated over a range of 0 °C to 70 °C [32 °F to 158 °F] but can be operated from -25 °C to 105 °C [-13 °F to 221 °F]. These sensors are intended for use with non-corrosive, non-ionic working fluids such as air and dry gases.

Devices are available to measure absolute, differential and gage pressures from 1 psi (ASCX01) up to 150 psi (ASCX150). The absolute devices (A) have an internal vacuum reference and an output voltage proportional to absolute pressure. Differential devices (D) allow application of pressure to either side of the sensing diaphragm and can be used for gage or differential pressure measurements.

The ASCX series devices feature an integrated circuit (IC) sensor element and laser trimmed thick film ceramic housed in a compact solvent resistant case. It provides excellent corrosion resistance and isolation to external packaging stresses. The package has convenient mounting holes and pressure ports for ease of use with standard plastic tubing for pressure connection.

All ASCX devices are calibrated for span to within ± 1 % (typically ± 0.2 %) of FSO. The devices are characterized for operation from a single 5 volt supply although sensitivity is ratiometric to the supply voltage and any dc supply from 5 Vdc to 16 Vdc is acceptable.

The ACSX series requires very low quiescent current compared to other signal conditioned pressure sensors, thus, this series is ideal for battery-powered applications.

The 100 microseconds response time makes this series an excellent choice for computer peripherals and pneumatic control applications.

Contact your local honeywell representative, or go to Honeywell's website at www.honeywell.com/sensing for additional details.

⚠ WARNING
PERSONAL INJURY
DO NOT USE these products as safety or emergency stop devices or in any other application where failure of the product could result in personal injury.
Failure to comply with these instructions could result in death or serious injury.

⚠ WARNING
MISUSE OF DOCUMENTATION
- The information presented in this product sheet is for reference only. Do not use this document as a product installation guide.
- Complete installation, operation, and maintenance information is provided in the instructions supplied with each product.
Failure to comply with these instructions could result in death or serious injury.

Microstructure Pressure Sensors

Signal Conditioned
0 psi to 1 psi up to 0 psi to 150 psi

ASCX PERFORMANCE CHARACTERISTICS [4]

Characteristic	Min.	Typ.	Max.	Unit
Offset [5]				
Models ASCX15/30/100/150xN	0.205	0.250	0.295	Volts
Models ASCX01/05DN	0.180	0.250	0.320	Volts
Output @ FS Pressure		4.750		Volts
Combined Pressure Linearity and Hysteresis [6]	–	±0.1	±0.5	% FSO
Temperature Effect on Span 0 °C to 70 °C [32 °F to 158 °F] [7]				
Models ASCX15/30/100/150xN	–	±0.2	±1.0	% FSO
Models ASCX01/05DN	–	±0.2	±1.5	% FSO
Temperature Effect on Offset 0 °C to 70 °C [32 °F to 158 °F] [7]				
Models ASCX15/30/100/150xN	–	±0.5	±1.0	% FSO
Models ASCX01/05DN		±0.5	±1.5	% FSO
Repeatability [8]	–	±0.2	±0.5	% FSO
Response Time [9]	–	100	–	Microsec.

SPECIFICATION NOTES

Note 1: Full-Scale Span is the algebraic difference between the output voltage at full-scale pressure and the output at zero pressure.

Note 2: Maximum pressure above which causes permanent sensor failure.

Note 3: Sensitivity is ratiometric to supply voltage.

Note 4: Performance specs are shown at reference conditions. Specifications apply for absolute pressure devices with pressure applied to Port A. For gage devices, pressure is applied to Port B and Port A is left open for ambient. For differential pressures, Port B is the high-pressure port. All differential devices feature dual pressure ports and can be used as gage or differential sensors. For absolute devices, Port B is inactive.

Note 5: Offset calibration is at the lowest pressure for each given device.

Note 6: Linearity refers to the best straight line fit as measured for offset, full scale and ½ full-scale pressure.

Note 7: Temperature errors are the maximum shift over 0 °C to 70 °C [32 °F to 158 °F], relative to the 25 °C [77 °F] reading.

Note 8: Maximum difference in output at any pressure within the operating pressure range and the temperature within 0 °C to 70 °C [32 °F to 158 °F] after:
        a) 100 temperature cycles, 0 °C to 70 °C [32 °F to 158 °F]
        b) 1.0 million pressure cycles, 0 psi to Full-Scale Span.

Note 9: Response time for a 0 psi to Full-Scale Span pressure step change, 10 % to 90 % rise time.

ELECTRICAL CONNECTION

Pinout	ASCX Series
	PIN 1) External Offset Adjustment PIN 2) V$_s$ PIN 3) + Output PIN 4) Ground PIN 5) N/C PIN 6) Do Not Use

Microstructure Pressure Sensors
Signal Conditioned
0 psi to 1 psi up to 0 psi to 150 psi

ASCX Series

PHYSICAL DIMENSIONS for Reference Only (mm/in)

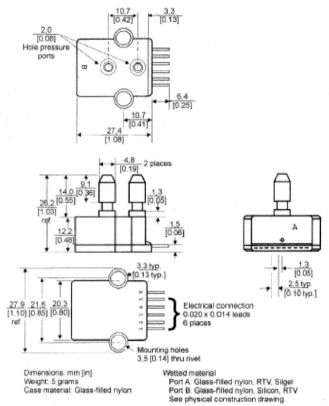

Dimensions: mm [in]
Weight: 5 grams
Case material: Glass-filled nylon

Wetted material
Port A: Glass-filled nylon, RTV, Silgel
Port B: Glass-filled nylon, Silicon, RTV
See physical construction drawing

WARRANTY/REMEDY

Honeywell warrants goods of its manufacture as being free of defective materials and faulty workmanship. Contact your local sales office for warranty information. If warranted goods are returned to Honeywell during the period of coverage, Honeywell will repair or replace without charge those items it finds defective. **The foregoing is Buyer's sole remedy and is in lieu of all other warranties, expressed or implied, including those of merchantability and fitness for a particular purpose.**

Specifications may change without notice. The information we supply is believed to be accurate and reliable as of this printing. However, we assume no responsibility for its use.

While we provide application assistance personally, through our literature and the Honeywell web site, it is up to the customer to determine the suitability of the product in the application.

For application assistance, current specifications, or name of the nearest Authorized Distributor, contact a nearby sales office. Or call:
1-800-537-6945 USA/Canada
1-815-235-6847 International
FAX
1-815-235-6545 USA
INTERNET
www.honeywell.com/sensing
info.sc@honeywell.com

Honeywell

Sensing and Control
www.honeywell.com/sensing
Honeywell
11 West Spring Street
Freeport, Illinois 61032

Index

Index

Index

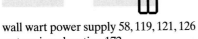
Index

About the Author

Tom Petruzzellis is an electronics engineer with 30 years' experience currently working with the geophysical field equipment department at the State University of New York—Binghamton. He is also an instructor at Binghamton. He has written extensively for industry publications, including *Electronics Now, Modern Electronics, QST, Microcomputer Journal,* and *Nuts & Volts,* and is the author of four earlier books: *Build Your Own Electronics Workshop; STAMP 2 Communications and Control Projects; Optoelectronics, Fiber Optics, and Laser Cookbook;* and *Alarm, Sensor, and Security Circuit Cookbook,* all from McGraw-Hill. Mr. Petruzzellis lives in Vestal, New York.

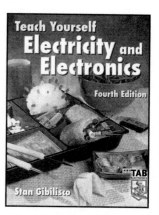